ECOTOXI
OF EXPL

ECOTOXICOLOGY OF EXPLOSIVES

Edited by
Geoffrey I. Sunahara • Guilherme Lotufo
Roman G. Kuperman • Jalal Hawari

CRC Press
Taylor & Francis Group
Boca Raton London New York

CRC Press is an imprint of the
Taylor & Francis Group, an **informa** business

CRC Press
Taylor & Francis Group
6000 Broken Sound Parkway NW, Suite 300
Boca Raton, FL 33487-2742

First issued in paperback 2019

© 2009 by Taylor and Francis Group, LLC
CRC Press is an imprint of Taylor & Francis Group, an Informa business

No claim to original U.S. Government works

ISBN-13: 978-0-8493-2839-8 (hbk)
ISBN-13: 978-0-367-38559-0 (pbk)

Library of Congress Cataloging-in-Publication Data

Ecotoxicology of explosives / editors, Geoffrey I. Sunahara ... [et al.].
    p. cm.
   Includes bibliographical references and index.
   ISBN 978-0-8493-2839-8 (hardcover : alk. paper)
   1. Explosives--Toxicology. 2. Explosives--Environmental aspects. I. Sunahara, Geoffrey I. (Geoffrey Isao), 1953- II. Title.

RA1270.E93E36 2009
363.17'98--dc22
                                                2009015403

Visit the Taylor & Francis Web site at
http://www.taylorandfrancis.com

and the CRC Press Web site at
http://www.crcpress.com

Contents

List of Figures vii
List of Tables ix
Preface xiii
The Editors xv
Contributors xix
Abbreviations xxiii

Chapter 1   Introduction 1

Geoffrey I. Sunahara, Roman G. Kuperman, Guilherme R. Lotufo, Jalal Hawari, Sonia Thiboutot, and Guy Ampleman

Chapter 2   Fate and Transport of Explosives in the Environment: A Chemist's View 5

Fanny Monteil-Rivera, Annamaria Halasz, Carl Groom, Jian-Shen Zhao, Sonia Thiboutot, Guy Ampleman, and Jalal Hawari

Chapter 3   Effects of Energetic Materials on Soil Organisms 35

Roman G. Kuperman, Michael Simini, Steven Siciliano, and Ping Gong

Chapter 4   Aquatic Toxicology of Explosives 77

Marion Nipper, R. Scott Carr, and Guilherme R. Lotufo

Chapter 5   Fate and Toxicity of Explosives in Sediments 117

Guilherme R. Lotufo, Marion Nipper, R. Scott Carr, and Jason M. Conder

Chapter 6   Bioconcentration, Bioaccumulation, and Biotransformation of Explosives and Related Compounds in Aquatic Organisms 135

Guilherme R. Lotufo, Michael J. Lydy, Gregory L. Rorrer, Octavio Cruz-Uribe, and Donald P. Cheney

Chapter 7   Toxicity of Energetic Compounds to Wildlife Species 157

Mark S. Johnson and Christopher J. Salice

Chapter 8   Genotoxicity of Explosives 177

Laura Inouye, Bernard Lachance, and Ping Gong

Chapter 9   Mechanisms of the Mammalian Cell Cytotoxicity of Explosives ... 211

*Narimantas Čėnas, Aušra Nemeikaitė-Čėnienė, Jonas
Šarlauskas, Žilvinas Anusevičius, Henrikas Nivinskas,
Lina Misevičienė, and Audronė Marozienė*

Chapter 10   Bioconcentration, Bioaccumulation, and Biomagnification of
Nitroaromatic and Nitramine Explosives in Terrestrial Systems 227

*Mark S. Johnson, Christopher J. Salice, Bradley E. Sample,
and Pierre Yves Robidoux*

Chapter 11   Habitat Disturbance at Explosives-Contaminated Ranges 253

*Rebecca A. Efroymson, Valerie Morrill, Virginia H. Dale,
Thomas F. Jenkins, and Neil R. Giffen*

Chapter 12   Ecological Risk Assessment of Soil Contamination with
Munition Constituents in North America 277

*Roman G. Kuperman, Ronald T. Checkai, Mark S. Johnson,
Pierre Yves Robidoux, Bernard Lachance, Sonia Thiboutot,
and Guy Ampleman*

Chapter 13   Closing Remarks 309

*Guilherme R. Lotufo, Geoffrey I. Sunahara, Jalal Hawari, and
Roman G. Kuperman*

Index 313

List of Figures

FIGURE 2.1 Abiotic transformation routes of TNT: (A) under photolytic conditions [23]; (B) in the presence of Fe(0) [24] or Fe(II) [25]; (C) under alkaline conditions [26].

FIGURE 2.2 Abiotic degradation pathways of RDX: (A) initiated by denitration (hydrolysis, photolysis, electrolysis, Fe(II), zero valent metals), or (B) initiated by reduction to the corresponding nitroso derivatives (2e-transfer) (photolysis, electrolysis, zero valent metals). (Data from Refs. 42, 48–50.)

FIGURE 2.3 Proposed denitration routes of CL-20 by photolysis [60], zero valent iron [61], and salicylate 1-monooxygenase from *Pseudomonas* sp. strain ATCC 29352 [62], and degradation products of A (or B) intermediates.

FIGURE 2.4 Biotransformation of TNT under aerobic and anaerobic conditions. TAT is only produced under strict anaerobic conditions.

FIGURE 2.5 Biodegradation of RDX initiated by denitration under aerobic (path A) and anaerobic (path B) conditions.

FIGURE 2.6 Biodegradation of RDX initiated by reduction to the nitroso derivatives followed by α-hydroxylation (path A), denitration (path B), or reduction (path C).

FIGURE 3.1 Schematic toxicological impacts of TNT on the soil microbial community.

FIGURE 3.2 The toxicity of TNT to the nitrogen and carbon soil biogeochemical cycles. Values are concentrations of TNT shown to inhibit the specified process by 50%. (For nitrification, the entire nitrification process is indicated but the assay only measures the production of nitrite and not nitrate. Numbers in brackets are references.)

FIGURE 6.1 Relationship between the log bioconcentration factors (BCFs) for explosive and related compounds reported in Table 6.1 and log K_{ow}. The solid line represents the best-fit linear equation (log BCF = 0.53 log K_{ow} – 0.23, r^2 = 0.37) for all values except those with log BCF <–0.9. The dashed line represents the linear equation from Meylan et al. [15].

FIGURE 6.2 Bioconcentration factors (BCF) for aquatic animals exposed to TNT determined for body residues for parent compound, sum concentration of extractable compounds, and for the sum concentration of extractable and nonextractable compounds.

FIGURE 7.1 Changes in body mass and feed consumption of northern bobwhites (*Colinus virginianus*) to dietary exposures of RDX for (a) 14 days and (b) 90 days. Egg production for (c) 14 days as other data are presented as means ±SEM.

FIGURE 8.1 Linking effects on an individual level to an ecological scale. Effects measured in individuals are commonly utilized as biomarkers of exposure and can range from premutagenic lesions (adducts, etc.) that can be repaired, to mutations and clastogenic effects. Depending on the cell type (somatic or germ) mutated, different effects may be observed at the population level, as shown in the gray box. Differential species sensitivity will contribute to effects observed at the ecosystem level; sensitive species are affected to a greater extent than resistant species, resulting in altered diversity and interspecies interactions.

FIGURE 9.1   Structural formulae of nitroaromatic explosives and their metabolites.

FIGURE 9.2   The dependence of LC_{50} of nitroaromatic explosives, their metabolites, and model nitroaromatic compounds in FLK cells on their calculated single-electron reduction potentials ($E^{1}_{7(calc)}$).

FIGURE 10.1   Microcosm design for assessing the accumulation of [^{14}C]-RDX or [^{14}C]-HMX in earthworms.

FIGURE 11.1   Vegetation removal due to various stressors during a single training, testing, or range maintenance event. Example relative extents of disturbance and uncertainties of spatial disturbance pattern of many of these stressors are depicted.

FIGURE 11.2   Aerial photograph of an impact area at Yuma Proving Ground. The small bright circles are ordnance impact scars on the dark desert pavement. Larger, circular light areas are scars in the pavement where plants are growing or have grown in the past. Four large, light gray desert washes cross the photo, with trees and shrubs seen as dark spots. (U.S. Army file photo, scale of photo unknown.)

FIGURE 11.3   Drought-stressed white bursage (*Ambrosia dumosa*) that has apparently colonized an impact crater at Yuma Proving Ground, Arizona. (U.S. Army file photo.)

FIGURE 12.1   Eight-step ecological risk assessment process for Superfund.

FIGURE 12.2   Derivation of Eco-SSL.

List of Tables

TABLE 2.1 Military Explosives Classification

TABLE 2.2 Most Commonly Used Physicochemical Properties of Some of the Explosives Needed for Environmental Risk Assessment

TABLE 2.3 Aerobic Transformation of TNT by Bacteria and Their Major Products

TABLE 2.4 Biodegradation of TNT with Fungi

TABLE 2.5 Biodegradation of RDX, HMX, and CL-20 by Microorganisms

TABLE 2.6 Adsorption Coefficients of Explosive Contaminants in Natural and Model Systems

TABLE 3.1 Summary of Effective Concentrations of TNT That Reduce Soil Microbial Endpoints by 50% (EC_{50})

TABLE 3.2 Selected Ecotoxicological Benchmarks for Energetic Materials Established in Standardized Single-Species Toxicity Tests with Terrestrial Plants

TABLE 3.3 Selected Ecotoxicological Benchmarks for Energetic Materials Established in Standardized Single-Species Toxicity Tests with Soil Invertebrates and in a Microcosm Test with Indigenous Soil Microinvertebrate Community

TABLE 4.1 Toxicity of Nitrobenzenes and Derivatives

TABLE 4.2 Toxicity of Nitrotoluenes

TABLE 4.3 Toxicity of TNT Reduction Products and Their Isomers

TABLE 4.4 Toxicity of Reduction Products of Several DNT Isomers

TABLE 4.5 Toxicity of Nitrophenols and Derivatives

TABLE 4.6 Toxicity of Various Ordnance Compounds

TABLE 4.7 Final and Interim Acute and Chronic Values Derived as Water Quality Criteria (μmoles L^{-1}) for the Protection of Aquatic Life

TABLE 5.1 Nominal Concentration of TNT and Concentration of the Sum of TNT and its Transformation Products (ΣTNT) Measured after Mixing into Spiked Sediments of Different Types

TABLE 5.2 Toxicity of Explosives and Their Transformation Products Determined for Aquatic Invertebrates Using Spiked Sediment Exposures

TABLE 5.3 Toxicity of Porewater Extracted from Sandy or Fine-Grained Sediments Spiked with 2,6-DNT, Tetryl, or Picric Acid

TABLE 6.1   Empirically Determined Bioconcentration Factor (BCF) Values for Fish and Aquatic Invertebrates

TABLE 6.2   Bioconcentration Factor (BCF) Values Predicted from log K_{ow} Values Using a Regression Model and Range of Experimentally Determined Values

TABLE 6.3   Bioconcentration Factor (BCF) Values for Aquatic Animals Determined Using the Sum Concentration of the Parent Compound and Major Transformation Products, or Radioactivity in Extract, or Total Radioactivity Surrogate for Body Residue

TABLE 6.4   Specific Rate Constants for TNT Removal from Seawater by Tissue Cultures of Marine Macroalgae

TABLE 6.5   Comparison of Metabolite Levels from TNT Biotransformation by Three Tissue Cultures of Marine Macroalgae

TABLE 7.1   Acute Oral Median Lethal Dose (LD_{50}) of Specific Explosive Compounds to Western Fence Lizards (*Sceloporus occidentalis*)

TABLE 7.2   Developmental Toxicity Values of Several Selected Substances Using the FETAX Assay

TABLE 8.1   In Vitro Microorganism-Based Systems Used to Assess Mutagenic Potential for Explosives and Explosives-Related Compounds

TABLE 8.2   In Vitro Mammalian Cell-Based Systems Used to Assess Mutagenic Potential for Explosives and Explosives-Related Compounds

TABLE 8.3   Summary for Genotoxicity Testing of Explosives Compounds

TABLE 9.1   Concentrations of Nitroaromatic Explosives and Their Metabolites for the 50% Cell Survival (LC_{50}) in Mammalian Cell Culture Cytotoxicity Studies

TABLE 10.1   Log K_{ow} and Water Solubility of Nitramine and Nitroaromatic Explosives

TABLE 10.2   Analyte Concentrations (mg kg^{-1} dry weight) of Yellow Nutsedge Grown in Hydroponic Media

TABLE 10.3   Properties of Tested Soils

TABLE 10.4   Concentration of Tissue TNT-Derived Radiocarbon and Distribution of Radiocarbon in Wheat and Blando Brome Grown in TNT-Amended Soil for 60 Days[a]

TABLE 10.5   Tissue Concentrations[a] of RDX from Plants Grown in RDX-Amended Soil of 10 mg kg^{-1} for 60 Days

TABLE 10.6   Bioaccumulation Factors and Associated Coefficients of Determination (*R^2*) for Sunflower

TABLE 10.7   RDX Concentrations in Plant Tissue (in mg kg^{-1})

TABLE 10.8   Calculated Bioaccumulation Factors (BAFs) from Plant Tests Using Selected Nitramines

TABLE 11.1   Effects of Explosives Ranges on Terrestrial Wildlife Habitat Variables

TABLE 12.1 Growth (Shoot Fresh Mass) Toxicity Benchmarks (EC_{20} mg kg^{-1}) for Munition Constituents (MC) Freshly Amended or Weathered and Aged in Sassafras Sandy Loam Soil for Selected Test Species

TABLE 12.2 Growth (Shoot Dry Mass) Toxicity Benchmarks (EC_{20} mg kg^{-1}) for Munition Constituents (MC) Freshly Amended or Weathered and Aged in Sassafras Sandy Loam Soil for Selected Test Species

TABLE 12.3 Reproduction Toxicity Benchmarks (EC_{20} mg kg^{-1}) for Munition Constituents (MC) Freshly Amended or Weathered and Aged in Sassafras Sandy Loam Soil for Selected Test Species

TABLE 12.4 Derivation of Proposed Soil Invertebrate-Based Ecotoxicological Tolerance Values (ETVs) for CL-20 in Freshly Amended Sassafras Sandy Loam Soil as the Geometric Mean of Reproduction Benchmarks for Selected Species

TABLE 12.5 Proposed Terrestrial Plant-Based Ecotoxicological Tolerance Values for Munition Constituents Determined Using Ecotoxicological Benchmarks Established in Similar Individual Definitive Toxicity Tests with Sassafras Sandy Loam Soil

TABLE 12.6 Proposed Soil Invertebrate-Based Ecotoxicological Tolerance Values for Munition Constituents Determined Using Ecotoxicological Benchmarks Established in Similar Individual Definitive Toxicity Tests with Sassafras Sandy Loam Soil

TABLE 12.7 Proposed Military-Specific Canadian Environmental Sustainability Indices (ESI) for Energetic Soil Contaminants at Department of National Defense Installations

Preface

An international meeting of defense ministries and department representatives from Australia, Canada, New Zealand, the United Kingdom, and the United States was held in Edinburgh, Australia, on February 23–25, 2004, under the auspices of The Technical Cooperation Program (TTCP). At this meeting, the Conventional Weapons Group (WPN) Technical Panel 4 (TP-4) for Energetic Materials and Propulsion agreed that a project "Development of Environmental Threshold Values for Defense Sites Contaminated with Energetic Materials" be conducted to promote the data acquisition and sharing of ecotoxicological information to address environmental problems related to energetic materials (EMs). The first objective of this international collaboration involved the development of environmental threshold concentrations for explosives and propellants. These criteria were needed for the ecological risk assessment (ERA) of sites contaminated with EMs, that is, to know "how clean is clean." Although the choice of approach and the use of ERA tools may differ from country to country, the assurance of quality and direction of ecotoxicological research were recognized as key concerns shared by the scientific community.

The second objective of this project was to review the accessible fate and ecotoxicological data for EMs and the methodologies for their development, and to make them available to interested parties through the publication of a book on the ecotoxicology of explosives. The book presented here is structured to first characterize the fate of explosives in the environment; then to provide information on the their ecological effects in key environmental media, including aquatic, sediment, and terrestrial habitats; and finally to describe the practical application of fate and ecotoxicological information to the environmental risk assessment of EM-contaminated sites. Information presented in this book follows the recognition that the ecotoxicological characterization of an EM-contaminated site can be accomplished through establishing a relationship between the measures of exposure to the EMs determined by chemical analyses and the measures of the effect of the EMs determined by toxicity testing. Approaches to assembling these lines of evidence for environmental risk assessment purposes may not be obvious to the nonspecialist and are discussed in relevant chapters of the book.

In this book, the reader is introduced to the fate and transport of EMs in the environment. The knowledge of the transport, transformation, and degradation pathways of these chemicals in the environment will help the reader to understand the potential hazardous impact and bioaccumulation of EMs in different terrestrial and aquatic ecological receptors. The genotoxic effects of EMs as well as the current understanding of the cellular and molecular mechanisms underlying the toxicity of these chemicals are reviewed. Integration of the preceding information is important for ERA of EM-contaminated sites.

The chapter authors focused primarily on the peer-reviewed publications in the open literature, although technical reports were also reviewed for this book. Each chapter of this book has been evaluated by at least two external peer reviewers familiar with the specified research area. This book is intended for readers at the graduate and undergraduate university levels, as well as for a wide range of environmental professionals, including scientists, engineers, consultants, site managers, regulators, and decision makers at EM-contaminated installations and ranges.

The editors and contributors wish to thank the many individuals who assisted us with the peer review process of this book, and Helen E. Blaho for proofreading and typing the manuscript. We acknowledge the U.S. Strategic Environmental R&D Program (SERDP) and the TTCP for their vision and support during the preparation of this book.

Geoffrey I. Sunahara

The Editors

Geoffrey I. Sunahara is a senior research scientist and the group leader of Applied Ecotoxicology at the Biotechnology Research Institute (National Research Council–Canada) in Montreal. He has more than 20 years of professional experience in the field of biochemical toxicology and environmental risk assessment, having gained this expertise in Canada, the United States, and Europe. He has more than 200 research publications, proceedings, and presentations. His current fields of research include the ecotoxicological characterization of recalcitrant soil contaminants such as the energetic substances (TNT, RDX, and HMX) and their metabolites using whole organisms (bacteria, plant, and earthworm toxicity tests) and cultured cell approaches (mutagenicity and cell proliferation). Sunahara has served on several editorial boards, and he was the lead editor of the recent ecotoxicology book *Environmental Analysis of Contaminated Sites*. He has participated on expert advisory committees for Environment Canada, Natural Sciences and Engineering Reesearch Council of Canada, U.S. Strategic Environmental Research and Development Program (SERDP), and U.S. EPA research projects. Sunahara received his M.Sc. in pharmacology from the University of Toronto and his Ph.D. in pharmacology and toxicology from the University of British Columbia in Vancouver, Canada. He was a Fogarty International Post-Doctoral Fellow at the National Institute of Environmental Health Sciences (National Institute of Health, North Carolina). Sunahara was a corecipient of the TTCP Frances Bérubé Award for Environmental Awareness (2005) and holds academic positions at McGill University and Concordia University.

Guilherme R. Lotufo was born in Brazil and earned a bachelor's degree in biology and a master's degree in zoology from the University of Sao Paulo, and a Ph.D. in zoology from Louisiana State University. He is the lead scientist for a wide diversity of ecotoxicology-related projects for the U.S. Army Research and Development Center in Vicksburg, Mississippi. Most of his research relates to sediment quality assessment, bioaccumulation of hydrophobic organic compounds in aquatic environments, and ecological risk assessment. He has expanded his expertise to numerous investigations of the aquatic toxicology of explosive compounds under funding from the Army and Navy. Lotufo has

maintained active collaborations with leading national and international research organizations. Lotufo has authored more than 40 scientific articles in renowned journals.

Roman G. Kuperman has more than 20 years experience in field and laboratory methods for assessment of the ecological effects of soil pollutants using invertebrate, microbial, and molecular toxicology methods. He earned his Ph.D. in entomology from the Ohio State University in 1993. As a principal investigator at Edgewood Chemical Biological Center, Kuperman leads U.S. Strategic Environmental Research and Development Program (SERDP)-funded research programs developing data on environmental fate and the effects of explosive materials. He leads the Key Technical Area (KTA 4-32-04) "Development of Environmental Tolerance Values for Defense Sites Contaminated with Energetic Materials" of The Technical Cooperation Program (TTCP) Weapons Technical Panel (WPN TP-4), an international organization that collaborates in defense scientific and technical information exchange activities for Australia, Canada, New Zealand, the United Kingdom, and the United States. Kuperman also serves as the Soil Ecology Society liaison on the U.S. National Academy of Sciences National Committee for Soil Science; the U.S. representative on several international work groups developing new international standardized testing methods for ecotoxicological research, and is a member of the U.S. Environmental Protection Agency-sponsored Ecological Soil Screening Levels (Eco-SSL) National Task Group. Kuperman chaired the Society of Environmental Toxicology and Chemistry (SETAC) Contaminated Soils Advisory Group and is a member of several international scientific advisory committees. Kuperman has published more than 200 scientific papers in peer-reviewed journals, symposia proceedings, and technical reports.

Jalal Hawari has been with the National Research Council of Canada since 1983. He is a principal research scientist and leader of the Environmental and Analytical Chemistry Group at the Biotechnology Research Institute (BRI). He obtained his Ph.D. in chemistry from the Christopher Ingold Laboratories at University College London. Since 1994, Hawari has served as an adjunct professor at McGill University. His current research is focused on detection, fate, and ecological impact; biotransformation pathways of pollutants and new substances; and development of green processes for the extraction and conversion of renewable feedstocks into value-added products such as fine

chemicals and biopolymers. Working at the interface of chemistry and microbiology, the group develops enabling bioanalytical tools to identify microorganisms and to measure the environmental fate, transformation pathways, and health risk associated with the use of nitramines and nitroaromatics in soil, groundwater, and marine sediments. In the last few years, the group received several prestigious research grants from Defence Research and Development Canada–Department of National Defence (DRDC-DND; Canada), the Office of Naval Research (U.S. Navy), and the Strategic Environmental Research and Development Program (SERDP; United States) to determine the microbial degradation and environmental fate and impact of emerging contaminants in marine sediments and coastal waterways. Hawari has won a number of Canadian (2002 NRC Outstanding Achievement Award for Industrial Partnership, 2002 Golden Medal–Queen Elizabeth II, 2004 NRC Outstanding Achievement Award for External Recognition, and four Director General awards) and international awards (2003 SERDP–Cleanup Project of the Year Award, 2004 ES&T–American Chemical Society Excellence in Review Award; 2005 TTCP Frances Bérubé Environmental Awareness Award). In addition, Hawari holds seven patents, shares a licensed soil bioremediation technology, and has published over 200 scientific articles and 150 proceedings and critical reviews.

Contributors

Guy Ampleman
Defence Research and Development
    Canada–Valcartier
National Defence Canada
Val-Bélair, Québec, Canada

Žilvinas Anusevičius
Institute of Biochemistry
Vilnius, Lithuania

R. Scott Carr
U.S. Geological Survey
Columbia Environmental Research
    Center
Marine Ecotoxicology Research Station
Corpus Christi, Texas

Narimantas Čėnas
Institute of Biochemistry
Vilnius, Lithuania

Ronald T. Checkai
Environmental Toxicology Branch
U.S. Army Edgewood Chemical
    Biological Center
Aberdeen Proving Ground, Maryland

Donald P. Cheney
Department of Biology
Northeastern University
Boston, Massachusetts

Jason M. Conder
ENVIRON International Corporation
Irvine, California

Octavio Cruz-Uribe
Department of Chemical Engineering
Oregon State University
Corvallis, Oregon

Virginia H. Dale
Environmental Sciences Division
Oak Ridge National Laboratory
Oak Ridge, Tennessee

Rebecca A. Efroymson
Environmental Sciences Division
Oak Ridge National Laboratory
Oak Ridge, Tennessee

Neil R. Giffen
Environmental Sciences Division
Oak Ridge National Laboratory
Oak Ridge, Tennessee

Ping Gong
SpecPro, Inc.
Vicksburg, Mississippi

Carl Groom
Biotechnology Research Institute
National Research Council of Canada
Montreal, Québec, Canada

Annamaria Halasz
Biotechnology Research Institute
National Research Council of Canada
Montreal, Québec, Canada

Jalal Hawari
Biotechnology Research Institute
National Research Council of Canada
Montreal, Québec, Canada

Laura Inouye
U.S. Army Engineer Research and
    Development Center
Waterways Experiment Station, EP-R
Vicksburg, Mississippi

Thomas F. Jenkins
Cold Regions Research & Engineering
  Laboratory
U.S. Army Corps of Engineers
Hanover, New Hampshire

Mark S. Johnson
Health Effects Research Program
U.S. Army Center for Health Promotion
  and Preventive Medicine
Aberdeen Proving Ground, Maryland

Roman G. Kuperman
U.S. Army Edgewood Chemical
  Biological Center
Environmental Toxicology Branch
Aberdeen Proving Ground, Maryland

Bernard Lachance
Biotechnology Research Institute
National Research Council of Canada
Montreal, Québec, Canada

Guilherme R. Lotufo
U.S. Army Corps of Engineers
U.S. Army Engineer Research and
  Development Center
Vicksburg, Mississippi

Michael J. Lydy
Fisheries and Illinois Aquaculture
  Center and
Department of Zoology
Southern Illinois University
Carbondale, Illinois

Audronė Marozienė
Institute of Biochemistry
Vilnius, Lithuania

Lina Misevičienė
Institute of Biochemistry
Vilnius, Lithuania

Fanny Monteil-Rivera
Biotechnology Research Institute
National Research Council of Canada
Montreal, Québec, Canada

Valerie Morrill
U.S. Army Yuma Proving Ground
Yuma, Arizona, USA

Aušra Nemeikaitė-Čėnienė
Institute of Immunology
Vilnius University
Vilnius, Lithuania

Marion Nipper
Center for Coastal Studies
Texas A&M University–Corpus Christi
Corpus Christi, Texas

Henrikas Nivinskas
Institute of Biochemistry
Vilnius, Lithuania

Pierre Yves Robidoux
Biotechnology Research Institute
National Research Council of Canada
Montreal, Québec, Canada

Gregory L. Rorrer
Department of Chemical Engineering
Oregon State University
Corvallis, Oregon

Christopher J. Salice
Health Effects Research Program
U.S. Army Center for Health Promotion
  and Preventive Medicine
Aberdeen Proving Ground, Maryland

Bradley E. Sample
CH2M HILL Inc.
Sacramento, California

Jonas Šarlauskas
Institute of Biochemistry
Vilnius, Lithuania

Steven Siciliano
University of Saskatchewan
Saskatoon, Saskatchewan, Canada

Michael Simini
Environmental Toxicology Branch
U.S. Army Edgewood Chemical
  Biological Center
Aberdeen Proving Ground, Maryland

Geoffrey I. Sunahara
Biotechnology Research Institute
National Research Council of Canada
Montreal, Québec, Canada

Sonia Thiboutot
Defence Research and Development
  Canada–Valcartier
National Defence Canada
Val-Bélair, Québec, Québec, Canada

Jian-Shen Zhao
Biotechnology Research Institute
National Research Council of Canada
Montreal, Québec, Canada

Abbreviations

[^{14}C]DDT	[^{14}C]-dichlorodiphenyltrichloroethane
[^{14}C]HMX	[^{14}C]-octahydro-1,3,5,7-tetranitro-1,3,5,7-tetrazocine
[^{14}C]RDX	[^{14}C]-hexahydro-1,3,5-trinitro-1,3,5-triazine
1,2-DNB	1,2-dinitrobenzene (o-DNB)
1,3,5-TNB	1,3,5-trinitrobenzene (TNB)
1,3-DNB	1,3-dinitrobenzene (m-DNB)
1,4-DNB	1,4-dinitrobenzene (p-DNB)
2,2'-AZT	4,4',6,6'-tetranitro-2,2'-azoxytoluene isomer
2,3,6-TNT	2,3,6-trinitrotoluene
2,3-DNT	2,3-dinitrotoluene
2,4,6-TNT	2,4,6-trinitrotoluene (TNT)
2,4'-AZT	2',4,6,6'-tetranitro-2,4'-azoxytoluene
2,4-DANT	2,4-diamino-6-nitrotoluene (2,4-diaminonitrotoluene)
2,4-DAP	2,4-diaminophenol
2,4-DAT	2,4-diaminotoluene
2,4-DNA	2,4-dinitiroaniline
2,4-DNP	2,4-dinitrophenol
2,4-DNT	2,4-dinitrotoluene
2,5-DNP	2,5-dinitrophenol
2,5-DNT	2,5-dinitrotoluene
2,6-DANT	2,6-diamino-4-nitrotoluene (2,6-diaminonitrotoluene)
2,6-DAT	2,6-diaminotoluene
2,6-DNP	2,6-dinitrophenol
2,6-DNT	2,6-dinitrotoluene
2',4-AZT	2,4',6,6'-tetranitro-2',4-azoxytoluene
2-A-3,6-DNT	2-amino-3,6-dinitrotoluene
2-A-3-NT	2-amino-3-nitrotoluene
2-A-4,6-DNT	2-amino-4,6-dinitrotoluene (2-ADNT)
2-A-4-NT	2-amino-4-nitrotoluene
2-A-5-NT	2-amino-5-nitrotoluene
2-A-6-NT	2-amino-6-nitrotoluene
2-ADNT	2-amino-4,6-dinitrotoluene (2-A-4,6-DNT)
2-NP	2-nitrophenol (o-NP)
2-NT	2-nitrotoluene (o-NT)
3,4-DNT	3,4-dinitrotoluene
3,5-DNA	3,5-dinitroaniline
3,5-DNT	3,5-dinitrotoluene
3-A-2,4-DNT	3-amino-2,4-dinitrotoluene
3-A-2,6-DNT	3-amino-2,6-dinitrotoluene
3-A-4-NT	3-amino-4-nitrotoluene
3-AP	3-aminophenol (m-AP)
3-NP	3-nitrophenol (m-NP)

3-NT	3-nitrotoluene (m-NT)
4,4'-AZT	2,2',6,6'-tetranitro-4,4'-azoxytoluene
4-A-2,6-DNT	4-amino-2,6-dinitrotoluene (4-ADNT)
4-A-2-NP	4-amino-2-nitrophenol
4-A-2-NT	4-amino-2-nitrotoluene
4-A-3,5-DNT	4-amino-3,5-dinitrotoluene
4-A-3-NT	4-amino-3-nitrotoluene
4-ADNT	4-amino-2,6-dinitrotoluene (4-A-2,6-DNT)
4-AP	4-aminophenol (p-AP)
4-NP	4-nitrophenol (p-NP)
4-NT	4-nitrotoluene (p-NT)
5-A-2,4-DNT	5-amino-2,4-dinitrotoluene
8-oxo-dG	8-oxo-7,8-dihydro-2'-deoxyguanosine
ADNT	aminodinitrotoluene
ADR	adrenodoxin reductase
ADX	adrenodoxin
AFLP	amplified fragment length polymorphism
ANTA	3-amino-5-nitro-1,2,4-triazole
ARAR	applicable or relevant and appropriate requirement
$ArNO_2$	nitroaromatics
$ArNO_2^{-\bullet}$	nitroaromatic anion-radical
ATCC	American Type Culture Collection
ATSDR	Agency for Toxic Substance Disease Registry
BAF	bioaccumulation factor
BAP	benzo[a]pyrene, (B[a]P)
BCF	bioconcentration factor
BCNU	1,3-bis-(2-chloroethyl)-1-nitrosourea
BD/DR	building demolition and debris removal
BERA	baseline ecological risk assessment
BR	basal respiration
CCME	Canadian Council of Ministers of the Environment
cDNA	complementary DNA
CERCLA	Comprehensive Environmental Response, Compensation, and Liability Act of 1980
CHO	Chinese hamster ovary
CL-14	4,6-diamino-5,7-dinitrobenzofuroxan, 5,7-diamino-4,6-dinitrobenzofurazane-1-oxide, diaminodinitrobenzofuroxan
CL-20	2,4,6,8,10,12-hexanitro-2,4,6,8,10,12-hexaazaisowurtzitane
COPC	contaminants of potential concern
C_{org}	organic carbon
CRE	Comprehensive Range Evaluations
DANT	diaminonitrotoluene
DAT	diaminotoluene
DEGDN	diethylene glycol dinitrate
DHA	dehydrogenase activity
DNT	dinitrotoluene
DNX	hexahydro-5-nitro-1,3-dinitroso-1,3,5-triazine
DoD	Department of Defense

DODI	Department of Defense Instruction
DPPD	N,N′-diphenyl-p-phenylene diamine
DQO	EPA Data Quality Objective
DU	depleted uranium
EC	effective concentration
Eco-SSL	ecological soil screening level
EDYS	ecological dynamics simulation modeling
EGDN	ethylene glycol dinitrate
EM	energetic material
EMH	extramedullary hematopoiesis
EOD	explosive ordnance disposal
EPA	Environmental Protection Agency
ERA	ecological risk assessment
ERAGS	Ecological Risk Assessment Guidance for Superfund
EROD	ethoxyresorufin-O-deethylase
ESI	Environmental Sustainability Index
EST	expressed sequence tag
ETV	ecotoxicological tolerance value
FAD	flavin adenine dinucleotide
FMN	flavin mononucleotide
FUDS	Formerly Used Defense Sites
GIS	geographic information system
GSH	reduced glutathione
GSSG	oxidized glutathione
GST	glutathione S-transferase
H_2O_2	hydrogen peroxide
Hexyl	2,2′,4,4′,6,6′-hexanitrodiphenylamine
HMX	octahydro-1,3,5,7-tetranitro-1,3,5,7-tetrazocine (high melting explosive)
HPLC	high performance liquid chromatography
HTRW	hazardous toxic radioactive waste
IRIS	USEPA Integrated Risk Information System
ISO	International Organization for Standardization
k_{cat}/K_m	bimolecular rate constants
k_e	elimination rate constant
K_{ow}	n-octanol/water partition coefficient
KTA	Key Technical Area
LC	lethal concentration
LC/MS	liquid chromatography–mass spectrometry
LC/MS (ES−)	liquid chromatography–electrospray (negative) ionization mass spectrometry
LC_{50}	median lethal concentration
LCTA	Land Condition Trend Analysis
LD_{50}	median lethal dose
LOAEC	lowest observed (observable) adverse effect concentration
LOEC	lowest observed (observable) concentration
LOEL	lowest observed (observable) effect level
$\log K_{ow}$ (or log P)	logarithm of the octanol/water partition coefficient

MC	munition constituents
MCN	micronucleus
MCOC	munition constituents of concern
MEDINA	methylenedinitramine
MIM	maneuver impact miles
MMRP	Military Munitions Response Program
MNX	hexahydro-3,5-dinitro-1-nitroso-1,3,5-triazine
MOU	memorandum of understanding
NA	nitroaniline
NAC	nitroaromatic compound
NADP	nicotinamide adenine dinucleotide phosphate
NB	nitrobenzene
NC	nitrocellulose
NDAB	4-nitro-2,4-diazabutanal
NFA	nitrogen fixation activity
NG	nitroglycerin(e)
nNOS	neuronal nitric oxide synthase
NOAEC	no observed adverse effect concentration
NOAEL	no observed adverse effect level
NOCs	nitro organic compounds
NOEC	no observed effect concentration
NOEL	no observed effect level
NQ	nitroguanidine
NTO	5-nitro-1,2,4-triazol-3-one
$O_2^{-\bullet}$	superoxide anion
OC	organic carbon
$OH^\bullet$	hydroxyl radical
OM	organic matter
ORAP	Operational Range Assessment Program
ORC	ordnance related compound
PA	2,4,6-trinitrophenol or picric acid
PAT	production or lot acceptance testing
PCR	polymerase chain reaction
pentryl	2,4,6-trinitrophenyl-N-nitramino-ethylnitrate
PETN	pentaerythritoltetranitrate
PICT	pollution induced community tolerance
PLFA	phospholipid fatty acid
PNA	potential nitrification activity
QAPP	quality assurance project plan
QSAR	quantitative structure–activity relationship
RAPD	random amplified polymorphic DNA
RCA	range condition assessments
RDX	hexahydro-1,3,5-trinitro-1,3,5-triazine (royal demolition explosive)
REVA	range environmental vulnerability assessment
RFLP	restriction fragment length polymorphism
RI/FS	remedial investigation and feasibility study
Rsim	regional simulator

SAGE	serial analysis of gene expression
SARA	Superfund Amendments and Reauthorization Act of 1986
SARs	small-arms ranges
SEEM	spatially explicit exposure model
SERDP	Strategic Environmental Research and Development Program
SIR	substrate-induced respiration
SLERA	screening level ecological risk assessment
SMDP	scientific management decision point
SPME	solid phase microextraction
SQGe	environmental soil quality guideline
SRO	sustainable range oversight
SSL	Sassafras sandy loam
SSR	simple sequence repeats
T&E	threatened and endangered
TAT	2,4,6-triaminotoluene
tetryl	2,4,6-trinitrophenyl-N-methylnitramine (trinitrophenylmethylnitramine)
TNB	1,3,5-trinitrobenzene (1,3,5-TNB)
TNBO	4,5,6,7-tetranitrobenzimidazol-2-one
TNC	1,3,6,8-tetranitrocarbazole
TNT	2,4,6-trinitrotoluene (2,4,6-TNT)
TNT⁻	2,4,6-trinitrobenzyl anion
TNX	hexahydro-1,3,5-trinitroso-1,3,5-triazine
Trad-MCN	Tradescantia micronucleus bioassay
TRV	toxicity reference value
Trx	thioredoxin
TTCP	The Technical Cooperation Program
UDS	unscheduled DNA synthesis assay
UFP	uniform federal policy
USACE	U.S. Army Corps of Engineers
USACHPPM	U.S. Army Center for Health Promotion and Preventive Medicine
USAF	U.S. Air Force
USEPA	U.S. Environmental Protection Agency (EPA)
UV	ultraviolet
UXO	unexploded ordnance
WCL	Webster clay loam
YPG	Yuma Proving Ground

1 Introduction

Geoffrey I. Sunahara, Roman G. Kuperman,
Guilherme R. Lotufo, Jalal Hawari,
Sonia Thiboutot, and Guy Ampleman

Managing sites contaminated with munitions constituents has become an international problem shared by many countries, but few have the technical expertise or have developed the methodology for the effective characterization of environmental impacts from site contamination with explosives and related energetic materials (EMs) [1]. Some countries, including Canada and the United States, are developing and testing new munition constituents that are designed to decrease environmental damage of military activities. These and other activities, including military base closures and demilitarization, rely on the use of ecotoxicological data for munitions constituents. The greatly increased need for such ecotoxicological data provided an impetus for extensive investigations of EM environmental impacts in recent years.

Explosives and related EMs still remain in the environment as a legacy of past munitions production and disposal, as well as testing and training practices. These activities cause different levels of contamination that can be toxic to ecological receptors that inhabit the impacted sites and surrounding areas exposed to the offsite migration of contaminants. Certain issues related to the release of EMs in the environment were reviewed in earlier publications [2–7]. Unexploded ordnance (UXO) at terrestrial and aquatic sites also presents an international environmental problem due to the release of EMs from the corroding ordnance, in addition to the risks associated with the potential for an accidental detonation. Various European countries are also dealing with legacy EM-contaminated sites following the wars of the 20th century [8–10], and sea dumping or land burying of many obsolete munitions [11]. The Technical Cooperation Program (TTCP) collaborates defense scientific and technical information exchange and shared research activities among Australia, Canada, New Zealand, the United Kingdom, and the United States. The TTCP countries have agreed on the need for ecotoxicological research on EMs, and that this information be made available to the environmental professionals, risk managers, and owners of EM-contaminated sites [1]. This book was produced as part of the activities of the Key Technical Area (KTA 4-32) "Development of Environmental Tolerance Values for Defense Sites Contaminated with Energetic Materials" of The Technical Cooperation Program (www.dtic.mil/ttcp/), Weapons Technical Panel (WPN TP-4). A group of scientists involved in this KTA have been contributing to the four-year effort of reviewing available ecotoxicological data for EMs and methodologies for their development, and establishing new ecotoxicological benchmarks for explosives and related EMs when data gaps were identified. The key findings of KTA 4-32 are included in several chapters of this book.

Many EM-contaminated sites contain explosives and propellant materials in soil, sediment, or groundwater at concentrations that span several orders of magnitude [1,12,13]. These EMs can have direct or indirect adverse effects on human health and the environment. Although some EMs are persistent and can be highly mobile in the environment, their effects on the biota at environmentally relevant concentrations have not been sufficiently investigated. This knowledge gap presents a challenge for site managers who have to distinguish those sites that pose significant environmental risks from those that do not, prioritize contaminated sites by their respective level of risk, quantify the risks, and develop appropriate remedial actions and cleanup goals. Therefore, the main objectives of this book are to (a) evaluate the peer-reviewed literature on environmental fate and ecotoxicological effects of explosives, propellants, and related EMs; and (b) provide the different stakeholders of EM-contaminated sites with a compilation of critically reviewed information that can enable them to better evaluate the strengths and limitations of the ecotoxicological data used for environmental risk assessment and management of these sites.

The readers of this book will discover that the ecotoxicology of explosives is a multidisciplinary environmental science that involves the contributions from analytical chemists, biochemists, biologists with expertise at different organizational scales ranging from molecular to ecosystem levels, environmental toxicologists, ecotoxicologists, soil scientists, and ecological risk assessors. The readers will be introduced to the state of the science of the ecotoxicology of explosives and propellants, including those that were used in the past, are currently in use, or have been developed recently. We have invited the experts in the field to present information on the fate and transport of these chemicals in the environment, their toxicological effects on different species, as well as some ecological issues associated with EM-contaminated sites. We have encouraged our chapter authors to critically examine the peer-reviewed literature, to identify and prioritize the knowledge gaps, and to recommend future areas of research. We hope that the information provided in this book will stimulate further research to develop and apply scientifically based ecotoxicological data that will advance the assessments and protection of military testing and training ranges, and will ensure their management as sustainable resources.

REFERENCES

1. The Technical Cooperation Program–Key Technical Area 4-28 (TTCP-KTA 4-28), *Characterization of Sites Contaminated with Energetic Materials (EM)*, 2005, http://www.em-guidelines.org/index.html#overview (accessed June 2008).
2. Rickert DE, *Toxicity of Nitroaromatic Compounds*, Hemisphere Publishing, New York, 1982.
3. Yinon J, *Toxicity and Metabolism of Explosives*, CRC Press, Boca Raton, FL, 1990.
4. Talmage SS et al., Nitroaromatic munition compounds: Environmental effects and screening values, *Rev. Environ. Contam. Toxicol.*, 161, 1, 1999.

5. Sunahara GI et al., Laboratory and field approaches to characterize the soil ecotoxicology of polynitro explosives, in *Environmental Toxicology and Risk Assessment: Science, Policy and Standardization–Implications for Environmental Decisions: Tenth Volume STP 140*, Greenberg BM, Hull RN, Roberts Jr. MH, Gensemer RW, Eds., American Society for Testing and Materials, West Conshohocken, PA, 2001, 293.

6. Pennington JC and Brannon JM, Environmental fate of explosives, *Thermochimica Act*, 384, 163, 2002.

7. Juhasz AL and Naidu R, Explosives: Fate, dynamics, and ecological impact in terrestrial and marine environments, *Rev. Environ. Contam. Toxicol.*, 191, 163, 2007.

8. Spain JC, Hughes JB, and Knackmuss H-J, Eds., *Biodegradation of Nitroaromatic Compounds and Explosives*, CRC Press, Boca Raton, FL, 2000.

9. Michels J, Ed., *Joint Research Group Processes for the Bioremediation of Soil Compilation of Current Projects of the Joint Research Group*, German Ministry of Education, Science, Research and Technology, 1998, http://www.epa.gov/tio/download/remed/procbior.pdf (accessed June 2008).

10. Michel J, Gehrke U, and Track T, German joint research group: Processes for the bioremediation of soil, in *Proceedings of the Seventh International FZK/TNO Conference on Contaminated Soil*, CONSOIL, Ed., Leipzig, Germany, 2000, 1101.

11. Castelo Branco P, Schubert H, and Campos J, Eds., *Defense Industries: Science and Technology Related to Security: Impact of Conventional Munitions on Environment and Population*. NATO Science Series: IV: Earth and Environmental Sciences Series, Vol. 44, 2004.

12. Jenkins TF et al., Identity and distribution of residues of energetic compounds at army live-fire training ranges, *Chemosphere*, 63, 1280, 2006.

13. Simini M et al., Evaluation of soil toxicity at Joliet army ammunition plant, *Environ. Toxicol. Chem.*, 14, 623, 1995.

2 Fate and Transport of Explosives in the Environment
A Chemist's View

Fanny Monteil-Rivera, Annamaria Halasz,
Carl Groom, Jian-Shen Zhao, Sonia Thiboutot,
Guy Ampleman, and Jalal Hawari

CONTENTS

2.1   Introduction 5
2.2   Transformation Pathways...8
    2.2.1   Abiotic Transformations 8
        2.2.1.1   Nitroaromatic Compounds..8
        2.2.1.2   Cyclic Nitramines 11
2.3   Biotic Transformations 13
    2.3.1   Nitroaromatic Compounds 13
        2.3.1.1   TNT Biotransformation 13
        2.3.1.2   DNT Biotransformation 18
        2.1.3.3   Picric Acid Biotransformation 18
    2.3.2   Cyclic Nitramines...18
        2.3.2.1   Anaerobic Degradation 18
        2.3.2.2   Aerobic Degradation...20
2.4   Sorption Processes and Bioavailablity..21
    2.4.1   Nitroaromatic Compounds 22
    2.4.2   Cyclic Nitramines 24
2.5.   Conclusions and Future Outlook 25
References...25

2.1   INTRODUCTION

Explosives are highly energetic chemicals that rapidly release large amounts of gaseous products and energy upon detonation. The most frequently manufactured and used energetic chemicals include 2,4,6-trinitrotoluene (TNT), dinitrotoluenes (DNTs), 1,3,5-trinitrobenzene (TNB), 2,4,6-trinitrophenylmethylnitramine (tetryl), 2,4,6-trinitrophenol (picric acid or PA), nitrocellulose (NC), pentaerythritoltetranitrate (PETN), nitroglycerine (NG), ethyleneglycoldinitrate (EGDN), hexahydro-1,3,5-trinitro-1,3,5-triazine (RDX), and octahydro-1,3,5,7-tetranitro-1,3,5,7-tetrazocine (HMX) (Table 2.1). The polycyclic nitramine, 2,4,6,8,10,12-hexanitro-2,4,6,8,10,12-hexaazaisowurtzitane, or CL-20, which was first synthesized in the late 1980s [1], is presently considered as a potential replacement for RDX and HMX [2]. Table 2.2 summarizes the physicochemical properties of the most often encountered energetic materials [3–9].

Manufacturing, training, open burning or detonation (OB/OD), and improper disposal in landfills and sea dumps have led to the contamination of both terrestrial and marine environments [10–13]. In the United States more than 2000 sites have been identified as potentially contaminated by energetic chemicals [14], and in Canada approximately 100 OB/OD and training sites are known to be associated with activities involving RDX, HMX, and TNT [15]. The degree of contamination at these sites is extremely varied, and the spatial contamination distribution is often highly heterogeneous [13,16,17].

For most of the world, reliable estimates of the extent of environmental contamination by energetic chemicals are not available. Presently, several U.S. and Canadian federal and private agencies support and conduct research on site characterization to determine the fate and environmental impact of energetic chemicals (e.g., Strategic Environmental Research and Development Program (SERDP), Arlington, Virgina [18]; U.S. Army Environmental Center, Aberdeen Proving Ground, Maryland [19]). The Canadian Department of Defense established an Energetic Materials Working Group in 1996 to help limit the effects of explosives on human health and the environment, and to supervise the research and development dedicated to the environmental impact of munitions [20]. In Germany, munition manufacturing sites that were demolished immediately after World War II are being characterized and remediated [11]. Similar programs exist in the United Kingdom and Australia.

Several literature reports and reviews are available that describe contamination, distribution, and fate of explosives in the environment [21,22]. The present review will specifically analyze the (bio)transformation pathways and availability of explosives to help understand what causes their toxicity. We will focus our discussion on the most frequently encountered energetic chemicals, including TNT, DNTs, aminodinitrotoluenes (ADNTs), diaminonitrotoluenes (DANTs); and cyclic nitramines, including RDX, HMX, and the emerging chemical CL-20. All of these nitro organic compounds (NOCs) possess the characteristic labile X-NO$_2$ functional groups (with X = C or N), which play a key role in determining their transport and transformation routes, and thereby their environmental impact. We will first analyze recent research reports on various chemical, microbial, and enzymatic transformation routes of the NOCs aforementioned. Then, the sorption of NOCs and their degradation products to soils will be reviewed and discussed. Knowing the transformation pathways and physical transport of explosives will help one understand the toxicity associated with these chemicals.

TABLE 2.1

Military Explosives Classification

Molecular Groups	Common Name	Chemical Name	Structure	Chemical Formula
Nitroaromatic compounds	TNT	2,4,6-trinitro-toluene		$C_7H_5N_3O_6$
	2,4-DNT	2,4-dinitrotoluene		$C_7H_6N_2O_4$
	Tetryl	2,4,6-trinitro-phenylmethyl-nitramine		$C_7H_5N_5O_8$
	TATB	1,3,5-triamino-trinitrobenzene		$C_6H_6N_6O_6$
	Picric Acid (PA)	2,4,6-trinitro-phenol		$C_6H_3N_3O_7$
Nitrate esters	NC	Nitrocellulose		$C_6H_7O_2(OH)_x(ONO_2)_y$ $x + y = 3$
	PETN	Pentaerythritol tetranitrate		$C_5H_8N_4O_{12}$
	NG	Nitroglycerin		$C_3H_5N_3O_9$
	EGDN	Ethyleneglycol dinitrate		$C_2H_4N_2O_6$

TABLE 2.1 (continued)
Military Explosives Classification

Nitramines	RDX	Hexahydro-1,3,5-trinitro-1,3,5-triazine		$C_3H_6N_6O_6$
	HMX	Octahydro-1,3,5,7-tetranitro-1,3,5,7-tetrazocine		$C_4H_8N_8O_8$
	CL-20	2,4,6,8,10,12-hexanitro-2,4,6,8,10,12-hexaazaisowurtzitane		$C_6H_6N_{12}O_{12}$
	NGu	Nitroguanidine		$CH_4N_4O_2$
Azides	GAP	Glycidyl azide polymer		$C_3H_5N_3O$

2.2 TRANSFORMATION PATHWAYS

2.2.1 ABIOTIC TRANSFORMATIONS

2.2.1.1 Nitroaromatic Compounds

Figure 2.1 summarizes various abiotic reactions that TNT, a typical nitroaromatic compound (NAC), may encounter in natural media [23–26].

2.2.1.1.1 Alkaline Hydrolysis

Nitroaromatic compounds (NACs) contain electron-withdrawing nitro groups ($-NO_2$) that reduce the electron density of the aromatic ring, thus favoring nucleophilic attack by bases such as OH^-. As a consequence, NACs are known to be unstable in alkaline media [27,28]. For instance, TNT [29–31], 2,4-dinitrotoluene (2,4-DNT), 2-amino-4,6-dinitrotoluene (2-ADNT), and 4-amino-2,6-dinitrotoluene (4-ADNT) [32] have been demonstrated to degrade in water or soils at pH > 11. At pH 12 and room temperature, nitroaromatics degraded in soil with rates decreasing in the following order: TNT ($t_{1/2}$ = 2.9 h) > 2-ADNT ($t_{1/2}$ = 6.2 h) ≈ 4-ADNT ($t_{1/2}$ = 6.5 h) > 2,4-DNT ($t_{1/2}$ = 8.4 h) [32]. The disappearance of TNT under alkaline conditions occurred via a rapid initial step in which a highly colored transient species (λ_{max} = 450 nm), identified as a Meisenheimer complex (σ-anionic), was reversibly formed (Figure 2.1)

TABLE 2.2

Most Commonly Used Physicochemical Properties of Some of the Explosives Needed for Environmental Risk Assessment

Common Name	Molecular Weight (g mol⁻¹)	Melting Point (°C)	Water Solubility at 25°C (mg L⁻¹)	Octanol/Water Partition Coefficient (log K_{ow})	Henry's Law Constant at 25°C (atm m³ mol⁻¹)	Vapor Pressure at 25°C (mm Hg)
TNT	227.13	80.1	130 [3]	1.6 [4]	4.57×10^{-7} [4]ᵃ	1.99×10^{-4} [4]ᵃ
2,4-DNT	182.15	71	270 [4]	1.98 [4]	1.86×10^{-7} [5]	1.47×10^{-4} [4]
Tetryl	287.17	129.5	75 [3]	2.04 [3]	2.69×10^{-11} [5]	5.69×10^{-9} [5]
TATB	258.15	ND	32 [3]	0.7 [3]	5.8×10^{-12} [3]	1.34×10^{-11} [3]
Picric Acid	229.10	121.8	12800 [4]	1.33 [4]	1.7×10^{-8} [4]	7.5×10^{-7} [4]
NC	10⁵–10⁶ [6]	206 [7]ᵇ	Insoluble [5]	NDᶜ	ND	ND
PETN	316.17	143.3	43 [4]	3.71 [5]	1.7×10^{-9} [5]	5.38×10^{-9} [5]
NG	227.11	13.5	1800 [4]	1.62 [4]	3.4×10^{-6} [3]ᵃ	2×10^{-4} [4]
EGDN	152.08	−22.3	5200 [4]	1.16 [4]	2.52×10^{-6} [5]	7.2×10^{-2} [4]
RDX	222.26	205	56.3 [8]	0.90 [8]	1.96×10^{-11} [5]	4.0×10^{-9} [5]
HMX	296.16	286	4.5 [8]	0.17 [8]	2.60×10^{-15} [5]	3.3×10^{-14} [5]
CL-20	438.19	260ᵇ	3.7 [8]	1.92 [8]	ND	ND
NGu	104.07	239	4400 [4]	−0.89 [4]	4.67×10^{-16} [5]	1.43×10^{-11} [5]
GAP	1700–2300 [9]	ND	ND	ND	ND	ND

Note: ND = Not determined.

ᵃ At 20°C.

ᵇ With decomposition.

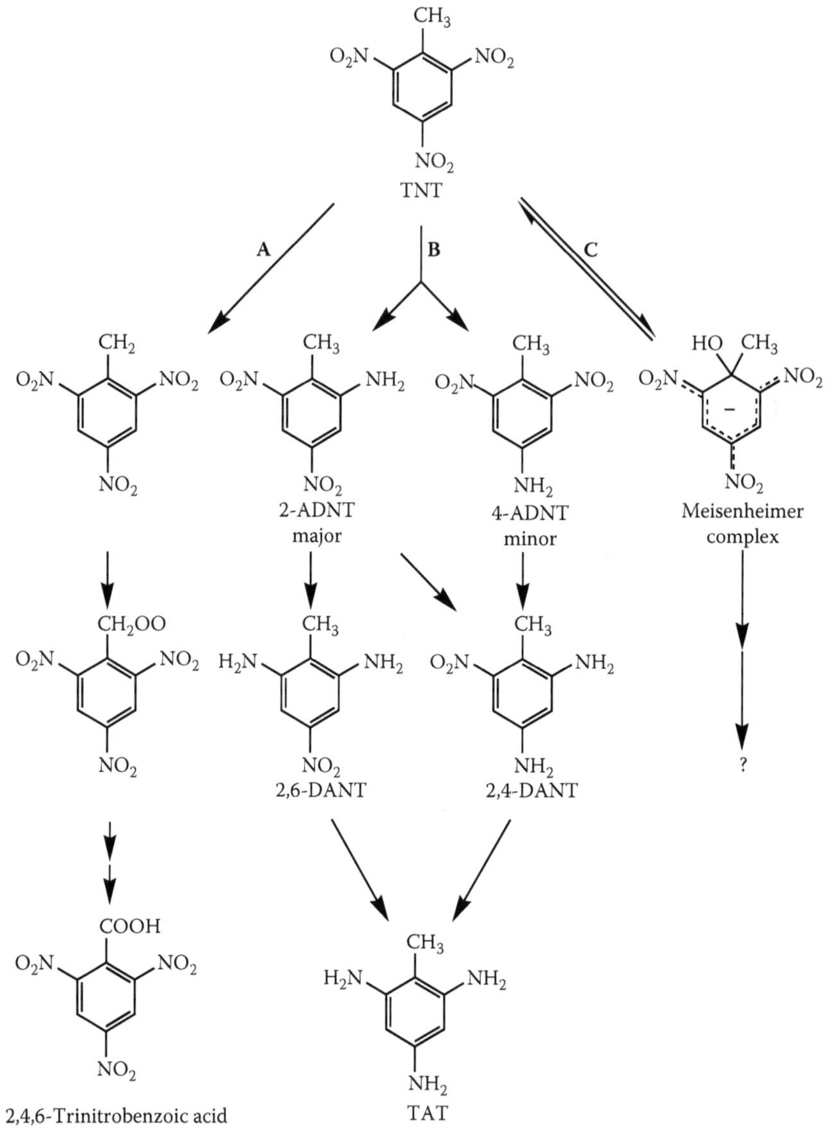

FIGURE 2.1 Abiotic transformation routes of TNT: (A) under photolytic conditions [23]; (B) in the presence of Fe(0) [24] or Fe(II) [25]; (C) under alkaline conditions [26].

[26,33]. This colored complex was used to develop colorimetric tests for many nitro aromatic compounds. Whether the Meisenheimer complex or 2,4,6-trinitrobenzyl anion (TNT⁻) resulting from proton abstraction reacted further with TNT (or with excess hydroxide) to form larger species is still not clear, but alkaline hydrolysis of TNT in aqueous media led to large molecular weight compounds (60% > 30 kDa [28]; 40% > 1 kDa [30]).

2.2.1.1.2 Photolysis

In addition to hydrolysis, photolysis is recognized as another environmentally important mechanism for the removal (attenuation) of organic pollutants in natural surface waters [34]. Several approaches including solar photolysis [35–39], UV (ultraviolet) photolysis [35], UV peroxide [40], UV ozone, and TiO_2 photocatalysis [23,41] have been employed to treat TNT-contaminated water. The rate of photolysis was pH-dependent, increasing threefold as pH increased from 4 to 8 [36]. In most cases, TNT photolysis yields a variety of products including 4-ADNT, 3,5-dinitroaniline, 2,4,6-trinitrophenol, 2,4,6-trinitrobenzyl alcohol, 2,4,6-trinitrobenzoic acid, and 1,3,5-trinitrobenzene (TNB), among which 2,4,6-trinitrobenzoic acid was the most often encountered [23]. Although the aromatic ring normally remains intact, ring cleavage and high mineralization rates could be achieved in the presence of a photocatalyst, such as TiO_2 [23].

2.2.1.1.3 Reduction by Iron

The presence of metals in soil can also influence the fate and environmental impact of NACs. Zero valent iron, Fe(0), reduces TNT [24,42,43], 1,3,5-TNB [44,45], and 2,4-DNT [46] to partially or completely reduced amines. Furthermore, Fe(II) at the surface of Fe(III) (hydr)oxides reduces TNT and other NACs to corresponding aromatic polyamines [25,47]. In the case of Fe(0), using high-purity iron allowed complete reduction of TNT to 2,4,6-triaminotoluene (TAT), while scrap iron gave only partial reduction [24]. Both Fe(0) and Fe(II) led to the predominant formation of 2-ADNT over 4-ADNT.

2.2.1.2 Cyclic Nitramines

The degradation of RDX under a variety of abiotic conditions (hydrolytic, photolytic, and reductive) is summarized in Figure 2.2 [42, 48–50].

2.2.1.2.1 Alkaline Hydrolysis

Cyclic nitramine explosives such as RDX and HMX degrade under alkaline conditions [49,51–54]. At pH 12 and 50°C, HMX hydrolysis ($t_{1/2}$ = 2.5 h) occurred at a rate approximately 14 times slower than that of RDX ($t_{1/2}$ = 10.9 min) [53]. The disappearance of RDX and HMX was concurrent with the formation of nitrite and corresponding imino intermediates (Figure 2.2, path A) [49,51], and both nitramines gave closely related product distributions (NO_2^-, N_2O, NH_3, N_2, HCHO, and $HCOO^-$). Using LC/MS (liquid chromatography–mass spectrometry), Balakrishnan et al. [49] were able to directly detect the ring cleavage product 4-nitro-2,4-diazabutanal (NDAB), and thus provide experimental evidence of hydrolytic ring cleavage and decomposition (Figure 2.2, path A). CL-20 is also a heterocyclic nitramine, which like RDX and HMX, contains the characteristic $N-NO_2$ functional groups (Table 2.1). However, in contrast to RDX and HMX, which are two-dimensional monocyclic nitramines, CL-20 is a polycyclic nitramine characterized by a strained three-dimensional cage structure. This structural difference may cause significant variations in the degradation pathways of CL-20 from those reported for RDX and HMX [55]. When challenged with alkaline conditions (pH 10), CL-20 degraded 1.5 times faster than RDX,

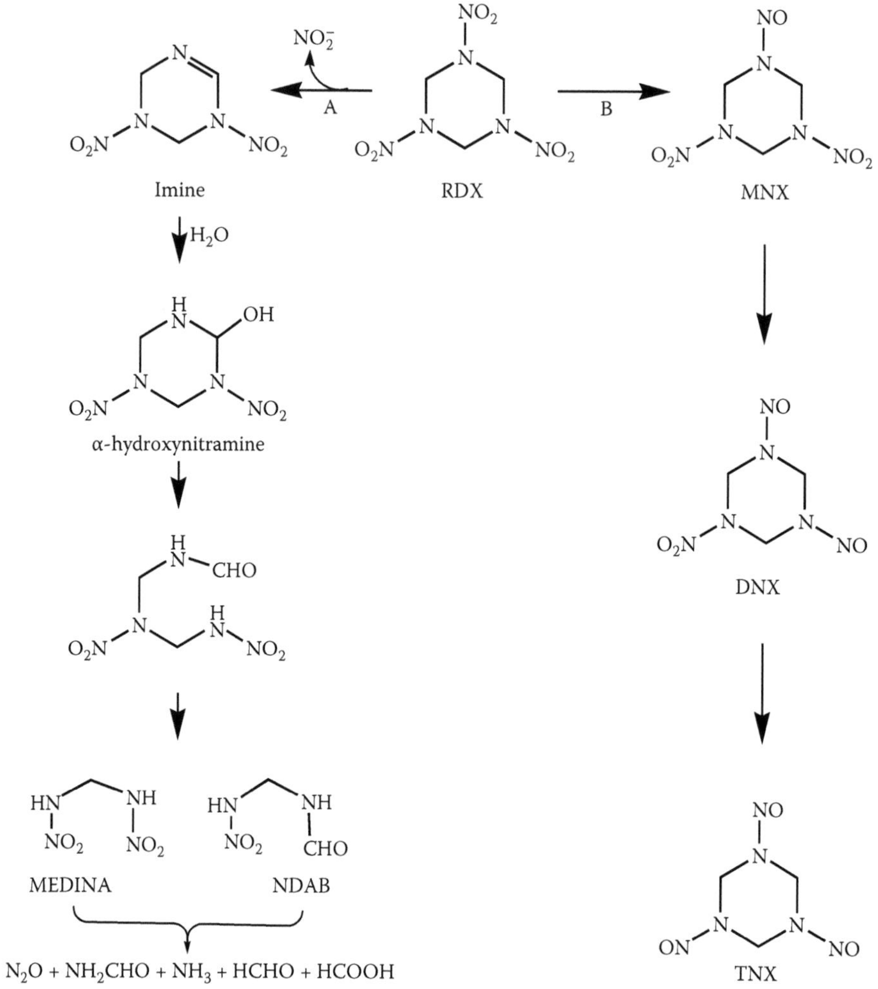

FIGURE 2.2 Abiotic degradation pathways of RDX: (A) initiated by denitration (hydrolysis, photolysis, electrolysis, Fe(II), zero valent metals), or (B) initiated by reduction to the corresponding nitroso derivatives (2e-transfer) (photolysis, electrolysis, zero valent metals). (Data from Refs. 42, 48–50.)

and led to the formation of NO_2^-, N_2O, NH_3, and $HCOO^-$, with no formaldehyde detected [49].

2.2.1.2.2 Photolysis

Photolysis is another abiotic process that could contribute to the attenuation of nitramines in natural media. Because HMX is more resistant to degradation than RDX [49,52,53], very few reports pertain to HMX photolysis. However, several reports regarding the photodegradation of RDX [48,56–58] suggested that photolysis was accomplished through the initial homolysis of the N-NO$_2$ bond. Such a bond scission ultimately yields products resembling those found in hydrolysis (HCHO, HCOOH, NH_3, NO_2^-, and N_2O). In addition, Hawari et al. [48] identified two key intermediates, NDAB and methylenedinitramine (MEDINA, $CH_2(NHNO_2)_2$), during the photolysis of RDX. The latter degrades spontaneously in water to form HCHO and N_2O [59]. Recently, the rigid molecule CL-20 was also photolyzed and produced NO_2^-, NO_3^-, NH_3, HCOOH, N_2, and N_2O, thereby suggesting a complete degradation of the chemical in water [60]. Several initial intermediates (A and B in Figure 2.3) involved in the degradation of CL-20 were detected, and tentatively identified using uniformly amino labeled [15]N-[CL-20] and LC/MS (ES–) (liquid chromatography–electrospray (negative) ionization mass spectrometry) [60–62].

2.2.1.2.3 Reduction by Iron

Like TNT, nitramines can be abiotically degraded in the presence of reduced iron. Fe(0) was reported to degrade RDX to produce NH_4^+ and unidentified water-soluble compounds [24,42,63]. The nitroso byproducts—hexahydro-3,5-dinitro-1-nitroso-1,3,5-triazine (MNX), hexahydro-5-nitro-1,3-dinitroso-1,3,5-triazine (DNX), and hexahydro-1,3,5-trinitroso-1,3,5-triazine (TNX; Figure 2.2, path B)—were usually detected within the first hours of reaction but never accumulated [63,64]. As the amount of TNX always remained small [63,64], it was suggested that nitro moiety reduction was not the sole pathway of RDX removal and that a second pathway involving initial denitration to give nitrite was occurring (Figure 2.2, path A). Using Fe(II) bound to magnetite to degrade RDX, Gregory et al. [50] recently observed similar reaction patterns and identified NH_4^+, N_2O, and HCHO as stable products with the transient formation of MNX, DNX, and TNX. In the presence of Fe(0), CL-20 underwent rapid denitration (losing two equivalents NO_2^-) to ultimately produce N_2O, ammonium, formate, glyoxal (H(O)C-C(O)H), and glycolic acid (CH_2OH-COOH) [61]. LC/MS (ES–) measurements of Fe(0)-treated [15]N-[CL-20] allowed identification of the denitrated imine intermediates detected earlier during the photolysis of CL-20 (Compounds A and B in Figure 2.3).

2.3 BIOTIC TRANSFORMATIONS

2.3.1 NITROAROMATIC COMPOUNDS

Biodegradation of nitroaromatics has been reviewed by several groups [15,21,65–68]. The presence of nitro ($-NO_2$) electron withdrawing groups on the aromatic ring protects nitroaromatics from initial attack by oxygenases and favors reduction. The lower number of $-NO_2$ groups in DNTs compared to TNT greatly affects the reactivity of these compounds under aerobic and anaerobic conditions.

2.3.1.1 TNT Biotransformation

Although TNT can be transformed under both aerobic and anaerobic conditions by bacteria and fungi (Table 2.3 [69–82] and Table 2.4 [83–88]), the presence of three nitro groups limits oxidative attack from aerobic organisms [89], and the reductive mechanism predominates. Reductive attack produces corresponding amino ($-NH_2$)

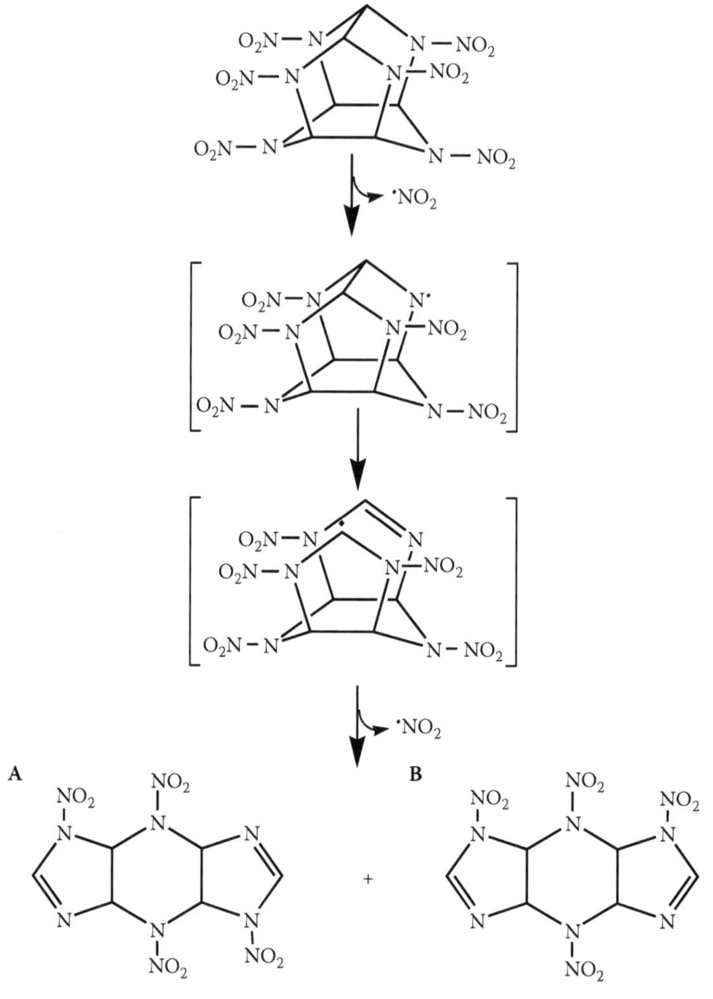

$$\mathbf{A}\ (\text{or}\ \mathbf{B}) \xrightarrow{H_2O} NO_2^- + N_2O + HCOOH + CHOCHO + NH_3$$

FIGURE 2.3 Proposed denitration routes of CL-20 by photolysis [60], zero valent iron [61], and salicylate 1-monooxygenase from *Pseudomonas* sp. strain ATCC 29352 [62], and degradation products of A (or B) intermediates.

electron donating groups [90,91]. TNT first reduces to produce ADNTs (preferentially 4-ADNT) [92,93], and DANTs (preferentially 2,4-DANT). Only under strict anaerobic conditions does the reduction continue to produce TAT [94,95] (Figure 2.4). The resulting amines are not further degraded by anaerobes, and hence persist in

TABLE 2.3
Aerobic Transformation of TNT by Bacteria and Their Major Products

Bacterial Strains	TNT (mg/L)	Degradation (%)	NO_2^-	Denitrated Products	Meisenheimer	ADNT	Reference
Bacillus sp.	100	95	+	2A4NT	ND	+	69
Enterobacter cloacae PB2	113	98	+	–	+	ND	70
Methylobacterium sp.	25	100	ND	ND	ND	+	71
Mycobacterium sp. HL 4NT-1	113	98	+	–	+	+	72
Mycobacterium vaccae JOB-5[b]	90	100	ND	ND	ND	+	73
Pseudomonas sp. clone A[b,c]	50	100	+	DNTs, 2-NT, T	+	+	74
Pseudomonas aeruginosa MA01[b,c]	100	98	ND	ND	ND	+	75
Pseudomonas aeruginosa	100	100	+	2A4NT	ND	+	69
Pseudomonas aeruginosa MX[c]	102	99	ND	ND	ND	+	76
Pseudomonas pseudoalcaligenes JS52	50	100	+	–	ND	+	77
Pseudomonas savastanoi	70	100	+	24DNT	ND	+	78
Pseudomonas fluorescens I-C[c,d]	ND	ND	+	–	+	+	79
Rhodococcus erythropolis HL PM-1			–	–	+	+	80
Serratia marcescens	50	100	ND	ND	ND	+	81
Streptomyces chromofuscus A11[b]	25	100	–	ND	ND	+	82
Staphylococcus sp.	100	98	+	2A4NT	ND	+	69

Note: +, produced; –, not produced; ND, no data; T, toluene; NT, nitrotoluene; DNT, dinitrotoluene; 2A4NT, 2-amino-4-nitrotoluene; ADNT, aminodinitrotoluene.

[a] Mineralization either not reported or <1%.
[b] Diaminonitrotoluene produced.
[c] Azoxynitrotoluene produced.
[d] Aminodimethyltetranitro-biphenyl produced.

TABLE 2.4
Biodegradation of TNT with Fungi

Fungal Strain	TNT (mg/l)	CO$_2$ (%)	Biotransformation (%)	Reference
Phanerochaete chrysosporium ATCC 1767	100.0	13.7	100	83
Phanerochaete chrysosporium BKM-F-1767	100.0	18.0	100	84
Phanerochaete chrysosporium BKM-F-1767	40.0	10.0	100	85
Phanerochaete chrysosporium BKM-F-1767	20.0	39.0	100	86
Phanerochaete chrysosporium	10.0	40.0	100	87
Fomes fomentarius MWF01-4	22.7	8.0	93	88
Trametes versicolor TM5	22.7	28.0	100	88
Agaricus eastivalis TMAest1	22.7	9.0	100	88
Agrocybe praecox TM70.3.1	22.7	14.5	100	88
Clitocybe odora TM3	22.7	5.1	88	88
Coprinus comatus TM6	22.7	0.8	82	88
Alternaria sp. TMRZ/WN2	22.7	0.4	74	88
Aspergillus terrus MWi458	22.7	0.2	100	88
Fusarium sp. TMS21	22.7	0.6	100	88
Mucor mucedo DSM810	22.7	0.1	95	88
Neurospora crassa TM	22.7	0.6	100	88
Penicillium frequentans ATCC96048	22.7	0.5	100	88
Rhizoctonia solani MWi5	22.7	0.2	90	88

the environment under anaerobic conditions. However, under aerobic conditions they can be oxidized or biotransformed [45,96]. The reduction to amines occurs through the formation of nitroso derivatives, ArNO (2-NO-DNT and 4-NO-DNT), and hydroxylamine derivatives, ArNHOH (2-HADNT and 4-HADNT; Figure 2.4). The latter are very reactive molecules that can undergo several secondary reactions such as rearrangement (Bamberger rearrangement) to phenol amines [97], formation of azoxy compounds [85,90,93,98], or irreversible binding to humic acids [89,92], and are thus rarely observed in natural systems [99].

Due to the presence of the stable aromatic ring, mineralization of TNT by bacteria is often very poor under both aerobic and anaerobic conditions. The highest mineralization rate (40%, measured as CO$_2$ liberated) was reported during TNT incubation with *Phanerochaete chrysosporium* in the presence of Tween 80 [86]. The reaction intermediates included ArNO, 2- and 4-HADNT, 2- and 4-ADNT, and 2,4- and 2,6-DANT, in addition to their formyl and acetyl derivatives [85]. Scheibner et al. [100] reported 52% mineralization of uniformly ring-labeled [^{14}C]-2-ADNT by manganese peroxidase (MnP) from the white-rot fungus *Nematoloma frowardii*, but TNT was not a suitable substrate for the MnP system. Further details on the role of fungi on the transformation of NACs can be found in Frische et al. [93].

In summary, ADNTs, DANTs, TAT, and ArNO are all products of TNT degradation in soil. Therefore, their toxicity and availability should be well defined to fully address the potential risk of TNT in soils and sediments.

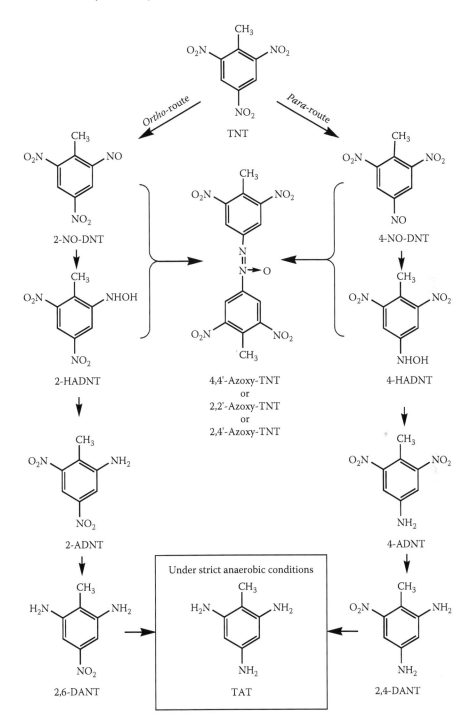

FIGURE 2.4 Biotransformation of TNT under aerobic and anaerobic conditions. TAT is only produced under strict anaerobic conditions.

2.3.1.2 DNT Biotransformation

Like TNT, DNTs can be reduced to the corresponding mono or diamino derivatives. Furthermore, with fewer electron withdrawing groups, DNTs are more susceptible than TNT to initial oxygenation under aerobic conditions. The oxidation reaction can occur: (a) directly on DNT with release of the nitro group as nitrite and production of hydroxyl groups necessary for ring cleavage [101–103]; (b) after partial or complete reduction of the nitro groups to hydroxylaminoaromatics or aminoaromatics; or (c) with formation of a hydride Meisenheimer complex [67].

2.1.3.3 Picric Acid Biotransformation

Like TNT, 2,4,6-trinitrophenol (picric acid, PA) is also degraded through reduction. The hydride Meisenheimer complex is recognized as a key intermediate of denitration for the microbial degradation of PA [104,105]. Lenke and Knackmuss [106] described an initial reduction of PA via the intermediate formation of a hydride–Meisenheimer complex to produce 2,4-dinitrophenol by the enzymatic elimination of nitrite. Under aerobic conditions, the mono- or dinitrophenol can then be oxygenated with subsequent ring cleavage.

2.3.2 CYCLIC NITRAMINES

Unlike NACs, cyclic nitramines undergo ring cleavage and mineralization under both aerobic and anaerobic conditions (Table 2.5) [107–131].

2.3.2.1 Anaerobic Degradation

McCormick et al. [132] found that a municipal sludge containing a consortium of bacteria degraded RDX (hexahydro-1,3,5-trinitro-1,3,5-triazine) to form the nitroso compounds MNX (hexahydro-1-nitroso-3,5-dinitro-1,3,5-triazine), DNX (hexahydro-1,3-dinitroso-5-nitro-1,3,5-triazine), and TNX (hexahydro-1,3,5-trinitroso-1,3,5-triazine) prior to ring cleavage to form methanol, hydrazine, and formaldehyde. Kitts et al. [110] also identified the nitroso byproducts of RDX and HMX with *Providencia rettgeri*, *Citrobacter freundii*, and *Morganella morganii* (soil isolates of the family Enterobacteriaceae) without identification of ring cleavage intermediates. Subsequent work by Hawari and co-workers revealed the identity of a key ring cleavage product, namely, methylenedinitramine (MEDINA), from both RDX and HMX, which allowed the elucidation of degradation routes for cyclic nitramines under anaerobic conditions [133,134]. Figure 2.5 and Figure 2.6 summarize the degradation pathways of RDX during incubation with *Klebsiella pneumoniae* (Figure 2.5, path B) and *Clostridium bifermentans* (Figure 2.6, path B) isolated from an anaerobic domestic sludge [107,113]. *Klebsiella pneumoniae* demonstrated an initial denitration [107] followed by ring cleavage and decomposition, whereas the second strain, *Clostridium bifermentans,* induced an initial reduction of RDX to MNX, followed by denitration and decomposition [113]. Both strains produced a similar product distribution including MEDINA, N_2O, HCHO, MeOH, and CO_2.

TABLE 2.5

Biodegradation of RDX, HMX, and CL-20 by Microorganisms

Substrate	Microorganisms	Concentration (µM)	Degradation (%)	CO$_2$ (%)	Reference
		Anaerobic Degradation			
RDX	*Klebsiella pneumoniae*	100–225	70–100	72–ND	107, 108
	Enterobacter cloacae	225	80	ND	109
	Escherichia coli	225	80	ND	108
	Serratia marcescens	225	100	ND	108
	Citrobacter freundii	330	85	9	110
	Morganella morganii	330	100	5	110
	Providencia rettgeri	330	100	8	110
	Shewanella sp.	100	100	25–45	111
	Desulfovibrio sp.	112	100	ND	112
	Clostridium sp.	20–225	80–100	3.5	113–116
	Acetobacterium malicum	30	100	NA	117
HMX	*Clostridium* sp.	20–36	100	2–15	115, 118
	Desulfovibrio sp.	85	100	ND	112
	Citrobacter freundii	50	15	ND	110
	Morganella morganii	50	85	ND	110
	Providencia rettgeri	50	15	ND	110
CL-20	*Clostridium* sp.	20	100	ND	115
		Aerobic Degradation			
RDX	*Rhodococcus* sp.	162–202	100	35-ND	119–122
	Stenotrophomonas maltophilia	203	100	ND	123
	Corynebacterium sp.	180	98–100	ND	124, 125
	Pseudomonas putida	225	50	ND	108
	Xanthomonas maltophilia	225	80	ND	108
CL-20	*Pseudomonas* sp.	25	100	ND	126
	Agrobacterium sp.	100	100	ND	127
		Aerobic Degradation by Fungi			
RDX	*Phanerochaete chrysosporium*	90–270	65–100	62–70	128–130
HMX	*Phanerochaete chrysosporium*	47	100	70	131

Note: ND, no data or not detected.

When CL-20 was anaerobically incubated, as the sole nitrogen source, with the denitrifying strain *Pseudomonas* sp. FA1, the energetic chemical biotransformed with the concurrent release of NO_2^-, N_2O, and HCOOH [126]. Once again initial denitration of the cyclic nitramine led to ring cleavage and gave a product distribution closely related to that observed following photolysis [60] or alkaline hydrolysis [49] (Figure 2.3). Subsequent work by Bhushan et al. [62] showed that enzymatic

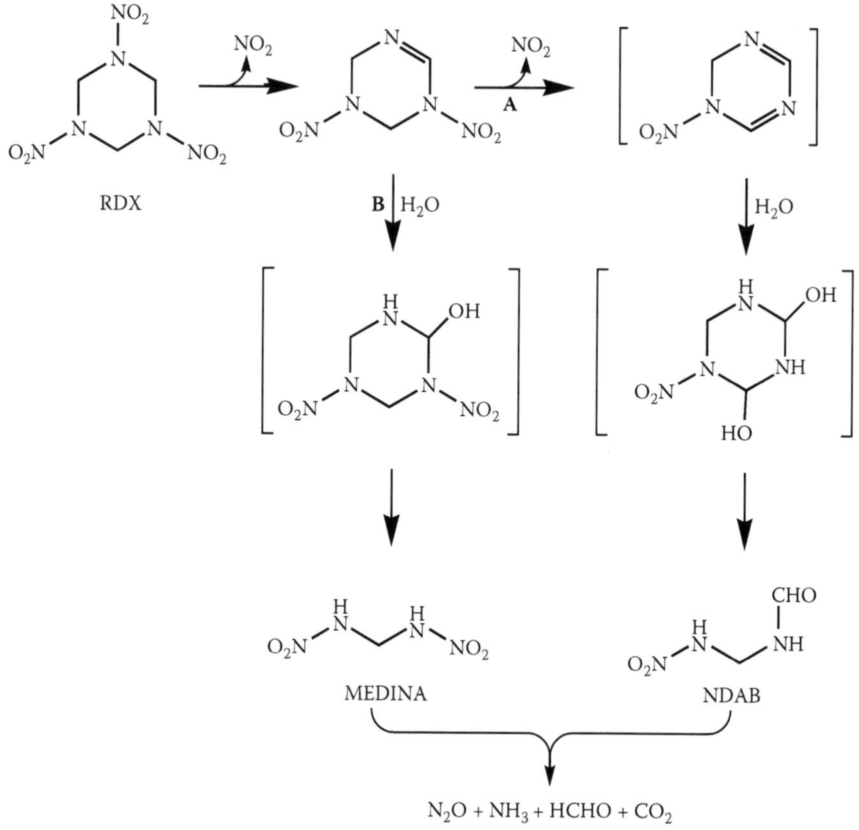

FIGURE 2.5 Biodegradation of RDX initiated by denitration under aerobic (path A) and anaerobic (path B) conditions.

denitration of CL-20 by salicylate 1-monooxygenase also yielded NO_2^-, N_2O, and HCOOH, through the transient formation of intermediates A and B (Figure 2.3).

2.3.2.2 Aerobic Degradation

In contrast to the multiple reaction pathways (reduction or denitration) observed under anaerobic conditions, the aerobic degradation of RDX using *Rhodococcus* sp. strain DN22 proceeded exclusively through denitration (Figure 2.5, path A) [119,120,122]. Initiation of the overall degradation process in *Rhodococcus* sp. Strain DN22 was attributed to the action of a cytochrome P450 enzyme [122,135,136]. The denitration products included NO_2^-, N_2O, NH_4^+, HCHO, and 4-nitro-2,4-diazabutanal (NDAB). In the presence of *Rhodococcus* sp. strain DN22, NDAB was stable and considered to be an end product of degradation [120], but recently the compound was shown to be aerobically degraded by *Phanerochaete chrysosporium* to produce N_2O and CO_2 [137]. Interestingly, NDAB has been observed in contaminated soils [137] and in plant tissues grown on RDX contaminated soil and subjected to photolysis [138].

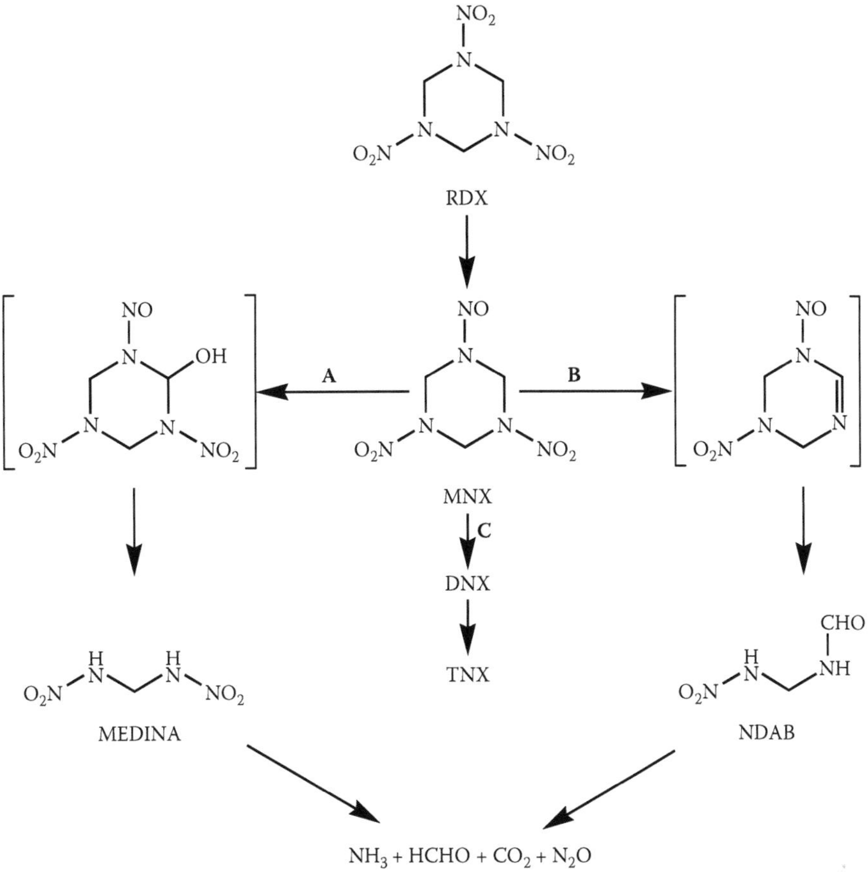

FIGURE 2.6 Biodegradation of RDX initiated by reduction to the nitroso derivatives followed by α-hydroxylation (path A), denitration (path B), or reduction (path C).

The preceding discussion revealed that once an initial attack, whether biotic or abiotic, disturbed the cyclic nitramine (RDX, HMX, or CL-20), the subsequent degradation reactions are largely governed by the interactions of the newly formed intermediates with water to eventually produce NO_2^-, N_2O, NH_3, HCHO, and HCOOH (Figures 2.2, 2.3, 2.5, and 2.6). Knowledge of the product distributions provides insight into the adverse effect of this family of explosives to various receptors in the environment.

2.4 SORPTION PROCESSES AND BIOAVAILABLITY

Table 2.2 summarizes the physicochemical properties (solubility [S], octanol/water partition coefficient [log K_{ow}], Henry's law constant, and vapor pressure) of several explosives. As evident in Table 2.2, the volatility of most explosives is relatively low, so that with the exception of NG, when released into the environment these chemicals do not generally migrate to the atmosphere [139]. The compounds can be solubilized in water and potentially migrate through subsurface soil at a rate dependent on their solubility in water, rate of dissolution, and affinity for stationary components of the soil matrix.

2.4.1 NITROAROMATIC COMPOUNDS

Several studies have dealt with sorption and transport of TNT and other related NACs in natural and model systems. These investigations showed that sorption of NACs and particularly TNT to minerals and natural soils is not linear, and can usually be well described by Freundlich (Equation 2.1) [140–144], Langmuir (Equation 2.2) [145,146], or multicomponent (Equation 2.3) [147] sorption models:

$$\frac{x}{m} = K_F C^{1/n} \tag{2.1}$$

$$\frac{x}{m} = \frac{KnC}{1+KC} \tag{2.2}$$

$$\frac{x}{m} = \frac{K_1 n_1 C}{1+K_1 C} + K_2 C \tag{2.3}$$

where x/m is the mass of solute sorbed per unit mass at equilibrium (mg kg^{-1}); C is the aqueous phase concentration of solute at equilibrium (mg L^{-1}); K_F is the Freundlich constant that gives a measure of the adsorbent capacity (mg$^{1-1/n}$ kg^{-1} L$^{1/n}$); 1/n gives a measure of the intensity of sorption; n is the maximum possible surface coverage corresponding to one monolayer; and K, K_1, n_1, and K_2 are constants.

In contrast to many other nonionic organic pollutants, sorption of TNT and related NACs is dominated by strong and specific interactions with certain matrix components, rather than by hydrophobic partitioning. Among all matrix components commonly found in soils and aquifers, including clays, carbonates, quartz, aluminum, iron (hydr)oxides, and natural organic matter, only clays were found to be strong sorbents for NACs [144,146,148–151]. As a consequence, the values of the distribution coefficient (K_d) measured for NACs are generally not correlated to the amount of organic carbon (see K_d values for TNT in Table 2.6 [152–155]), and estimating the adsorption of NACs using a general K_{oc} value may lead to erroneous predictions. Soil-specific adsorption coefficient measurements for NACs are therefore recommended when predicting contaminant migration in a given environment.

Although K_d values as high as 21,500 L kg^{-1} are reported for TNT with pure homoionic K$^+$- or NH$_4^+$-clays, K_d values for TNT sorption onto soils are typically low (<10 L kg^{-1}; Table 2.6). As a result, TNT is predicted to be mobile in soils and aquifers, and its mobility should be even greater at high concentrations where specific binding sites are saturated. However, although highly mobile, TNT may not be available for ecological receptors due to its transformation to break down products such as

TABLE 2.6

Adsorption Coefficients of Explosive Contaminants in Natural and Model Systems

Sorbent	Clay (%)	Total Organic Carbon (%)	K_d (L kg^{-1})				
			TNT	2,4-DNT	RDX	HMX	CL-20
Soils							
Sassafras sand	11	0.33	—	—	0.3 [152]	0.7 [152]	2.4 [153]
Topsoil	4	2.3[a]	4.2 [143]	—	1.9 [154]	2.5 [152]	15 [153]
Newport	5.6	3.5	2.3 [145]	—	—	—	—
Lonestar	10.0	0.06	2.5 [145]	—	—	—	—
Cornhuskers	20.0	0.83	4.1 [145]	—	—	—	—
Crane	20.6	2.8	3.7 [145]	—	—	—	—
Joliet	23.8	3.6	6.8 [145]	—	—	—	—
Holston B	43.8	1.2	3.0 [145]	—	—	—	—
Sharkey clay	54.4	2.4	11 [145]	—	—	—	—
K$^+$- LAAP D	32	0.20	167 [155]	—	0.66 [155]	1.73 [155]	—
Aqua-gel	>87	ND	130 [147]	130 [147]	6.6 [147]	8.9 [147]	—
Minerals							
K$^+$-mont.	NA	ND	414 [155]		3.17 [155]	22.1 [155]	—
K$^+$-mont.	NA	ND	21500 [146]	7400 [146]	1.2 [146]	—	—
H$^+$-mont.						15.6 [153]	0.6 [153]
K$^+$-kaolinite	NA	ND	1800 [146]	690 [146]			
K$^+$-illite	NA	ND	12500 [146]	3650 [146]			

Note: NA, not applicable; ND, not determined.

[a] Only organic matter (8.4%) was mentioned in the referred papers but total organic carbon (TOC) was also measured and found to be 2.3 %.

4-ADNT, 2-ADNT, 2,4-DANT, and 2,6-DANT [140,143,145,156–159]. Under aerobic conditions, the resulting aromatic (poly)amines of TNT are irreversibly bound to mineral surfaces [160] and natural organic matter [159–168]. Under anaerobic conditions, aromatic (poly)amines are stable and reversibly bound to soil [25,159]. The irreversible binding of aromatic amines occurs through binding between the amino groups, and the carbonyl [165] or phenolic [163] functional groups of humic acids, which probably explains why in natural soil, the distribution coefficient for TNT and transformation products increase with the number of amino groups (2,4-DANT > 4-ADNT > TNT) [143], whereas in pure clays, the coefficients increase with the number of nitro groups (TNT > 2-ADNT > 4-ADNT > 2,6-DANT > 2,4-DANT) [147]. Under strictly anaerobic conditions, TNT can continue its transformation to eventually produce TAT. The latter irreversibly binds to the soil matrix, thereby rendering the chemical unavailable to potential receptors [89,92].

2.4.2 Cyclic Nitramines

Significantly less data is found in the literature on the sorption of nitramines. Unlike those of TNT, sorption isotherms of RDX and HMX are usually well described by a linear equation (Equation 2.4), thus indicating either a nonspecific or a specific but nonsaturated sorption process [142,147,152,154,169]:

$$\frac{x}{m} = K_d C \tag{2.4}$$

where x/m and C have the same meaning as described earlier and K_d is the distribution coefficient (L kg^{-1}).

As demonstrated by the low K_d values measured in various soil systems (Table 2.6), RDX and HMX are only poorly immobilized by soils. The slightly higher adsorption coefficients measured for HMX despite a K_{ow} inferior to that of RDX (Table 2.2) suggests that as for TNT, adsorption of RDX and HMX to soils is governed by interactions with minerals rather than by association with soil organic matter. Examining the results of several investigations, Monteil-Rivera et al. [152] drew a close connection between the content of clay, and the level of retention for HMX. The authors thus concluded that the immobilization of nitramines was favored by association to clay minerals. Surprisingly, the sorption of CL-20 to clay and other minerals appeared negligible (Table 2.6) [153]. Relatively high K_d values (300 L kg^{-1}) were measured in soils with high organic content, suggesting that sorption of CL-20 was dominated by interactions with natural organic matter. The larger log K_{ow} measured for CL-20 compared to those of RDX and HMX (Table 2.2) is probably responsible for the stronger sorption of CL-20 to organic rich soils.

In general, monocyclic nitramines, and especially HMX, are degraded in soils at a much slower rate than TNT. As discussed earlier, RDX and HMX can be degraded via initial denitration to form intermediates (MEDINA, NDAB) [59,120] that are further degraded in soil to give CO_2, N_2O, and NH_3, or via sequential reduction of the N-NO_2 groups to the corresponding nitroso products [108,152,154]. No data are presently available on the fate of nitroso derivatives of HMX, while MNX, DNX, and TNX resulting from in situ reduction of RDX were recently reported to be relatively stable in soils and to exhibit transport potential consistent with that of RDX (0.29 < K_d < 0.85 L kg^{-1} with the Louisiana Army Ammunition Plant [LAAP D] soil to be compared to 0.33 L kg^{-1} for RDX) [13]. With a water solubility exceeding that of the parent nitramine, nitroso compounds may be of greater mobility and bioavailability than RDX and HMX. The toxicity of nitroso derivatives is not well established and should be therefore carefully evaluated.

CL-20 is weakly soluble in water and exhibits a higher affinity for soils than do RDX and HMX. As a result, the polycyclic nitramine should be less available for receptors. It was shown to degrade under several chemical and biochemical conditions to give NO_2^-, N_2O, $HCOO^-$, NH_3, and CH(O)CH(O) (Figure 2.3). Glyoxal is known to be toxic [170], and its presence in soils and sediments could cause greater adverse effects than the parent compound. The persistence and mobility of glyoxal in soils and sediments may thus be a determining factor when assessing the environmental impact of CL-20.

2.5. CONCLUSIONS AND FUTURE OUTLOOK

The wide use of nitro organic based energetic chemicals (NOCs), such as the aromatic TNT, and the nonaromatic cyclic nitramines RDX and HMX has resulted in the contamination of terrestrial and aquatic systems. Several reports (see Chapters 3–5 and 7–9 of this book) described the toxic and carcinogenic effects of explosives and their degradation products to various terrestrial, aquatic, and avian receptors. However, to determine the true identity of the chemicals that cause toxicity, the transport and transformation mechanisms of these chemicals must be understood.

The migration of explosives from surface soil to the water table is governed by their physicochemical properties and the mechanisms of their interaction with the soil matrix (mainly clay, and soil organic matter). The frequent detection of TNT (S = 140 mg/L) and RDX (S = 45 mg/L) in groundwater of contaminated sites is confirmed. As for HMX, its limited solubility in water (<5 mg/L) restricts its presence close to the soil surface. No field data is yet available on CL-20.

On the other hand, TNT is a labile molecule that (bio)transforms under both aerobic and anaerobic conditions to produce ADNTs and DANTs, with TAT only produced under strictly anaerobic conditions. Thus far only poor mineralization has been observed from reported laboratory experiments undertaken to biodegrade TNT. In field samples contaminated with RDX or HMX (no data on the use of CL-20 is available), both parent compounds were detected together with some of their corresponding nitroso derivatives and the ring cleavage product 4-nitro-2,4-diazabutanal, thus providing experimental evidence on the occurrence of in situ natural attenuation. Recent laboratory experiments demonstrated that once the nonaromatic nitramines—RDX, HMX, or CL-20—are disrupted by a successful initial attack such as denitration, the molecule undergoes complete decomposition in water to produce N_2O, NO_2^-, HCHO, and CO_2 (from RDX and HMX); and N_2O, NO_2^-, NH_4^+, HCOOH, and CH(O)CH(O) (from CL-20).

Therefore, to unequivocally relate the impact (e.g., toxic effects) of an explosive on a selective receptor, the identity of its transformed product(s) and the mechanisms of interaction with soil should be known. Very sensitive and specific detection techniques (e.g., LC/MS) capable of direct detection of trace amounts of explosives and their derivatives should be employed.

REFERENCES

1. Nielsen AT, U.S. Department of Navy, U.S. Patent Office Application Case No. 70631, June 24, 1987.
2. Wardle RB et al., Proceedings of the 27th International Annual Conference on ICT, ADPA, Arlington, VA, June 25–28, 1996, 27-1.
3. Chemical Properties Database, Groundwater Services, Inc., http://www.gsi-net.com/UsefulTools/ChemPropDatabaseHome.asp (accessed July 2008).

4. HSDB Hazardous Substances Data Bank, National Library of Medicine, Bethesda, MD, http://toxnet.nlm.nih.gov/ (accessed July 2008).

5. Rosenblatt DH et al., in *The Handbook of Environmental Chemistry*, Vol. 3, Part G, Hutzinger O, Ed., Springer-Verlag, Heidelberg, 1991, 195.

6. Fifer RA, Chemistry of nitrate ester and nitramine propellants, in *Fundamentals of Solid-Propellant Combustion*, Vol. 90, Kuo KK, and Summerfield M, Eds., American Institute of Aeronautics and Astronautics Inc., New York, 1984, chap. 4.

7. Binke N et al., Study on the melting process of nitrocellulose by thermal analysis method, *J. Therm. Anal. Calorimetry,* 58, 249, 1999.

8. Monteil-Rivera F et al., Physico-chemical measurements of CL-20 for environmental applications. Comparison with RDX and HMX, *J. Chromatogr. A*, 1025, 125, 2004.

9. Sahu SK et al., Thermal and photodegradation of glycidyl azide polymers, *Polym. Degradation Stab.,* 62, 495, 1998.

10. Thiboutot S et al., Characterization of Antitank Firing Ranges at CFB Valcartier, WATC Wainwright and CFAD Dundurn, Report DREV-R-9809, Valcartier, QC, 1998.

11. Spain JC, Introduction, in *Biodegradation of Nitroaromatic Compounds and Explosives*, Spain JC, Hughes JB, and Knackmuss H-J, Eds., CRC Press, Boca Raton, FL, 2000, 1.

12. Pennington JC et al., Distribution and Fate of Energetics on DoD Test and Training Ranges: Interim Report 1, ERDC TR-01-13, Engineer Research and Development Center, prepared for US Army Corps of Engineers, Washington, DC, 2001.

13. Pennington JC et al., Distribution and Fate of Energetics on DoD Test and Training Ranges: Interim Report 2, ERDC TR-02-08, Engineer Research and Development Center, prepared for US Army Corps of Engineers, Washington, DC, 2002.

14. United States General Accounting Office, Report GAO-04-147 Military Munitions: DOD Needs to Develop a Comprehensive Approach for Cleaning Up Contaminated Sites, December 2003.

15. Hawari J and Halasz A, Microbial degradation of explosives, in *The Encyclopedia of Environmental Microbiology*, Bitton G, Ed., John Wiley & Sons, Amsterdam, Netherlands, 2002, 1979.

16. Jenkins TF et al., Sampling error associated with collection and analysis of soil samples at TNT-contaminated sites, *Field Anal. Chem. Technol.*, 1, 151, 1997.

17. Thiboutot S et al., Environmental Conditions of Surface Soils and Biomass Prevailing in the Training Area at CFB Gagetown, New Brunswick, Report DRDC-TR 2003-152, Valcartier, QC, 2003.

18. Strategic Environmental Research Development Program (SERDP), http://www.serdp.org/ (accessed July 2008).

19. U.S. Army Environmental Center (USAEC), http://aec.army.mil/usaec/ (accessed July 2008).

20. Office of the Auditor General of Canada, April 2003 Report, Chap. 7: National Defense—Environmental Stewardship of Military Training and Test Areas, 2003.

21. Spain JC, Hughes JB, and Knackmuss H-J, Eds., *Biodegradation of Nitroaromatic Compounds and Explosives,* CRC Press, Boca Raton, FL, 2000.

22. Pennington JC and Brannon JM, Environmental fate of explosives, *Thermochimica Acta*, 384, 163, 2002.

23. Schmelling DC and Gray KA, Photocatalytic transformation and mineralization of 2,4,6-trinitrotoluene (TNT) in TiO_2 slurries, *Water Res.,* 29, 2651, 1995.

24. Oh S-Y et al., Effect of adsorption to elemental iron on the transformation of 2,4,6-trinitrotoluene and hexahydro-1,3,5-trinitro-1,3,5-triazine in solution, *Environ. Toxicol. Chem.*, 21, 1384, 2002.

25. Hofstetter TB et al., Complete reduction of TNT and other (poly)nitroaromatic compounds under iron-reducing subsurface conditions, *Environ. Sci. Technol.*, 33, 1479, 1999.

26. Mills A, Seth A, and Peters G, Alkaline hydrolysis of trinitrotoluene, TNT, *Phys. Chem. Chem. Phys.*, 5, 3921, 2003.

27. Urbanski T, *Chemistry and Technology of Explosives*, Vol. 1, Jeczalikowa I and Laverton S, trans., Pergamon Press, New York, 1964, 230.

28. Saupe A, Garvens HJ, and Heinze L, Alkaline hydrolysis of TNT and TNT in soil followed by thermal treatment of the hydrolysates, *Chemosphere*, 36, 1725, 1998.

29. Emmrich M, Kinetics of the alkaline hydrolysis of 2,4,6-trinitrotoluene in aqueous solution and highly contaminated soils, *Environ. Sci. Technol.*, 33, 3802, 1999.

30. Felt DR, Larson SL, and Hansen LD, Molecular Weight Distribution of the Final Products of TNT–Hydroxide Reaction, Technical Report ERDC/EL TR-01-16, US Army Engineer Research and Development Center, Vicksburg, MS, 2001.

31. Felt DR, Larson SL, and Hansen LD, Kinetics of Base-Induced 2,4,6-Trinitrotoluene Transformation, Technical Report ERDC/EL TR-01-17, US Army Engineer Research and Development Center, Vicksburg, MS, 2001.

32. Emmrich M, Kinetics of the alkaline hydrolysis of important nitroaromatic co-contaminants of 2,4,6-trinitrotoluene in highly contaminated soils, *Environ. Sci. Technol.*, 35, 874, 2001.

33. Buncel E, Dust JM, and Terrier F, Rationalizing the regioselectivity in polynitroarene anionic sigma-adduct formation. Relevance to nucleophilic aromatic substitution, *Chem. Rev.*, 95, 2261, 1995.

34. Hwang H-M, McCullum D, and Slaughter L, Phototransformation of 2,4,-dichloroaniline in a surface freshwater environment: Effects on microbial assemblages, *Bull. Environ. Contam. Toxicol.*, 60, 81, 1998.

35. Andrews CC and Osmon JL, The Effects of Ultraviolet Light on TNT in Aqueous Solutions, Report WQEC/C 75-197 (AD-B008175), Weapons Quality Engineering Center, Naval Weapons Support Center, Crane, IN, 1975.

36. Mabey WR et al., Photolysis of nitroaromatics in aquatic systems. I. 2,4,6-trinitrotoluene, *Chemosphere,* 12, 3, 1983.

37. Spanggord RJ et al., Environmental Fate Studies on Certain Munitions Wastewater Constituents—Phase IV. Lagoon Model Studies, Report ADA 138550, SRI International, Menlo Park, CA, 1983.

38. Hwang H-M et al., Photochemical and microbial degradation of 2,4,6-trinitrotoluene (TNT) in a freshwater environment, *Bull. Environ. Contam. Toxicol.*, 65, 228, 2000.

39. Cui H et al., Effect of photo sensitizer riboflavin on the fate of 2,4,6-trinitrotoluene in a freshwater environment, *Chemosphere*, 44, 621, 2001.

40. Andrews CC, Photooxidative treatment of TNT contaminated wastewaters. Report WQEC/C 80-137 (AD-A084684), Weapons Quality Engineering Center, Naval Weapons Support Center, Crane, IN, 1980.

41. Schmelling DC, Gray KA, and Kamat PV, The influence of solution matrix on the photocatalytic degradation of TNT in TiO_2 slurries, *Water Res.*, 31, 1439, 1997.

42. Hundal LS et al., Removal of TNT and RDX from water and soil using iron metal, *Environ. Pollut.*, 97, 55, 1997.

43. Tratnyek PG, Miehr R, and Bandstra JZ, Kinetics of reduction of TNT by iron metal, in *Ground Water Quality: Natural and Enhanced Restoration of Groundwater,* Thornton SF and Oswald SE, Eds., International Association of Hydrological Science, Sheffield, UK, 2002, 427.

44. Agrawal A and Tratnyek PG, Reduction of nitro aromatic compounds by zero-valent iron metal, *Environ. Sci. Technol.*, 30, 153, 1996.

45. Bell LS et al., A sequential zero valent iron and aerobic biodegradation treatment system for nitrobenzene, *J. Contam. Hydrol.*, 66, 201, 2003.

46. Oh S-Y, Cha DK, and Chiu PC, Graphite-mediated reduction of 2,4-dinitrotoluene with elemental iron, *Environ. Sci. Technol.*, 36, 2178, 2002.

47. Rügge K et al., Characterization of predominant reductants in an anaerobic leachate-contaminated aquifer by nitroaromatic probe compounds, *Environ. Sci. Technol.*, 32, 23, 1998.
48. Hawari J et al., Photodegradation of RDX in aqueous solution: a mechanistic probe for biodegradation with *Rhodococcus* sp., *Environ. Sci. Technol.*, 36, 5117, 2002.
49. Balakrishnan VK, Halasz A, and Hawari J, The alkaline hydrolysis of the cyclic nitramine explosives RDX, HMX and CL-20: New insights into degradation pathways obtained by the observation of novel intermediates, *Environ. Sci. Technol.*, 37, 1838, 2003.
50. Gregory KB et al., Abiotic transformation of hexahydro-1,3,5-trinitro-1,3,5-triazine by Fe^{II} bound to magnetite, *Environ. Sci. Technol.*, 38, 1408, 2004.
51. Hoffsommer JC, Kubose DA, and Glover DJ, Cationic micellar catalysis of the aqueous alkaline hydrolyses of 1,3,5-triaza-1,3,5-trinitrocyclohexane and 1,3,5,7-tetraaza-1,3,5,7-tetranitrocyclooctane, *J. Phys. Chem.*, 81, 380, 1977.
52. Croce M and Okamoto Y, Kinetic isotope effects and intermediate formation for the aqueous alkaline homogeneous hydrolysis of 1,3,5-triaza-1,3-5-trinitrocyclohexane (RDX), *J. Org. Chem.*, 44, 2100, 1979.
53. Heilmann HM, Wiesmann U, and Stenstrom MK, Kinetics of the alkaline hydrolysis of high explosives RDX and HMX in aqueous solution and adsorbed to activated carbon, *Environ. Sci. Technol.*, 30, 1485, 1996.
54. Bishop RL et al., Base hydrolysis of HMX and HMX-based plastic-bonded explosives with sodium hydroxide between 100 and 155°C, *Ind. Eng. Chem. Res.,* 38, 2254, 1999.
55. Hawari J, Biodegradation of RDX and HMX: From basic research to field application, in *Biodegradation of Nitroaromatic Compounds and Explosives*, Spain JC, Hughes JB, and Knackmuss H-J, Eds., CRC Press, Boca Raton, FL, 2000, 277.
56. Glover DJ and Hoffsommer JC, Photolysis of RDX in aqueous solution, with and without ozone, Technical Report 77-20 (ADA 042199), Naval Surface Weapons Center, White Oak Laboratory, Silver Spring, MD, 1977.
57. Bose P, Glaze WH, and Maddox S, Degradation of RDX by various advanced oxidation processes: II. Organic by-products, *Water Res.*, 32, 1005, 1998.
58. Peyton GR, Lefaivre MH, and Maloney SW, Verification of RDX photolysis mechanism, *CERL Technical Report 99/93,* Champaign, IL, 1999.
59. Halasz A et al., Insights into the formation and degradation mechanisms of methylene-dinitramine during the incubation of RDX with anaerobic sludge, *Environ. Sci. Technol.*, 36, 633, 2002.
60. Hawari J et al., Photodegradation of CL-20: Insights into the mechanisms of initial reactions and environmental fate, *Water Res.*, 38, 4055, 2004.
61. Balakrishnan VK et al., Decomposition of the polycyclic nitramine explosive, CL-20, by Fe^0, *Environ. Sci. Technol.*, 38, 6861, 2004.
62. Bhushan B et al., Initial reaction(s) in biotransformation of CL-20 is catalyzed by salicylate 1-monooxygenase from *Pseudomonas* sp. strain ATCC 29352, *Appl. Environ. Microbiol.,* 70, 4040, 2004.
63. Singh J, Comfort SD, and Shea PJ, Remediating RDX-contaminated water and soil using zero-valent iron, *J. Environ. Qual.*, 27, 1240, 1998.
64. Oh BT, Craig LJ, and Alvarez PJJ, Hexahydro-1,3,5-trinitro-1,3,5-triazine mineralization by zerovalent iron and mixed anaerobic cultures, *Environ. Sci. Technol.*, 35, 4341, 2001.
65. Hawari J et al., Mini-review: Microbial degradation of explosives: biotransformation versus mineralization, *Appl. Microbiol. Biotechnol.*, 54, 605, 2000.
66. Esteve-Núñez A, Caballero A, and Ramos JL, Biological degradation of 2,4,6-trinitro-toluene, *Microbiol. Mol. Biol. Rev.*, 65, 335, 2001.
67. Ye J, Singh A and Ward OP, Biodegradation of nitroaromatics and other nitrogen-containing xenobiotics, *World J. Microbiol. Biotechnol.*, 20, 117, 2004.

68. Lewis TA, Newcombe DA, and Crawford RL, Bioremediation of soils contaminated with explosives, *J. Environ. Manage.* 70, 291, 2004.
69. Kalafut T et al., Biotransformation pattern of 2,4,6-trinitrotoluene by aerobic bacteria, *Current Microbiol.*, 36, 45, 1998.
70. French CE, Nicklin S, and Bruce NC, Aerobic degradation of 2,4,6-trinitrotoluene by *Enterobacter cloacae* PB2 and pentaerythritol tetranitrate reductase, *Appl. Environ. Microbiol.*, 64, 2864, 1998.
71. Van Aken B, Yoon JM, and Schnoor JL, Biodegradation of nitro-substituted explosives 2,4,6-trinitrotoluene, hexahydro-1,3,5-trinitro-1,3,5-triazine, and octahydro-1,3,5,7-tetra-nitro-1,3,5,7-tetrazocine by a phytosymbiotic *Methylobacterium* sp. associated with pop-lar tissues (*Populus deltoides* × nigra DN34), *Appl. Environ. Microbiol.*, 70, 508, 2004.
72. Vorbeck CV et al., Identification of a hydride-Meisenheimer complex as a metabolite of 2,4,6-trinitrotoluene by *Mycobacterium* strain, *J. Bacteriol.*, 176, 932, 1994.
73. Vanderberg LA, Perry JJ, and Unkefer PJ, Catabolism of 2,4,6-trinitrotoluene by *Mycobacterium vaccae*, *Appl. Microbiol. Biotechnol.*, 43, 937, 1995.
74. Duque E et al., Construction of a *Pseudomonas* hybrid strain that mineralizes 2,4,6-trinitrotoluene, *J. Bacteriol.*, 175, 2278, 1993.
75. Alvarez MA et al., *Pseudomonas aeruginosa* strain MA01 aerobically metabolizes the aminodinitrotoluenes produced by 2,4,6-trinitrotoluene nitro-group reduction, *Can. J. Microbiol.*, 41, 984, 1995.
76. Vasilyeva GK et al., Aerobic TNT reduction via 2-hydroxylamino-4,6-dinitrotoluene by *Pseudomonas aeruginosa* strain MX isolated from munitions-contaminated soil, *Biorem. J.*, 4, 111, 2000.
77. Fiorella PD and Spain JC, Transformation of 2,4,6-trinitrotoluene by *Pseudomonas pseudoalcaligenes* JS52, *Appl. Environ. Microbiol.*, 63, 2007, 1997.
78. Martin JL et al., Denitration of 2,4,6-trinitrotoluene by *Pseudomonas savastanoi*, *Can. J. Microbiol.*, 43, 447, 1997.
79. Pak JW et al., Transformation of 2,4,6-trinitrotoluene by purified xenobiotic reductase B from *Pseudomonas fluorescens* I-C, *Appl. Environ. Microbiol.*, 66, 4742, 2000.
80. Vorbeck CV et al., Initial reductive reactions in aerobic microbial metabolisms of 2,4,6-trinitrotoluene, *Appl. Environ. Microbiol.*, 64, 246, 1998.
81. Montpas S et al., Degradation of 2,4,6-trinitrotoluene by *Serratia marcescens*, *Bio-technol. Lett.*, 19, 291, 1997.
82. Pasty-Grigsby MB et al., Transformation of 2,4,6-trinitrotoluene (TNT) by actino-mycetes isolated from TNT-contaminated and uncontaminated environments, *Appl. Environ. Microbiol.*, 62, 1120, 1996.
83. Fernando T, Bumpus JA, and Aust SD, Biodegradation of TNT (2,4,6-trinitrotoluene) by *Phanerochaete chrysosporium*, *Appl. Environ. Microbiol.*, 56, 1666, 1990.
84. Hess TF et al., Combined photocatalytic and fungal treatment for the destruction of 2,4,6-trinitrotoluene (TNT), *Water Res.*, 32, 1481, 1998.
85. Hawari J et al., Biotransformation of 2,4,6-trinitrotoluene with *Phanaerochaete chrys-osporium* in agitated cultures at pH 4.5, *Appl. Environ. Microbiol.*, 65, 2977, 1999.
86. Hodgson J et al., Tween 80 enhanced TNT mineralization by *Phanerochaete chrysospo-rium*, *Can. J. Microbiol.*, 46, 110, 2000.
87. Stahl JD and Aust SD, Metabolism and detoxification of TNT by *Phanerochaete chrys-osporium*, *Biochem. Biophys. Res. Commun.*, 192, 477, 1993.
88. Scheibner K et al., Screening for fungi intensively mineralizing 2,4,6-trinitrotoluene, *Appl. Microbiol. Biotechnol.*, 47, 452, 1997.
89. Lenke H, Achtnich C, and Knackmuss H-J, Perspectives of bioelimination of polyni-troaromatic compounds, in *Biodegradation of Nitroaromatic Compounds and Explo-sives*, Spain JC, Hughes JB, and Knackmuss H-J, Eds., CRC Press, Boca Raton, FL, 2000, 92.

90. McCormick NG, Feeherry FF, and Levinson HS, Microbial transformation of 2,4,6-trinitrotoluene and other nitroaromatic compounds, *Appl. Environ. Microbiol.*, 31, 949, 1976.
91. Liu D, Thomson K, and Anderson C, Identification of nitroso compounds from biotransformation of 2,4-dinitrotoluene, *Appl. Environ, Microbiol.*, 47, 1295, 1984.
92. Rieger P-G and Knackmuss H-J, Basic knowledge and perspectives on biodegradation of 2,4,6-trinitrotoluene and related nitroaromatic compounds in contaminated soil, in *Biodegradation of Nitroaromatic Compounds*, Spain JC, Ed., Plenum Press, New York, 1995, 1.
93. Fritsche W et al., Fungal degradation of explosives: TNT and related nitroaromatic compounds, in *Biodegradation of Nitroaromatic Compounds and Explosives*, Spain JC, Hughes JB, and Knackmuss H-J, Eds., CRC Press, Boca Raton, FL, 2000, 213.
94. Preuss A, Fimpel J, and Diekert G, Anaerobic transformation of 2,4,6-trinitrotoluene (TNT), *Arch. Microbiol.*, 159, 345, 1993.
95. Hawari J et al., Characterization of metabolites in the biotransformation of 2,4,6-trinitrotoluene with anaerobic sludge: Role of triaminotoluene, *Appl. Environ. Microbiol.*, 64, 2200, 1998.
96. Peres CM, Naveau H, and Agathos SN, Biodegradation of nitrobenzene by its simultaneous reduction into aniline and mineralization of the aniline formed, *Appl. Microbiol. Biotechnol.*, 49, 343, 1998.
97. Corbett MD and Corbett BR, Bioorganic chemistry of the arylhydroxylamine and nitrosoarene functional groups, in *Biodegradation of Nitroaromatic Compounds*, Spain JC, Ed., Plenum Press, New York, 1995, 151.
98. Spanggord RJ, Stewart KR, and Riccio ES, Mutagenicity of tetranitroazoxytoluenes: A preliminary screening in *Salmonella typhimurium* strains TA100 and TA100R, *Mutation Res.*, 335, 207, 1995.
99. Cattaneo MV et al., Natural attenuation of explosives, in *Remediation of Engineering of Contaminated Soil*, Wise LD, Ed., Marcel Dekker, New York, 2000, 949.
100. Scheibner K, Hofrichter M, and Fritsche W, Mineralization of 2-amino-4,6-dinitrotoluene by manganese peroxidase of the white-rot fungus *Nematoloma frowardii*, *Biotechnol. Lett.*, 19, 835, 1997.
101. Suen WC and Spain JC, Cloning and characterization of *Pseudomonas* sp. strain dinitrotoluene genes for 2,4-dinitrotoluene degradation, *J. Bacteriol.*, 175, 1831, 1993.
102. Nishino SF et al., Mineralization of 2,4- and 2,6-dinitrotoluene in soil slurries, *Environ. Sci. Technol.*, 33, 1060, 1999.
103. Nishino SF, Paoli GC, and Spain JC, Aerobic degradation of dinitrotoluenes and pathway for bacterial degradation of 2,6-dinitrotoluene, *Appl. Environ. Microbiol.*, 66, 2139, 2000.
104. Behrend C and Heesche-Wagner K, Formation of hydride-Meisenheimer complexes of picric acid (2,4,6-trinitrophenol) and 2,4-dinitrophenol during mineralization of picric acid by *Nocardioides* sp. strain CB 22-2, *Appl. Environ, Microbiol.*, 65, 1372, 1999.
105. Rieger P-G et al., Hydride-Meisenheimer complex formation and protonation as key reactions of 2,4,6-trinitrophenol biodegradation by *Rhodococcus erythropolis*, *J. Bacteriol.*, 181, 1189, 1999.
106. Lenke H and Knackmuss H-J, Initial hydrogenation during catabolism of picric acid by *Rhodococcus erythropolis* HL-24-2, *Appl. Environ. Microbiol.*, 58, 2933, 1992.
107. Zhao J-S et al., Biodegradation of hexahydro-1,35-trinitro-1,3,5-triazine and its mononitroso derivative hexahydro-1-nitroso-3,5-dinitro-1,3,5-triazine by *Klebsiella peneumoniae* strain SCZ-1 isolated from an anaerobic sludge, *Appl. Environ. Microbiol.*, 68, 5336, 2002.
108. Young DM, Unkefer PJ, and Ogden KL, Biotransformation of hexahydro-1,3,5-trinitro-1,3,5-triazine (RDX) by a prospective consortium and its most effective isolate *Serratia marcescens*, *Biotechnol. Bioeng.*, 53, 515, 1997.

109. Pudge IB, Daugulis AJ, and Dubois CT, The use of *Enterobacter cloacae* ATCC 43560 in the development of a two-phase partitioning bioreactor for the destruction of hexahydro-1,3,5-trinitro-1,3,5-triazine (RDX), *J. Biotechnol.*, 100, 65, 2003.

110. Kitts CL, Cunningham DP, and Unkefer PJ, Isolation of three hexahydro-1,3,5-trinitro-1,3,5-triazine degrading species of the Family *Enterobacteriaceae* from nitramine explosive-contaminated soil, *Appl. Environ. Microbiol.*, 60, 4608, 1994.

111. Zhao J-S et al., Phylogeny of cyclic nitramine-degrading psychrophilic bacteria in marine sediment and their potential role in the natural attenuation of explosives, *FEMS Microbiol. Ecol.*, 49, 349, 2004.

112. Boopathy R et al., Metabolism of explosive compounds by sulfate-reducing bacteria, *Curr. Microbiol.*, 37, 127, 1998.

113. Zhao J-S et al., Metabolism of hexahydro-1,3,5-trinitro-1,3,5-triazine through initial reduction to hexahydro-1-nitroso-3,5-dinitro-1,3,5-triazine followed by denitration in *Clostridium bifermentans* HAW-1, *Appl. Microbiol. Biotechnol.*, 63, 187, 2003.

114. Zhao J-S, Spain J and Hawari J, Phylogenetic and metabolic diversity of hexahydro-1,3,5-trinitro-1,3,5-triazine (RDX)-transforming bacteria in strictly anaerobic mixed cultures enriched on RDX as nitrogen source, *FEMS Microbiol. Ecol.*, 46, 189, 2003.

115. Bhushan B et al., Chemotaxis-mediated biodegradation of cyclic nitramine explosives RDX, HMX, and CL-20 by *Clostridium* sp. EDB2, *Biochem. Biophys. Res. Commun.*, 316, 816, 2004.

116. Regan KM and Crawford RL, Characterization of *Clostridium bifermentans* and its biotransformation of 2,4,6-trinitrotoluene (TNT) and 1,3,5-triaza-1,3,5-trinitrocyclohexane (RDX), *Biotechnol. Lett.*, 16, 1081, 1994.

117. Adrian NR and Arnett CM, Anaerobic biodegradation of hexahydro-1,3,5-trinitro-1,3,5-triazine (RDX) by *Acetobacterium malicum* strain HAAP-1 isolated from a methanogenic mixed culture, *Curr. Microbiol.*, 48, 332, 2004.

118. Zhao J-S et al., Metabolism of octahydro-1,3,5,7-tetranitro-1,3,5,7-tetrazocine by *Clostridium bifermentants* strain HAW-1 and other several H_2-producing fermentative anaerobic bacteria, *FEMS Microbiol. Lett.*, 237, 65, 2004.

119. Coleman NV, Nelson DR, and Duxbury T, Aerobic biodegradation of hexahydro-1,3,5-trinitro-1,3,5-triazine (RDX) as a nitrogen source by a *Rhodococcus* sp. strain DN22, *Soil Biol. Biochem.*, 30, 1159, 1998.

120. Fournier D et al., Determination of key metabolites during biodegradation of hexahydro-1,3,5-trinitro-1,3,5-triazine with *Rhodococcus* sp. strain DN22, *Appl. Environ. Microbiol.*, 68, 166, 2002.

121. Jones AM et al., Biodegradability of selected highly energetic pollutants under aerobic conditions, Third International In Situ and On Site Bioreclamation Symposium, Battelle Press, 1995, 251.

122. Seth-Smith HMB et al., Cloning, sequencing, and characterization of the hexahydro-1,3,5-trinitro-1,3,5-triazine degradation gene cluster from *Rhodococcus rhodochrous*, *Appl. Environ. Microbiol.*, 68, 4764, 2002.

123. Binks PR, Nicklin S, and Bruce NC, Degradation of hexahydro-1,3,5-trinitro-1,3,5-triazine (RDX) by *Stenotrophomonas maltophilia* PB1, *Appl. Environ. Microbiol.*, 61, 1318, 1995.

124. Brenner A et al., Use of hexahydro-1,3,5-trinitro-1,3,5-triazine as a nitrogen source in biological treatment of munitions wastes, *Water Environ. Res.*, 72, 469, 2000.

125. Yang YX et al., Studies on three strains of *Corynebacterieum* degrading cyclotrimethylene-triamine (RDX), *Acta Microbiol. Sin.*, 23, 251, 1983.

126. Bhushan B et al., Biotransformation of 2,4,6,8,10,12-hexanitro-2,4,6,8,10,12-hexaazaisowurtzitane (CL-20) by denitrifying *Pseudomonas* sp. strain FA1, *Appl. Environ. Microbiol.*, 69, 5216, 2003.

127. Trott S et al., Biodegradation of the nitramine explosive CL-20, *Appl. Environ. Microbiol.*, 69, 1871, 2003.

128. Sheremata TW and Hawari J, Mineralization of RDX by white rot fungus *Phanerochaete chrysosporium* to carbon dioxide and nitrous oxide, *Environ. Sci. Technol.*, 34, 3384, 2000.

129. Fernando T and Aust SD, Biodegradation of munition waste TNT (2,4,6-trinitrotoluene) and RDX (hexahydro-1,3,5-trinitro-1,3,5-triazine) by *Phanerochaete chrysosporium*, *Industr. Eng. Chem. ACS Symp. Ser.*, 486, 214, 1991.

130. Stahl JD et al., Hexahydro-1,3,5-trinitro-1,3,5-triazine (RDX) biodegradation in liquid and solid-state matrices by *Phanerochaete crhysosporium*, *Biorem. J.*, 5, 13, 2001.

131. Fournier D et al., Biodegradation of octahydro-1,3,5,7-tetranitro-1,3,5,7-tetrazocine (HMX) by *Phanerochaete chrysosporium*: New insight into the degradation pathway, *Environ. Sci. Technol.*, 38, 4130, 2004.

132. McCormick NG, Cornell JH, and Kaplan AM, Biodegradation of hexahydro-1,3,5-trinitro-1,3,5-triazine, *Appl. Environ. Microbiol.*, 42, 817, 1981.

133. Hawari J et al., Characterization of metabolites during biodegradation of hexahydro-1,35-trinitro-1,3,5-triazine (RDX) with municipal anaerobic sludge, *Appl. Environ. Microbiol.*, 66, 2652, 2000.

134. Hawari J et al., Biotransformation routes of octahydro-1,3,5,7-tetranitro-1,3,5,7-tetrazocine by municipal anaerobic sludge, *Environ. Sci. Technol.*, 35, 70, 2001.

135. Coleman NV, Spain JC, and Duxbury T, Evidence that RDX biodegradation by *Rhodococcus* strain DN22 is plasmid-borne and involves a cytochrome p-450, *J. Appl. Microbiol.*, 93, 463, 2002.

136. Bhushan B et al., Biotransformation of hexahydro-1,3,5-trinitro-1,3,5-triazine (RDX) by a rabbit liver cytochrome P450: Insights into the mechanism of RDX biodegradation by *Rhodococcus* sp. strain DN22, *Appl. Environ. Microbiol.*, 69, 1347, 2003.

137. Fournier D et al., Biodegradation of the hexahydro-1,3,5-trinitro-1,3,5-triazine ring cleavage product 4-nitro-2,4-diazabutanal by *Phanaerochaete chrysosporium*, *Appl. Environ. Microbiol.*, 70, 1123, 2004.

138. Just CL and Schnoor JL, Phytophotolysis of hexahydro-1,3,5-trinitro-1,3,5-triazine (RDX) in leaves of reed canary grass, *Environ. Sci. Technol.*, 38, 290, 2004.

139. Spanggord RJ et al., Environmental Fate Studies of HMX Screening Studies, Final Report, Phase I, SRI Project No. LSU-4412, SRI International, Menlo Park, CA, 1982.

140. Comfort SD et al., TNT transport and fate in contaminated soil, *J. Environ. Qual.*, 24, 1174, 1995.

141. Selim HM, Xue SK, and Iskandar IK, Transport of 2,4,6-trinitrotoluene and hexahydro-1,3,5-trinitro-1,3,5-triazine in soils, *Soil Sci.*, 160, 328, 1995.

142. Xue SK, Iskandar IK, and Selim HM, Adsorption-desorption of 2,4,6-trinitrotoluene and hexahydro-1,3,5-trinitro-1,3,5-triazine in soils, *Soil Sci.*, 160, 317, 1995.

143. Sheremata TW et al., Fate of 2,4,6-trinitrotoluene and its metabolites in natural and model soil systems, *Environ. Sci. Technol.*, 33, 4002, 1999.

144. Boyd SA et al., Mechanisms for the adsorption of substituted nitrobenzenes by smectite clays, *Environ. Sci. Technol.*, 35, 4227, 2001.

145. Pennington JC and Patrick WH, Adsorption and desorption of 2,4,6-trinitrotoluene by soils, *J. Environ. Qual.*, 19, 559, 1990.

146. Haderlein SB, Weissmahr KW, and Schwarzenbach RP, Specific adsorption of nitroaromatic explosives and pesticides to clay minerals, *Environ. Sci. Technol.*, 30, 612, 1996.

147. Leggett DC, Sorption of Military Explosive Contaminants on Bentonite Drilling Muds, CRREL Report 85-18, U.S. Army Cold Regions Research and Engineering Laboratory, Hanover, NH, 1985.

148. Haderlein SB and Schwarzenbach RP, Adsorption of substituted nitrobenzenes and nitrophenols to mineral surfaces, *Environ. Sci. Technol.*, 27, 316, 1993.

149. Weissmahr KW et al., In situ spectroscopic investigations of adsorption mechanisms of nitroaromatic compounds at clay minerals, *Environ. Sci. Technol.*, 31, 240, 1997.
150. Weissmahr KW, Haderlein SB, and Schwarzenbach RP, Complex formation of soil minerals with nitroaromatic explosives and other π-acceptors, *Soil Sci. Soc. Am. J.*, 62, 369, 1998.
151. Weissmahr KW et al., Laboratory and field scale evaluation of geochemical controls on groundwater transport of nitroaromatic ammunition residues, *Environ. Sci. Technol.*, 33, 2593, 1999.
152. Monteil-Rivera F, Groom C, and Hawari J, Sorption and degradation of octahydro-1,3,5,7-tetranitro-1,3,5,7-tetrazocine in soil, *Environ. Sci. Technol.*, 37, 3878, 2003.
153. Balakrishnan VK et al., Sorption and stability of the polycyclic nitramine explosive CL-20 in soil, *J. Environ. Qual.*, 33, 1362, 2004.
154. Sheremata TW et al., The fate of the cyclic nitramine explosive RDX in natural soil, *Environ. Sci. Technol.*, 35, 1037, 2001.
155. Brannon JM et al., Aquifer soil cation substitution and adsorption of TNT, RDX, and HMX, *Soil Sediment Contam.*, 11, 327, 2002.
156. Krumholz LR et al., Transformations of TNT and related aminotoluenes in groundwater aquifer slurries under different electron-accepting conditions, *J. Ind. Microbiol. Biotechnol.*, 18, 161, 1997.
157. Price CB, Brannon JM, and Hayes C, Effect of redox potential and pH on TNT transformation in soil-water slurries, *J. Environ. Eng.*, 123, 988, 1997.
158. Brannon JM, Price CB, and Hayes C, Abiotic transformation of TNT in montmorillonite and soil suspensions under reducing conditions, *Chemosphere,* 36, 1453, 1998.
159. Elovitz MS and Weber EJ, Sediment-mediated reduction of 2,4,6-trinitrotoluene and fate of the resulting aromatic polyamines, *Environ. Sci. Technol.*, 33, 2617, 1999.
160. Daun G et al., Biological treatment of TNT-contaminated soil. 1. Anaerobic cometabolic reduction and interaction of TNT and metabolites with soil components, *Environ. Sci. Technol.*, 32, 1956, 1998.
161. Thorn KA, Pettigrew PJ, and Goldenberg WS, Covalent binding of aniline to humic substances 2. N-15 NMR studies of nucleophilic addition reactions, *Environ. Sci. Technol.*, 30, 2764, 1996.
162. Weber EJ, Spidle DL, and Thorn KA, Covalent binding of aniline to humic substances: 1. Kinetics studies, *Environ. Sci. Technol.*, 30, 2755, 1996.
163. Dawel G et al., Structure of a laccase-mediated product of coupling of 2,4-diamino-6-nitrotoluene (2,4-DANT) to guaiacol, a model for coupling of 2,4,6-trinitrotoluene (TNT) metabolites to humic organic soil matrix, *Appl. Environ. Microbiol.,* 63, 2560, 1997.
164. Li H and Lee LS, Sorption and abiotic transformation of aniline and alpha-naphthylamine by surface soils, *Environ. Sci. Technol.,* 33, 1864, 1999.
165. Thorn KA and Kennedy KR, ^{15}N NMR investigation of the covalent binding of reduced TNT amines to soil humic acid, model compounds, and lignocellulose, *Environ. Sci. Technol.*, 36, 3787, 2002.
166. Thorn KA, Pennington JC, and Hayes CA, ^{15}N NMR investigation of the reduction and binding of TNT in an aerobic bench scale reactor simulating windrow composting, *Environ. Sci. Technol.*, 36, 3797, 2002.
167. Weiß M et al., Fate and metabolism of [^{15}N] 2,4,6-trinitrotoluene in soil, *Environ. Toxicol. Chem.*, 23, 1852, 2004.
168. Weiß M et al., Fate and stability of ^{14}C-labeled 2,4,6-trinitrotoluene in contaminated soil following microbial bioremediation processes, *Environ. Toxicol. Chem.*, 23, 2049, 2004.
169. Myers TE et al., Laboratory studies of soil sorption/transformation of TNT, RDX, and HMX, Technical Report IRRP-98-8. U.S. Army Corps of Engineers, Waterways Experiment Station, Vicksburg, MS, 1998.
170. Shangari N et al., Toxicity of glyoxals-role of oxidative stress, metabolic detoxification and thiamine deficiency, *Biochem. Soc. Trans.*, 31, 1390, 2003.

3 Effects of Energetic Materials on Soil Organisms

Roman G. Kuperman, Michael Simini,
Steven Siciliano, and Ping Gong

CONTENTS

3.1 Introduction 36
3.2 Effects on Soil Microorganisms 37
    3.2.1 Effects of Cyclic Nitramines on Microbial Activity in Soil 37
    3.2.2 Microbial Toxicity of Nitroaromatic Compounds 38
        3.2.2.1 Effects of Nitroaromatic Compounds on the Soil
                Microbial Community 39
        3.2.2.2 Effects of Nitroaromatic Compounds on the Nitrogen
                Cycle 41
        3.2.2.3 Effects of Nitroaromatic Compounds on the Carbon
                Cycle 43
        3.2.2.4 Ecological Consequences of Soil Contamination with
                Nitroaromatic Compounds 44
3.3 Effects on Terrestrial Plants 45
    3.3.1 Effects of Cyclic Nitramines 45
    3.3.2 Phytotoxicity of Nitroaromatic Compounds 46
3.4 Effects on Soil Invertebrates 52
    3.4.1 Effects of Cyclic Nitramines 52
    3.4.2 Effects of Nitroaromatic Compounds 62
        3.4.2.1 Acute Toxicity of Nitroaromatic Compounds 62
        3.4.2.2 Chronic Toxicity of Nitroaromatic Compounds 64
    3.4.3 Effects of Perchlorate 64
3.5 Effects of Weathering and Aging Energetic Materials in Soil on
    Toxicity to Soil Organisms 65
3.6 Stimulating Effects of Energetic Materials 66
3.7 Mechanisms of Toxicity 67
3.8 Conclusions and Future Outlook 70
References 72

3.1   INTRODUCTION

Many sites associated with military operations that involve munitions manufacturing, disposal, testing, and training can contain elevated levels of energetic materials (EMs) and related compounds in soil. Understanding the impacts of EMs and their products on soil quality, fertility, and structure is essential to protecting and sustaining the ecological integrity of terrestrial ecosystems at these sites. Additionally, it is important to understand how soil physical and chemical properties affect the exposure of soil organisms to EM contaminants. Central to achieving these goals is the need to increase our knowledge of the effects of EMs on soil organisms. These compounds can exert their effects directly through toxicity to terrestrial plants, soil invertebrates and microorganisms, or indirectly by altering specific interactions within the soil biota community, or by disrupting the soil food webs. The intensity and duration of the environmental effects of EMs may depend upon those processes that influence the fate, persistence, and movement of contaminants through the soil and into soil organisms. Ultimately, these effects can interfere with the regulation, flow, and internal cycling of carbon and nutrients in terrestrial ecosystems and undermine the sustainable use of testing and training ranges at defense installations. This chapter reviews the available information on the effects of EM soil contaminants on the three major components of soil ecosystems, including the soil microbial community and soil processes it regulates, the terrestrial plants, and the soil invertebrates. The fate and biotransformation of explosives in soil (reviewed in Chapter 2) and the effects on organisms inhabiting the aqueous phase of soil ecosystems, such as microalgae and protozoans (reviewed in Chapter 4), are only briefly discussed in this chapter.

Literature on the effects of EMs on soil organisms is scant, and discrepancies are often found regarding the toxicity of the same chemical to different organisms. A majority of ecotoxicological data was established in studies that investigated the effects of the cyclic nitramine explosives hexahydro-1,3,5-trinitro-1,3,5-triazine (RDX) and octahydro-1,3,5,7-tetranitro-1,3,5,7-tetrazocine (HMX); or the nitroaromatic explosive 2,4,6-trinitrotoluene (TNT) and its production, use, and transformation/degradation products, including 2,4-dinitrotoluene (2,4-DNT), 2,6-dinitrotoluene (2,6-DNT), aminodinitrotoluenes (ADNTs), and 1,3,5-trinitrobenzene (TNB; also manufactured for vulcanizing natural rubber and other uses). Limited ecotoxicological data are available for perchlorate and for the recently developed explosive and propellant material, polycyclic nitramine, 2,4,6,8,10,12-hexanitro-2,4,6,8,10,12-hexaazaisowurtzitane (China Lake 20 or CL-20). The application of ecotoxicological benchmarks established in reported studies for use in the ecological risk assessment (ERA) process at contaminated sites is discussed in Chapter 12.

A preponderance of toxicity data reported to date was generated in studies using artificial soil (similarly formulated Organization for Economic Cooperation and Development [OECD] artificial soil or U.S. Environmental Protection Agency [USEPA] standard artificial soil). These investigations did not consider the effects of soil physical and chemical properties, which can vary widely at contaminated sites, on the bioavailability and subsequent toxicity of EM. Ecotoxicological data established in such studies may have limited relevance for site-specific ERAs. Therefore, special consideration was given to experimental data from studies that used natural soils and to data from studies that examined the effects of weathering and aging EM in soil on toxicity to soil organisms. Such studies more closely approximated the exposure conditions in the field and are more relevant for an ERA at locations where contaminants have been historically present.

3.2   EFFECTS ON SOIL MICROORGANISMS

Soil bacteria and fungi play an important role in the biodegradation and transformation of nitramine and nitroaromatic explosives (as reviewed in Chapter 2). Soil contamination with explosives at concentrations that exceed tolerance levels of soil microorganisms can adversely affect the capacity of soil for natural biodegradation or detoxification at contaminated sites. Perhaps even more important are the impacts of explosive soil contaminants on the critical ecosystem functions mediated by soil microorganisms involved in nutrient and carbon cycling. Notwithstanding the need for assessing the effects of soil contamination with EMs on microbially mediated processes, very few studies were conducted to ascertain such effects. The sparse ecotoxicological data pertinent to direct exposures in amended soils established from laboratory toxicity studies as well as available data from studies that investigated the effects on microbial communities at contaminated sites are reviewed in this section.

3.2.1   EFFECTS OF CYCLIC NITRAMINES ON MICROBIAL ACTIVITY IN SOIL

A review by Sunahara et al. [1] indicated that RDX and HMX have small to moderate effects on microbial activity as demonstrated by a 10% to 15% inhibition at the greatest tested nominal concentration in soil of 10,000 mg kg^{-1} for either EM. The effects of RDX on the indigenous soil microbial communities were investigated by Gong et al. [2] in two soil types with contrasting texture (sandy loam and silty clay loam) and organic carbon (C_{org}) content (11.2% and 3.5%). In these studies, several ecologically relevant assays, including dehydrogenase activity (DHA), potential nitrification activity (PNA), nitrogen fixation activity (NFA), basal respiration (BR), and substrate-induced respiration (SIR), were used. Based on nominal RDX concentrations, these studies established no significant inhibitory effects at approximately 6,000 mg kg^{-1} (no observed effect concentration, NOEC) and the lowest observed effect concentration (LOEC) of approximately 12,000 mg kg^{-1} for DHA in the two soil types after 4 weeks. The NOEC/LOEC values for PNA were 567/2,623 and 1,194/5,952 mg kg^{-1} in sandy loam soil after 1 and 4 weeks, respectively; and 975/9,829 and 6,143/12,237 mg kg^{-1} in silty clay loam soil after 1 and 12 weeks, respectively. The NOEC/LOEC values for NFA ranged from 975/6,143 to 1,235/12,237 mg kg^{-1} in silty clay loam soil with no inhibition up to 12,237 mg kg^{-1} (NOEC) in sandy loam soil during 12 weeks of incubation. The soil respiration assays established the NOEC/LOEC values of 248/1,235 and 1,235/6,143 mg kg^{-1} for SIR and BR, respectively, in silty clay loam; and no inhibition up to 12,237 mg kg^{-1} (NOEC) in sandy loam soil after the 12-week incubation.

Overall, the results of these studies [2] showed that RDX has a relatively low toxicity to the selected indicators of soil microbial activity. Inhibition of microbial activity by RDX appeared greater in silty clay loam soil, which had lower C_{org} content, although this trend would require confirmation in studies with multiple soil types. Generally, RDX is resistant to aerobic degradation in soil (Chapter 2), which limits metabolic activation of nitro groups and contributes to the low level of microbial toxicity. However, the authors [2] hypothesized that RDX can have long-term deleterious impacts on the soil microbial communities due, in part, to a gradual increase with time in the RDX concentration in soil solution. Such longer-term effects of RDX require further investigation.

Gong et al. [3] investigated the microbial toxicity of HMX in soil using the same soil types, exposure periods, and measurement endpoints as in the study of RDX [2]. These studies showed no significant effects of HMX on any of the five microbial indicators up to and including approximately 12,500 mg kg^{-1} nominal concentration in either soil type. In contrast with the results of Gong et al. [3], microbial activity measured as DHA, PNA, BR, and SIR was lower (compared with activity in the reference site soil) in samples collected from an antitank firing range where HMX was a principal contaminant. In these studies, trace quantities of co-contaminants including RDX at ≤2.8 mg kg^{-1}, TNT at ≤0.49 mg kg^{-1}, and ADNTs at ≤6.7 mg kg^{-1} were also detected in some samples [4]. However, concentrations of HMX ranging from 14 to 696 mg kg^{-1} in these soil samples did not correlate with any of the microbial endpoints assessed in that study [4]. Although HMX is structurally similar to RDX, it has lower aqueous solubility (see Chapter 2) and greater chemical stability compared with RDX, which can explain, in part, the contrasting effects of these two nitramines on the soil microbial activity.

Similar to HMX, no adverse effects on DHA or PNA were reported for CL-20 in a silty clay loam soil comparable to the one used in the studies by Gong et al. [2,3] and in Sassafras sandy loam (SSL) from the limit test with a single CL-20 concentration of 10,000 mg kg^{-1} during a two-week exposure [5]. The limit test is a variant of a definitive test that is performed when statistical analysis of the range-finding test data shows no significant effect at all treatment concentrations. In contrast to a multi-concentration definitive test, the limit test consists of only two treatments, including a control (0 mg kg^{-1}) and the selected greatest concentration of a chemical. The limit test with CL-20 showed that PNA was actually stimulated in the 10,000 mg kg^{-1} treatment, which was attributed to the release of nitrite during degradation of CL-20 in soil and further oxidation of nitrite to nitrate [5].

3.2.2 MICROBIAL TOXICITY OF NITROAROMATIC COMPOUNDS

In contrast with cyclic nitramines, the nitroaromatic compounds such as TNT and its transformation products can adversely affect basic soil processes, including denitrification and decomposition of organic matter (OM). The rapid transformation of TNT to the amino-nitro intermediates (see Chapter 2), including 2-amino-4,6-dinitrotoluene (2-ADNT) and 4-amino-2,6-dinitrotoluene (4-ADNT) following soil amendments and incubation, presented a challenge for partitioning the effects of the parent material (TNT) and its transformation products (ADNTs) on soil microorganisms. In

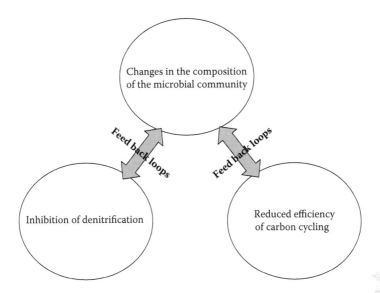

FIGURE 3.1   Schematic toxicological impacts of TNT on the soil microbial community.

most studies reviewed in this chapter, the exposure effects were assessed on the basis of acetonitrile-extractable concentrations of TNT or calculated as the sum of recovered TNT and ADNT concentrations. These effects were partially mediated by the differential toxicity of nitrotoluenes on Gram-negative compared to Gram-positive microorganisms. This differential impact not only affected the basic biogeochemical cycles of carbon and nitrogen but also significantly shifted the structural composition of the soil microbial community. Several studies [6–11] indicated that these effects were also reflected by the functional diversity and resiliency of the microbial communities. Studies reviewed in this chapter showed that TNT is a cytotoxic and genotoxic compound, and can affect the soil microbial community. The impact can be differentiated into distinct modes of actions resulting in specific and long-term changes in the soil microbial community. Studies have shown that TNT may alter the biogeochemical cycle of nitrogen and suggested that the denitrification portion of the nitrogen cycle can be affected most severely. Soil contamination with TNT can adversely affect the soil carbon cycle resulting in ecological changes in the entire ecosystem. These modes of action do not operate in isolation and can result in negative feedback loops, where changes in the composition of the microbial community could alter the efficiency of carbon cycling and denitrification. The change in carbon cycling can, in turn, alter microbial communities inhabiting the soil and thus alter the microbial community composition further (Figure 3.1).

3.2.2.1 Effects of Nitroaromatic Compounds on the Soil Microbial Community

The majority of ecotoxicological data for individual strains of soil microorganisms was derived from studies with amended growth media or soil slurries; therefore, these data cannot be used directly to infer the exposure effects in aerobic soil because the differences in the bioavailability and the fate of EMs in amended media were substantial. The average EC_{50} (effect median concentration) value of 7.8 µg TNT ml^{-1} was reported by Fuller and Manning [6] for cell growth inhibition of 14 different Gram-positive bacterial isolates. This inhibition was transitory; a decrease in TNT concentrations in soil slurries during bioremediation from 80 to 10 µg ml^{-1} resulted in a 1000-fold increase in number of culturable aerobic Gram-positive bacteria [8]. In contrast, most of the 17 Gram-negative organisms examined grew well in the presence of initial TNT concentrations up to 66 µg ml^{-1} [6]. Actinomycetes did not recover from TNT toxicity in soil slurries even when TNT concentrations were substantially decreased by bioremediation [8]. Furthermore, actinomycetes were reported to have a substantially lower EC_{50} of 362 mg kg^{-1} compared to Gram-positive organisms having EC_{50} of 4,177 mg kg^{-1}, as assessed by the prevalence of phospholipid fatty acids (PLFAs) under in situ soil conditions at the Joliet Army Ammunition Plant (Joliet, Illinois) [7]. The difference in TNT toxicity has long been thought to be due to differences in the uptake of TNT by Gram-positive as opposed to Gram-negative organisms. Fuller and Manning [6] hypothesized that the cell wall of Gram-positive bacteria can be more permeable to TNT compared to Gram-negative bacteria, or that Gram-negative bacteria can possess either active transport systems to move TNT out of the cell or enzymes that can detoxify TNT. The authors [6] based these hypotheses on microbial responses to the antimicrobial agent chloramphenicol, which they considered to be structurally similar to TNT. Chloramphenicol was reported to lose its effectiveness against strains possessing chloramphenicol impermeable cell membranes, chloramphenicol pumps, or deactivating enzymes. This can be the case for Gram-positive and Gram-negative bacteria but the precise reason why TNT is more toxic to actinomycetes is unclear at the present time. The soil ecological significance of selective impacts on Gram-positive organisms and actinomycetes is linked to the predominant role that these organisms play in the soil carbon cycle, especially in bulk soil. As will be discussed later, the inhibition of Gram-positive organisms can result in a dramatic inhibition of soil carbon cycling in TNT-contaminated soils.

The sensitivity of Gram-positive organisms to TNT does not mean that the effect of TNT on the soil microbial community is limited to only that group of organisms. In fact, exposure to TNT can lead to widespread changes in microbial community composition [9] and decreased diversity [10]. For example, an increase in TNT concentration from 10 to 80 µg ml^{-1} resulted in lower diversity as measured by a decrease in the number of identifiable PLFAs from 34 to 14 [8]. This decreased diversity often is not reflected in decreased total culturable heterotrophic bacterial numbers, which have often remained constant across TNT concentrations of several orders of magnitude [7,10]. The apparent lack of response is likely an artifact of the plate count method used to assess bacterial populations. This method is known to assess between 1% and 5% of the microbial community present in soil. Thus, this method is quite insensitive to bacterial community changes. In contrast, more sophisticated and sensitive techniques routinely detect differences in generic microbial community functions at low TNT concentrations. A study of functional diversity of microbial community based on utilization of 32 different substrates by a cultured indigenous community established the EC_{10} value of 0.2 µg ml^{-1} [11]. The EC_{20} values for inhibition of microbial respiration ranged from 70 mg kg^{-1} in a forest soil (4.1% C_{org}) to 530 mg kg^{-1} in a garden soil (8.7% C_{org}) [12]. These selective impacts have been assessed using the principle of pollution-induced community tolerance (PICT) in several investigations. The PICT principle states that a community exposed to a toxicant will become tolerant to that toxicant. As a result, when a tolerant (i.e., exposed) community is exposed to that toxicant it will retain more of its functionality at a given concentration of a toxicant compared to a nonexposed community. Exposure to TNT resulted in exactly this pattern of tolerance acquisition, as assessed by in vitro [11] or in situ [12] techniques. Recently, a non-PICT-based study has suggested that not only does TNT cause selective disruption of the microbial community but that these changes are irreversible [8].

3.2.2.2 Effects of Nitroaromatic Compounds on the Nitrogen Cycle

Selective TNT toxicity is readily evident in the nitrogen cycle (Figure 3.2). Soils contaminated with TNT often contain elevated concentrations of ammonia and nitrate [7,13] as a result of TNT degradation pathways described in Chapter 2. However, Fuller and Manning [7] found that ammonia oxidation is relatively insensitive to TNT concentrations with regression coefficients (r^2) between 0.15 and 0.39 ($p > 0.087$). In contrast, these authors hypothesized that the denitrification portion of the nitrogen cycle was easily disrupted by TNT, resulting in the accumulation of denitrification

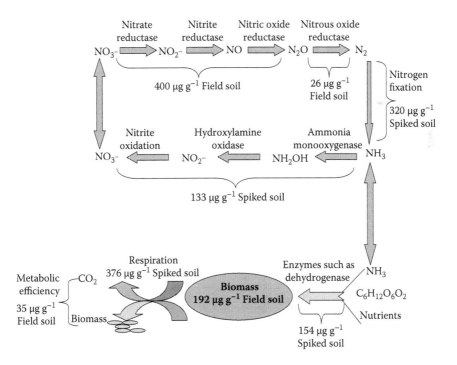

FIGURE 3.2 The toxicity of TNT to the nitrogen and carbon soil biogeochemical cycles. Values are concentrations of TNT shown to inhibit the specified process by 50%. (For nitrification, the entire nitrification process is indicated but the assay only measures the production of nitrite and not nitrate.)

substrates such as nitrate in the presence of high concentrations of munitions [7]. In support of this hypothesis, denitrification was reported to be very sensitive to TNT with EC_{50} of 26 mg kg^{-1} field soil [10]. Denitrification is the respiratory reduction of nitrate or nitrite to nitrous oxide. Nitrate is sequentially reduced to nitrous oxide by a set of four enzymes—nitrate reductase, nitrite reductase, nitric oxide reductase, and nitrous oxide reductase.

Typical assessments of denitrification activity only assess the activity of the first three enzymes and do not consider the sensitivity of nitrous oxide reductase to toxic compounds. However, as hypothesized by others [14,15], it appears that nitrous oxide reductase is much more sensitive to TNT than the other three enzymes based on corresponding EC_{50} values of 400 mg kg^{-1} for the first three enzymes and 26 mg kg^{-1} for nitrous oxide reductase [10].

Despite an observed decrease in denitrification activity, the percentage of the total microbial community comprising denitrifiers increased in the presence of TNT [10]. It should be emphasized that this effect is not an increase in the total numbers of denitrifiers but rather as a group; denitrifiers make up a larger proportion of the surviving community. The majority of denitrifiers are thought to be Gram-negative organisms and as discussed earlier, Gram-positive organisms are more sensitive to TNT compared with Gram-negative organisms. Thus, the increase in proportion is likely the result of a large decrease in the Gram-positive community. There are also selective changes in the denitrification community. Organisms possessing the copper-containing nitrite reductase are three times more resistant to TNT compared to organisms possessing the cytochrome-containing nitrite reductase for reasons that are not known at this time [10]. Given that no organism contains both forms of the nitrite reductase [16], it is not clear if this differential resistance to TNT is the result of differences in nitrite reductase or some other physiological process. Overall, these toxicological impacts on the nitrogen cycle are readily evident in many field sites and explain, in part, the elevated concentrations of nitrogenous compounds often associated with contaminated sites.

Other components of the nitrogen cycle, such as nitrogen fixation, are not as sensitive to TNT as denitrification. Nitrogen fixation is the conversion of nitrogen gas to an ionic species of nitrogen and is an important process that is carried out by free-living or symbiotic microorganisms. The commonly used nitrogen fixation toxicity assay assesses the ability of heterotrophic organisms in the soil to fix atmospheric nitrogen, that is, free-living nitrogen fixation. It does not assess the impact on the rhizobium–legume symbiosis. Under these conditions, the EC_{50} values for nitrogen fixation in soil ranged from 103 to 474 mg TNT kg^{-1} soil [2,17]. This relatively insensitive parameter is also of questionable ecological importance because, although free-living nitrogen-fixing organisms are present in all ecosystems, their contribution is commonly dwarfed by organic nitrogen mineralization or symbiotic nitrogen fixation activities. The sensitivity of free-living nitrogen fixers is comparable to that of actinomycetes, but it is not known if the nitrogen fixers are capable of recovering from exposure to TNT, or like the actinomycetes, will be permanently impaired.

Nitrification is the commonly used term that combines the activity of two distinctly different groups of microorganisms: ammonia oxidizing bacteria and nitrite oxidizing bacteria. In combination, these two groups of organisms convert ammonia to nitrate and derive energy from this process. These organisms are typically highly sensitive to changes in soil pH and their activity is often thought to be minimal in forest soils [18,19]. Certainly, the sensitivity of nitrifiers to TNT is similar to other functional groups based on reported EC_{50} values of 39 [2] and 227 [17] mg kg^{-1} for PNA, but greater than that for the nitrous oxide reductase found in the denitrification cycle.

3.2.2.3   Effects of Nitroaromatic Compounds on the Carbon Cycle

The carbon biogeochemical cycle can also be disrupted by soil contamination with TNT. An estimate of the number of microorganisms present in soil as microbial biomass is commonly assessed by fumigating the soil with a cell-lysing chemical such as chloroform, and then measuring how much carbon dioxide is released by organisms mineralizing dead microorganisms. Under these conditions, the EC_{50} and EC_{20} values for microbial biomass of 192 and 5 mg TNT kg^{-1} field soil, respectively, were reported by Frische and Hoper [20]. With the exception of nitrous oxide reductase, microbial biomass is more sensitive to TNT contamination than generic components of the nitrogen cycle. This may, in part, be due to how the nitrogen cycle is commonly assessed in toxicity assays. Typically, soil is assessed under slurry conditions, which are not representative of in situ conditions of the aerobic vadose zone of soil. In contrast, most estimators of the carbon cycle are assessed by fumigating a soil brought to 50%–60% of the water-holding capacity [20]. Under these conditions, different estimates of the impact of TNT on the carbon cycle performed in different laboratories are quite similar. For example, the EC_{50} values for inhibition of dehydrogenase activity ranging from 139 to 493 mg TNT kg^{-1} soil [17] compare favorably with the sensitivity of the microbial biomass observed by Frische and Hoper [20]. However, using the more sensitive indicators of carbon cycle impairment, such as the metabolic quotient, the EC_{20} and EC_{50} values of 3 and 35 mg TNT kg^{-1} field soil, respectively, have been established [20]. Both the metabolic quotient and the ratio of microbial biomass to soil C_{org} are estimates of microbial efficiency. Microbial efficiency is the proportion of a mineralized compound that can be devoted to growth and reproduction as opposed to cellular maintenance. It is commonly thought that when microorganisms are stressed by toxicants, their metabolic efficiency degrades because more of their cellular energy is devoted to cellular maintenance. Results of studies with TNT support the latter hypothesis by demonstrating that the metabolic quotient and the microbial biomass to soil C_{org} ratio are much more sensitive to TNT exposure compared to dehydrogenase activity or microbial biomass. This decreased metabolic efficiency was reflected by the high correlation ($r = -0.98$, $p < 0.001$) between adenosine triphosphate and TNT concentrations [21]. In this study, the soil was co-contaminated with heavy metals and it was difficult to derive an EC_{50} for TNT under mixed contamination conditions. However, the study by Lee et al. [21] confirms that microbial efficiency is negatively impacted by TNT [20]. This also has an impact on the nitrogen cycle because although the microorganisms are not incorporating as much of the carbon substrate, there is a corresponding decrease in the amount of nitrogen required by the microbial communities. Thus, there is likely a net mineralization occurring of nitrogen present in the soil OM.

Substrate-induced respiration appears to be a less sensitive indicator of TNT impacts on the carbon biogeochemical cycle compared with the measurement

TABLE 3.1

Summary of Effective Concentrations of TNT That Reduce Soil Microbial Endpoints by 50% (EC_{50})

Endpoint	EC_{50}	Reference
Gram-positive eubacteria	7.8 mg L^{-1} (growth media)	6
	4,177 mg kg^{-1} (field soil)	7
Actinomycetes	362 mg kg^{-1} (field soil)	7
Microbial biomass	192 mg kg^{-1} (field soil)	20
Metabolic quotient	35 mg kg^{-1} (field soil)	20
Substrate-induced respiration	376 mg kg^{-1} (silty clay loam)	2
Microbial biomass: Soil organic carbon	123 mg kg^{-1} field soil	19
Dehydrogenase activity	139 mg kg^{-1} (silty clay loam)	2, 17
	493 mg kg^{-1} (sandy loam)	
	168 mg kg^{-1} (silty clay loam)	
Nitrogen fixation	474 mg kg^{-1} (silty clay loam)	2, 17
	103 mg kg^{-1} (sandy loam)	
	165 mg kg^{-1} (silty clay loam)	
Nitrification	39 mg kg^{-1} (silty clay loam)	2, 17
	533 mg kg^{-1} (sandy loam)	
	227 mg kg^{-1} (silty clay loam)	
Nitrous oxide reductase	26 mg kg^{-1} (field soil)	10
Nitrate reductase + nitrite reductase + nitric oxide reductase	400 mg kg^{-1} (field soil)	10

endpoints discussed earlier (Table 3.1), based on the reported EC_{50} of 376 mg kg^{-1} [2]. Substrate-induced respiration is assessed by amending a soil with glucose and measuring the amount of carbon dioxide released by the in situ microbial community. The relative insensitivity of this parameter is not surprising given that TNT directly affects the metabolic efficiency of microorganisms. If that assumption is correct, then assessing the carbon dioxide released in response to a pulse of glucose would not reflect a change in microbial respiration of a diverse soil heterotrophic community. Thus, substrate-induced respiration is likely the most insensitive parameter of the carbon biogeochemical cycle as is reflected by the established EC_{50} values (Table 3.1).

3.2.2.4   Ecological Consequences of Soil Contamination with Nitroaromatic Compounds

The ecotoxicological data reviewed in this section show that TNT contamination of soil can disrupt the ecological functioning of a soil system and inhibit natural attenuation processes. This occurs because TNT can destroy a portion of the soil microbial community involved in OM decomposition and reduces the amount of soil C_{org} produced by this community. Consequently, the amount of bioavailable TNT is relatively high and it interferes with colonization of impacted areas by plants, resulting in toxicity of TNT to soil invertebrates and adversely affecting site remediation and restoration efforts. These adverse impacts of TNT on soil biota can lead to decreased primary productivity within the ecosystem, further limiting microbial activity. Thus, the EC_{50} of 35 mg TNT kg^{-1} soil reported by Frische and Hoper [20] to inhibit microbial efficiency appears particularly important in quantifying the overall impacts of TNT on soil microorganisms. This TNT concentration is similar to that of 26 mg kg^{-1} found to inhibit the activity of nitrous oxide reductase [10]. Once the carbon and nitrogen cycles are disrupted, negative feedback loops can develop and adversely affect the sustainability of contaminated terrestrial ecosystem. Based on these ecotoxicological benchmarks, a value of approximately 30 mg kg^{-1} appears to be a reasonable estimate of the TNT concentration in soil likely to impair critical microbially mediated functions in the soil ecosystems by 50%.

As discussed earlier, nitroaromatic compounds introduced into soil can undergo rapid transformation to the amino-nitro intermediates. Frequent co-occurrence of TNT, TNB, DNTs, and ADNTs in soils of contaminated sites or in experimentally contaminated soil treatments precluded investigators from partitioning the effects of the parent materials and their transformation products on soil microorganisms [7,12,17]. As a result, the established toxicity values for TNT reported in previous studies should not be accepted unequivocally. Additional studies will be required to definitively resolve the toxicity of individual nitroaromatic EMs to the soil microbial community and to critical processes in the soil ecosystem regulated by this community.

3.3 EFFECTS ON TERRESTRIAL PLANTS

This section reviews available information on the toxicity of energetic compounds to higher terrestrial plants, including visual injury, metabolism, and growth. Other photosynthetic organisms, such as algae and blue-green bacteria were not considered in this chapter because their growth, habitat, absorption mechanisms, and metabolic processes are often quite different from higher terrestrial plants, and therefore cannot be compared to higher plants. Detailed discussion of uptake and bioaccumulation of EMs is covered in Chapter 10 of this book.

3.3.1 EFFECTS OF CYCLIC NITRAMINES

Cyclic nitramine compounds such as RDX, HMX, and CL-20 have exhibited little or no phytotoxicity [5,22–24]. Among these nitramines, RDX and HMX are highly mobile within the plant and concentrate in leaf and flower tissue [25–27] despite their relatively low water solubility (Chapter 2), whereas accumulation of CL-20 occurred mainly in the roots [24]. Consequently, phytoaccumulation of nitramine explosives can pose a low risk of exposure to RDX and HMX for grazers of above-ground vegetation [24,25,28–30], and to CL-20 for grazers of below-ground vegetation [Chapter 10, this volume]. Studies using several trophic levels (plants–invertebrates–mammals–birds) are needed to assess the potential for biomagnification of RDX, HMX, and CL-20 across the food chain.

Robidoux et al. [23] found no reduction in seedling emergence or biomass of lettuce (*Lactuca sativa* L.) and barley (*Hordeum vulgare* L.) at measured soil concentrations up to and including 3320 mg kg^{-1} and 1866 mg kg^{-1} HMX in artificial soil or in forest soil (3.8% OM, pH 7.6), respectively (Table 3.2). Winfield [27] examined uptake and phytotoxic responses in 16 plant species (10 wild/cover crop species and 6 agronomic species) to short-term (<12 days) RDX exposure and found that dicotyledonous plants were more sensitive to RDX exposure than were monocotyledonous plants. Sainfoin (*Onobrychis viciifolia* Scop.) and sunflower (*Helianthus annuus* L.) were by far the most sensitive of the species tested. Toxicity symptoms included several adverse developmental effects indicative of teratogenicity, dysfunctions in cellular mainte-nance, repair mechanisms, or metabolic pathways. The author concluded that the observed symptoms were indicative of DNA–toxicant interactions. In contrast with the effects on sunflower, the only detrimental effect of RDX to perennial ryegrass (*Lolium perenne* L.) was the production of yellow spots on the leaves [27]. Vila et al. [31] showed that the emergence and growth of rice were not affected by RDX at concentrations up to 2000 mg kg^{-1} but that total chlorophyll content decreased at RDX concentrations greater than 500 mg kg^{-1}.

Rocheleau et al. [22] found that exposures to RDX or HMX either freshly amended or weathered and aged in SSL soil were not toxic to alfalfa (*Medicago sativa* L.), Japanese millet (*Echinochloa crusgalli* L.), and perennial ryegrass. Based on analytically determined concentrations, no phytotoxicity was observed at 9740 or 9537 mg kg^{-1} for RDX, and 10,410 or 9340 mg kg^{-1} for HMX in freshly amended or weathered-and-aged treatments, respectively. Furthermore, significant ($p < 0.0001$) growth stimulation was observed in Japanese millet and ryegrass exposed to these high concentrations of RDX or HMX. Similar growth stimulation was reported by Gong et al. [5] and Rocheleau et al. [24] for ryegrass exposed to approximately 10,000 mg kg^{-1} CL-20 in SSL for 21 days, whereas the growth of alfalfa was not affected at this concentration in SSL soil [5]. Extending ryegrass exposure to 42 days revealed inhibition of shoot growth up to 24% (compared with carrier control) at CL-20 concentrations ranging from 9 to 960 mg kg^{-1} and inhibition of root growth by 58% at 9604 mg kg^{-1} [24].

Cyclic nitramine compounds such as RDX and HMX appear to have different modes of action in plants compared to the nitroaromatic compounds. Their basic structure and behavior in plants are similar to those of the triazine herbicides such as atrazine and simazine, which inhibit photosystem II in photosynthesis [32]. The inhi-bition of photosynthesis may also be one of the modes of toxicity of cyclic nitramines based on symptoms such as chlorosis, necrosis, yellow leaf spots, and anthocyanin expression [27]. However, the mechanism(s) of phytotoxicity of cyclic nitramines have not been investigated thoroughly.

3.3.2 PHYTOTOXICITY OF NITROAROMATIC COMPOUNDS

Research on the effects of nitroaromatic EMs on terrestrial plants has taken a long and winding path that has yet to reach its destination. Early studies were mainly designed to demonstrate that newly developed explosives were not highly phytotoxic or focused on assessing the effects of TNT and related compounds associated with "pink water" waste on aquatic plants [33–35]. Schott and Worthley [34] found that the growth of *Lemna perpusilla* Torr. was reduced at TNT concentrations of 1 mg L^{-1}

TABLE 3.2

Selected Ecotoxicological Benchmarks for Energetic Materials Established in Standardized Single-Species Toxicity Tests with Terrestrial Plants

EM	Species	Soil Type	Soil pH	OM (%)	Clay (%)	Toxicity Benchmark	Value (mg kg^{-1})	Exposure	ERE	Ref.
RDX	A, JM, PR	SSL	5.2	1.2	17.0	LOAEC	>9740	Fresh	SE, SFM, SDM	22
RDX	A, JM, PR	SSL	5.2	1.2	17.0	LOAEC	>9537	W/A	SE, SFM, SDM	22
HMX	A, JM, PR	SSL	5.2	1.2	17.0	LOAEC	>10411	Fresh	SE, SFM, SDM	22
HMX	A, JM, PR	SSL	5.2	1.2	17.0	LOAEC	>9341	W/A	SE, SFM, SDM	22
CL-20	A, PR	SSL	5.2	1.2	17.0	LOAEC	>9832	Fresh	SE, SFM, SDM	5
TNT	L	OECD	6.5	10.0	20.0	IC_{20}	3113[a]	Fresh	SE	23
TNT	L	OECD	6.5	10.0	20.0	IC_{50}	>3120[a]	Fresh	SE	23
TNT	B	OECD	6.5	10.0	20.0	IC_{20}	8133	Fresh	SE	23
TNT	B	OECD	6.5	10.0	20.0	IC_{50}	8133	Fresh	SE	23
TNT	B	OECD	6.5	10.0	20.0	IC_{20}	134	Fresh	SFM	23
TNT	B	OECD	6.5	10.0	20.0	IC_{50}	8133	Fresh	SFM	23
TNT	B	OECD	6.5	10.0	20.0	IC_{20}	1200	Fresh	SDM	23
TNT	B	OECD	6.5	10.0	20.0	IC_{50}	8133	Fresh	SDM	23
TNT	B	OECD	6.5	10.0	20.0	IC_{20}, IC_{50}	<56	Fresh	RFM RDM	23
TNT	B	SFS	7.6	3.8	8.0	IC_{20}	398	Fresh	SE	23
TNT	B	SFS	7.6	3.8	8.0	IC_{50}	>7859	Fresh	SE	23
TNT	B	SFS	7.6	3.8	8.0	IC_{20}	139	Fresh	SFM	23
TNT	B	SFS	7.6	3.8	8.0	IC_{50}	256	Fresh	SFM	23
TNT	B	SFS	7.6	3.8	8.0	IC_{20}	272	Fresh	SDM	23
TNT	B	SFS	7.6	3.8	8.0	IC_{50}	>7859	Fresh	SDM	23
TNT	B	SFS	7.6	3.8	8.0	IC_{20}	<91	Fresh	RFM	23
TNT	B	SFS	7.6	3.8	8.0	IC_{50}	140	Fresh	RFM	23
TNT	B	SFS	7.6	3.8	8.0	IC_{20}	<91	Fresh	RDM	23
TNT	B	SFS	7.6	3.8	8.0	IC_{50}	137	Fresh	RDM	23
TNT	A	SSL	5.2	1.2	17.0	EC_{20}	95	Fresh	SE	40
TNT	A	SSL	5.2	1.2	17.0	EC_{50}	142	Fresh	SE	40
TNT	A	SSL	5.2	1.2	17.0	EC_{20}	41	Fresh	SFM	40
TNT	A	SSL	5.2	1.2	17.0	EC_{50}	77	Fresh	SFM	40
TNT	A	SSL	5.2	1.2	17.0	EC_{20}	43	Fresh	SDM	40
TNT	A	SSL	5.2	1.2	17.0	EC_{50}	93	Fresh	SDM	40
TNT	A	SSL	5.2	1.2	17.0	EC_{20}	119	W/A	SE	40
TNT	A	SSL	5.2	1.2	17.0	EC_{50}	277	W/A	SE	40
TNT	A	SSL	5.2	1.2	17.0	EC_{20}	8	W/A	SFM	40
TNT	A	SSL	5.2	1.2	17.0	EC_{50}	19	W/A	SFM	40

(continued on next page)

TABLE 3.2 (continued)
Selected Ecotoxicological Benchmarks for Energetic Materials Established in Standardized Single-Species Toxicity Tests with Terrestrial Plants

EM	Species	Soil Type	Soil pH	OM (%)	Clay (%)	Toxicity Benchmark	Value (mg kg⁻¹)	Exposure	ERE	Ref.
TNT	A	SSL	5.2	1.2	17.0	EC_{20}	4	W/A	SDM	40
TNT	A	SSL	5.2	1.2	17.0	EC_{50}	10	W/A	SDM	40
TNT	JM	SSL	5.2	1.2	17.0	EC_{20}	40	Fresh	SE	40
TNT	JM	SSL	5.2	1.2	17.0	EC_{50}	210	Fresh	SE	40
TNT	JM	SSL	5.2	1.2	17.0	EC_{20}	33	Fresh	SFM	40
TNT	JM	SSL	5.2	1.2	17.0	EC_{50}	99	Fresh	SFM	40
TNT	JM	SSL	5.2	1.2	17.0	EC_{20}	52	Fresh	SDM	40
TNT	JM	SSL	5.2	1.2	17.0	EC_{50}	159	Fresh	SDM	40
TNT	JM	SSL	5.2	1.2	17.0	EC_{20}	105	W/A	SE	40
TNT	JM	SSL	5.2	1.2	17.0	EC_{50}	673	W/A	SE	40
TNT	JM	SSL	5.2	1.2	17.0	EC_{20}	6	W/A	SFM	40
TNT	JM	SSL	5.2	1.2	17.0	EC_{50}	33	W/A	SFM	40
TNT	JM	SSL	5.2	1.2	17.0	EC_{20}	11	W/A	SDM	40
TNT	JM	SSL	5.2	1.2	17.0	EC_{50}	52	W/A	SDM	40
TNT	PR	SSL	5.2	1.2	17.0	EC_{20}	90	Fresh	SE	40
TNT	PR	SSL	5.2	1.2	17.0	EC_{50}	137	Fresh	SE	40
TNT	PR	SSL	5.2	1.2	17.0	EC_{20}	94	Fresh	SFM	40
TNT	PR	SSL	5.2	1.2	17.0	EC_{50}	129	Fresh	SFM	40
TNT	PR	SSL	5.2	1.2	17.0	EC_{20}	61	Fresh	SDM	40
TNT	PR	SSL	5.2	1.2	17.0	EC_{50}	86	Fresh	SDM	40
TNT	PR	SSL	5.2	1.2	17.0	EC_{20}	106	W/A	SE	40
TNT	PR	SSL	5.2	1.2	17.0	EC_{50}	246	W/A	SE	40
TNT	PR	SSL	5.2	1.2	17.0	EC_{20}	15	W/A	SFM	40
TNT	PR	SSL	5.2	1.2	17.0	EC_{50}	48	W/A	SFM	40
TNT	PR	SSL	5.2	1.2	17.0	EC_{20}	13	W/A	SDM	40
TNT	PR	SSL	5.2	1.2	17.0	EC_{50}	48	W/A	SDM	40
TNT	C, T	Lufa 2.3	5.6	5.0	7.0	LOAEC	136	Fresh	SFM	39
TNT	C, T	BBA	6.6	5.0	5.3	LOAEC	54	Fresh	SFM	39
TNT	O, W	Lufa 2.3	5.6	5.0	7.0	LOAEC	311	Fresh	SFM	39
TNT	O	BBA	6.6	5.0	5.3	LOAEC	355	Fresh	SFM	39
TNT	W	BBA	6.6	5.0	5.3	LOAEC	158	Fresh	SFM	39
TNB	A	SSL	5.2	1.2	17.0	EC_{20}	145	Fresh	SE	40
TNB	A	SSL	5.2	1.2	17.0	EC_{50}	172	Fresh	SE	40
TNB	A	SSL	5.2	1.2	17.0	EC_{20}	38	Fresh	SFM	40
TNB	A	SSL	5.2	1.2	17.0	EC_{50}	107	Fresh	SFM	40
TNB	A	SSL	5.2	1.2	17.0	EC_{20}	62	Fresh	SDM	40
TNB	A	SSL	5.2	1.2	17.0	EC_{50}	129	Fresh	SDM	40
TNB	A	SSL	5.2	1.2	17.0	EC_{20}	109	W/A	SE	40
TNB	A	SSL	5.2	1.2	17.0	EC_{50}	114	W/A	SE	40
TNB	A	SSL	5.2	1.2	17.0	EC_{20}	20	W/A	SFM	40
TNB	A	SSL	5.2	1.2	17.0	EC_{50}	63	W/A	SFM	40

TABLE 3.2 (continued)
Selected Ecotoxicological Benchmarks for Energetic Materials Established
in Standardized Single-Species Toxicity Tests with Terrestrial Plants

EM	Species	Soil Type	Soil pH	OM (%)	Clay (%)	Toxicity Benchmark	Value (mg kg^{-1})	Exposure	ERE	Ref.
TNB	A	SSL	5.2	1.2	17.0	EC$_{20}$	46	W/A	SDM	40
TNB	A	SSL	5.2	1.2	17.0	EC$_{50}$	92	W/A	SDM	40
TNB	JM	SSL	5.2	1.2	17.0	EC$_{20}$	109	Fresh	SE	40
TNB	JM	SSL	5.2	1.2	17.0	EC$_{50}$	204	Fresh	SE	40
TNB	JM	SSL	5.2	1.2	17.0	EC$_{20}$	16	Fresh	SFM	40
TNB	JM	SSL	5.2	1.2	17.0	EC$_{50}$	36	Fresh	SFM	40
TNB	JM	SSL	5.2	1.2	17.0	EC$_{20}$	43	Fresh	SDM	40
TNB	JM	SSL	5.2	1.2	17.0	EC$_{50}$	89	Fresh	SDM	40
TNB	JM	SSL	5.2	1.2	17.0	EC$_{20}$	139	W/A	SE	40
TNB	JM	SSL	5.2	1.2	17.0	EC$_{50}$	163	W/A	SE	40
TNB	JM	SSL	5.2	1.2	17.0	EC$_{20}$	0.3	W/A	SFM	40
TNB	JM	SSL	5.2	1.2	17.0	EC$_{50}$	0.9	W/A	SFM	40
TNB	JM	SSL	5.2	1.2	17.0	EC$_{20}$	0.7	W/A	SDM	40
TNB	JM	SSL	5.2	1.2	17.0	EC$_{50}$	2.0	W/A	SDM	40
TNB	PR	SSL	5.2	1.2	17.0	EC$_{20}$	28	Fresh	SE	40
TNB	PR	SSL	5.2	1.2	17.0	EC$_{50}$	95	Fresh	SE	40
TNB	PR	SSL	5.2	1.2	17.0	EC$_{20}$	45	Fresh	SFM	40
TNB	PR	SSL	5.2	1.2	17.0	EC$_{50}$	75	Fresh	SFM	40
TNB	PR	SSL	5.2	1.2	17.0	EC$_{20}$	56	Fresh	SDM	40
TNB	PR	SSL	5.2	1.2	17.0	EC$_{50}$	89	Fresh	SDM	40
TNB	PR	SSL	5.2	1.2	17.0	EC$_{20}$	107	W/A	SE	40
TNB	PR	SSL	5.2	1.2	17.0	EC$_{50}$	150	W/A	SE	40
TNB	PR	SSL	5.2	1.2	17.0	EC$_{20}$	46	W/A	SFM	40
TNB	PR	SSL	5.2	1.2	17.0	EC$_{50}$	83	W/A	SFM	40
TNB	PR	SSL	5.2	1.2	17.0	EC$_{20}$	51	W/A	SDM	40
TNB	PR	SSL	5.2	1.2	17.0	EC$_{50}$	86	W/A	SDM	40
2,4-DNT	A	SSL	5.2	1.2	17.0	EC$_{20}$	>47	Fresh	SE	40
2,4-DNT	A	SSL	5.2	1.2	17.0	EC$_{50}$	>47	Fresh	SE	40
2,4-DNT	A	SSL	5.2	1.2	17.0	EC$_{20}$	11	Fresh	SFM	40
2,4-DNT	A	SSL	5.2	1.2	17.0	EC$_{50}$	38	Fresh	SFM	40
2,4-DNT	A	SSL	5.2	1.2	17.0	EC$_{20}$	34	Fresh	SDM	40
2,4-DNT	A	SSL	5.2	1.2	17.0	EC$_{50}$	56	Fresh	SDM	40
2,4-DNT	A	SSL	5.2	1.2	17.0	EC$_{20}$	104	W/A	SE	40
2,4-DNT	A	SSL	5.2	1.2	17.0	EC$_{50}$	115	W/A	SE	40
2,4-DNT	A	SSL	5.2	1.2	17.0	EC$_{20}$	7	W/A	SFM	40
2,4-DNT	A	SSL	5.2	1.2	17.0	EC$_{50}$	30	W/A	SFM	40
2,4-DNT	A	SSL	5.2	1.2	17.0	EC$_{20}$	15	W/A	SDM	40
2,4-DNT	A	SSL	5.2	1.2	17.0	EC$_{50}$	42	W/A	SDM	40
2,4-DNT	JM	SSL	5.2	1.2	17.0	EC$_{20}$	55	Fresh	SE	40
2,4-DNT	JM	SSL	5.2	1.2	17.0	EC$_{50}$	70	Fresh	SE	40

(continued on next page)

TABLE 3.2 (continued)
Selected Ecotoxicological Benchmarks for Energetic Materials Established in Standardized Single-Species Toxicity Tests with Terrestrial Plants

EM	Species	Soil Type	Soil pH	OM (%)	Clay (%)	Toxicity Benchmark	Value (mg kg^{-1})	Exposure	ERE	Ref.
2,4-DNT	JM	SSL	5.2	1.2	17.0	EC_{20}	4	Fresh	SFM	40
2,4-DNT	JM	SSL	5.2	1.2	17.0	EC_{50}	10	Fresh	SFM	40
2,4-DNT	JM	SSL	5.2	1.2	17.0	EC_{20}	25	Fresh	SDM	40
2,4-DNT	JM	SSL	5.2	1.2	17.0	EC_{50}	34	Fresh	SDM	40
2,4-DNT	JM	SSL	5.2	1.2	17.0	EC_{20}	>32	W/A	SE	40
2,4-DNT	JM	SSL	5.2	1.2	17.0	EC_{50}	86	W/A	SE	40
2,4-DNT	JM	SSL	5.2	1.2	17.0	EC_{20}	4	W/A	SFM	40
2,4-DNT	JM	SSL	5.2	1.2	17.0	EC_{50}	7	W/A	SFM	40
2,4-DNT	JM	SSL	5.2	1.2	17.0	EC_{20}	6	W/A	SDM	40
2,4-DNT	JM	SSL	5.2	1.2	17.0	EC_{50}	10	W/A	SDM	40
2,4-DNT	PR	SSL	5.2	1.2	17.0	EC_{20}	90	Fresh	SE	40
2,4-DNT	PR	SSL	5.2	1.2	17.0	EC_{50}	137	Fresh	SE	40
2,4-DNT	PR	SSL	5.2	1.2	17.0	EC_{20}	94	Fresh	SFM	40
2,4-DNT	PR	SSL	5.2	1.2	17.0	EC_{50}	129	Fresh	SFM	40
2,4-DNT	PR	SSL	5.2	1.2	17.0	EC_{20}	61	Fresh	SDM	40
2,4-DNT	PR	SSL	5.2	1.2	17.0	EC_{50}	86	Fresh	SDM	40
2,4-DNT	PR	SSL	5.2	1.2	17.0	EC_{20}	>8	W/A	SE	40
2,4-DNT	PR	SSL	5.2	1.2	17.0	EC_{50}	>8	W/A	SE	40
2,4-DNT	PR	SSL	5.2	1.2	17.0	EC_{20}	5	W/A	SFM	40
2,4-DNT	PR	SSL	5.2	1.2	17.0	EC_{50}	7	W/A	SFM	40
2,4-DNT	PR	SSL	5.2	1.2	17.0	EC_{20}	2	W/A	SDM	40
2,4-DNT	PR	SSL	5.2	1.2	17.0	EC_{50}	8	W/A	SDM	40
2,6-DNT	A	SSL	5.2	1.2	17.0	EC_{20}	11	Fresh	SE	40
2,6-DNT	A	SSL	5.2	1.2	17.0	EC_{50}	19	Fresh	SE	40
2,6-DNT	A	SSL	5.2	1.2	17.0	EC_{20}	1.3	Fresh	SFM	40
2,6-DNT	A	SSL	5.2	1.2	17.0	EC_{50}	5	Fresh	SFM	40
2,6-DNT	A	SSL	5.2	1.2	17.0	EC_{20}	3	Fresh	SDM	40
2,6-DNT	A	SSL	5.2	1.2	17.0	EC_{50}	10	Fresh	SDM	40
2,6-DNT	A	SSL	5.2	1.2	17.0	EC_{20}	26	W/A	SE	40
2,6-DNT	A	SSL	5.2	1.2	17.0	EC_{50}	55	W/A	SE	40
2,6-DNT	A	SSL	5.2	1.2	17.0	EC_{20}	2	W/A	SFM	40
2,6-DNT	A	SSL	5.2	1.2	17.0	EC_{50}	7	W/A	SFM	40
2,6-DNT	A	SSL	5.2	1.2	17.0	EC_{20}	0.4	W/A	SDM	40
2,6-DNT	A	SSL	5.2	1.2	17.0	EC_{50}	5	W/A	SDM	40
2,6-DNT	JM	SSL	5.2	1.2	17.0	EC_{20}	40	Fresh	SE	40
2,6-DNT	JM	SSL	5.2	1.2	17.0	EC_{50}	57	Fresh	SE	40
2,6-DNT	JM	SSL	5.2	1.2	17.0	EC_{20}	13	Fresh	SFM	40
2,6-DNT	JM	SSL	5.2	1.2	17.0	EC_{50}	16	Fresh	SFM	40
2,6-DNT	JM	SSL	5.2	1.2	17.0	EC_{20}	11	Fresh	SDM	40
2,6-DNT	JM	SSL	5.2	1.2	17.0	EC_{50}	18	Fresh	SDM	40
2,6-DNT	JM	SSL	5.2	1.2	17.0	EC_{20}	>15	W/A	SE	40

TABLE 3.2 (continued)

Selected Ecotoxicological Benchmarks for Energetic Materials Established in Standardized Single-Species Toxicity Tests with Terrestrial Plants

EM	Species	Soil Type	Soil pH	OM (%)	Clay (%)	Toxicity Benchmark	Value (mg kg⁻¹)	Exposure	ERE	Ref.
2,6-DNT	JM	SSL	5.2	1.2	17.0	EC_{50}	>15	W/A	SE	40
2,6-DNT	JM	SSL	5.2	1.2	17.0	EC_{20}	5	W/A	SFM	40
2,6-DNT	JM	SSL	5.2	1.2	17.0	EC_{50}	9	W/A	SFM	40
2,6-DNT	JM	SSL	5.2	1.2	17.0	EC_{20}	6	W/A	SDM	40
2,6-DNT	JM	SSL	5.2	1.2	17.0	EC_{50}	11	W/A	SDM	40
2,6-DNT	PR	SSL	5.2	1.2	17.0	EC_{20}	29	Fresh	SE	40
2,6-DNT	PR	SSL	5.2	1.2	17.0	EC_{50}	38	Fresh	SE	40
2,6-DNT	PR	SSL	5.2	1.2	17.0	EC_{20}	18	Fresh	SFM	40
2,6-DNT	PR	SSL	5.2	1.2	17.0	EC_{50}	39	Fresh	SFM	40
2,6-DNT	PR	SSL	5.2	1.2	17.0	EC_{20}	26	Fresh	SDM	40
2,6-DNT	PR	SSL	5.2	1.2	17.0	EC_{50}	39	Fresh	SDM	40
2,6-DNT	PR	SSL	5.2	1.2	17.0	EC_{20}	42	W/A	SE	40
2,6-DNT	PR	SSL	5.2	1.2	17.0	EC_{50}	54	W/A	SE	40
2,6-DNT	PR	SSL	5.2	1.2	17.0	EC_{20}	24	W/A	SFM	40
2,6-DNT	PR	SSL	5.2	1.2	17.0	EC_{50}	39	W/A	SFM	40
2,6-DNT	PR	SSL	5.2	1.2	17.0	EC_{20}	21	W/A	SDM	40
2,6-DNT	PR	SSL	5.2	1.2	17.0	EC_{50}	34	W/A	SDM	40

Note: A = alfalfa (*Medicago sativa* L.), B = barley (*Hordeum vulgare* L.), BBA = German loamy sand, C = cress (*Lepidium sativum* L.), EM = energetic material, ERE = ecologically relevant endpoint, JM = Japanese millet (*Echinochloa crusgalli* L.), L = lettuce (*Lactuca sativa* L.), O = oat (*Avena sativa* L), OECD = Organization for Economic Co-operation and Development standard artificial soil, OM = organic matter, PR = perennial ryegrass (*Lolium perenne* (L.) Beauv.), RDM = root dry mass, RFM = root fresh mass, RE = root elongation, SDM = shoot dry mass, SE = seedling emergence, SFM = shoot fresh mass, SFS = sandy forest soil, SSL = Sassafras sandy loam soil, T = turnip (*Brassica rapa* Metzg.), W = wheat (*Triticum aestivum* L.), W/A = weathered and aged EM soil treatment.

[a] Nominal concentration. All remaining concentrations are based on chemical analysis.

and greater, and that plant death was observed at 5 mg L⁻¹ and greater. Palazzo and Leggett [36] reported leaf chlorosis and reduced plant yields of hydroponically grown forage grasses and legumes exposed to TNT-contaminated pink water. Wastewater containing 140 mg L⁻¹ was used in this study at full, half, and quarter strength. Severity of plant injury was positively correlated with increasing TNT concentration. In a controlled study [36], yellow nutsedge (*Cyperus esculentus* L.) was grown in hydroponic cultures containing TNT concentrations of 0, 5, 10, and 20 mg L⁻¹. Effects were rapid and occurred at all TNT treatments. Growth reduction was in the order of: root > leaf > rhizome. TNT and its metabolites, 4-ADNT and 2-ADNT, were recovered throughout the plant, with the greatest concentrations in the roots. Subsequent studies with the lower TNT concentrations showed that plant yield and growth rate decreased as a result of exposure to concentrations ranging from 0.5 to

5 mg L^{-1} [37]. The authors also found that faster growing plants were more tolerant to TNT injury and yield reduction.

Studies employing the more ecologically relevant soil exposures showed that nitroaromatic EMs were more toxic to terrestrial plants compared to cyclic nitramines [22,23,26,36–40]. Soil concentrations of TNT and related nitroaromatic compounds that cause phytotoxicity varied with the soil type and plant species (Table 3.2). Cataldo et al. [28] reported 50% reduction in plant (*Phaseolus vulgaris* L. [bean], *Triticum aestivum* L. [wheat], and *Bromus mollis* L. [blando broom grass]) height in two soils (1.7% and 7.2% OM) at 60 mg kg^{-1}; a 25% reduction in the plant height for wheat and grass at 30 mg kg^{-1}; and no effects at 10 mg kg^{-1}. Peterson et al. [38] observed significant linear reduction in *Festuca arundinacea* Schreb. (tall fescue) seed germination on agar plates amended with EM concentrations ranging from 7.5 to 60 mg L^{-1} for TNT and from 3 to 15 mg L^{-1} for 4-ADNT ($p < 0.005$; $r^2 = 0.37$). Gong et al. [39] studied toxicity of TNT to *Lepidium sativum* L. (cress), *Brassica rapa* Metzg (turnip), *Avena sativa* L. (oat), and wheat seedlings in two German soils. Fresh shoot mass of cress was significantly ($p < 0.01$) decreased at TNT concentration of 50 mg kg^{-1}, whereas fresh shoot mass of oat or wheat was decreased at 150 mg kg^{-1} and 350 mg kg^{-1}, respectively. The authors suggested that TNT phytotoxicity could be affected by properties of the two soils used in the study [39].

Using standardized phytotoxicity test methods, Rocheleau et al. [40] investigated the effects of TNT, TNB, 2,4-DNT, and 2,6-DNT freshly amended (following a 24-h moisture equilibration period) or weathered and aged (13 weeks) in SSL soil. Tests [40] showed that dinitrotoluenes were more toxic for plant species in freshly amended treatments compared with TNB or TNT based on the EC$_{20}$ values for shoot growth (dry mass), which ranged from 3 to 24 mg kg^{-1} for dinitrotoluenes, and from 43 to 62 mg kg^{-1} for TNB or TNT (Table 3.2). Among the plant species tested in this study, Japanese millet was the most sensitive, followed by alfalfa and ryegrass. Exposure of the three plant species to relatively low concentrations of the four compounds initially stimulated plant growth before the onset of inhibition at greater concentrations (hormesis; see Section 3.6). A decrease in tall fescue seed germination and seedling growth inhibition at 15 mg L^{-1} were reported for 4-ADNT by Peterson et al. [38].

3.4   EFFECTS ON SOIL INVERTEBRATES

3.4.1   EFFECTS OF CYCLIC NITRAMINES

The effects of nitramine EM on soil invertebrates are summarized in Table 3.3. Studies with RDX or HMX showed no adverse effect on survival of adult earthworms *Eisenia fetida* Savigny up to 500 mg kg^{-1} in artificial soil or in natural soils [41,42]. Survival of adult *E. andrei* Bouché was unaffected up to 756 mg kg^{-1} RDX in artificial soil in a study by Robidoux et al. [43]. These authors [43] observed significant adverse effects of RDX on the reproduction (productivity of hatched cocoons and juveniles; number of juveniles per hatched cocoon) at 189 mg kg^{-1} and on production of juveniles (total number, mass, and number per cocoon) at 95 mg kg^{-1}. Schäfer and Achazi [44] reported no effects on mortality and reproduction of enchytraeid worm (potworm) *Enchytraeus albidus* Henle and collembola *Folsomia candida* Willem

TABLE 3.3

Selected Ecotoxicological Benchmarks for Energetic Materials Established in Standardized Single-Species Toxicity Tests with Soil Invertebrates and in a Microcosm Test with Indigenous Soil Microinvertebrate Community

EM	Species	Soil Type	Soil pH	OM (%)	Clay (%)	Toxicity Benchmark	Value (mg kg^{-1})	Concentration	Exposure	Endpoint	Reference
RDX	F. candida	Lufa 2.2	5.6	4.3	6.7	LOEC	>1000	Measured	30-d aged	Mort	44
RDX	Ench. crypticus	Lufa 2.2	5.6	4.3	6.7	LOEC	>1000	Measured	30-d aged	Mort	44
RDX	E. andrei	OECD	6.0	10.0	20.0	LOEC	189	Nominal	Fresh	Repr	43
RDX	E. andrei	OECD	6.0	10.0	20.0	LOEC	95.0	Nominal	Fresh	Juv wt	43
RDX	E. andrei	OECD	6.0	10.0	20.0	NOEC	756	Nominal	Fresh	Mort	43
RDX	E. andrei	OECD	6.0	10.0	20.0	EC$_{20}$	160.1	Measured	Fresh	Cocoons	47
RDX	E. andrei	SFS	7.6	3.8	8.0	EC$_{20}$	117.4	Measured	Fresh	Ad wt	47
RDX	E. andrei	SFS	7.6	3.8	8.0	LOEC	14.5	Measured	Fresh	Repr	47
RDX	E. andrei	SFS	7.6	3.8	8.0	LOEC	90.8	Measured	Fresh	Cocoons	47
RDX	Ench. crypticus	Rac50-50	7.9	23.0	2.0	LOEC	>658	Measured	Fresh	Mort	45
RDX	Ench. crypticus	Rac50-50	7.9	23.0	2.0	LOEC	>658	Measured	Fresh	Repr	45
RDX	Ench. albidus	Rac50-50	7.9	23.0	2.0	LOEC	>918	Measured	Fresh	Repr	45
RDX	Ench. albidus	Rac50-50	7.9	23.0	2.0	EC$_{20}$	161	Measured	Fresh	Repr	45
RDX	Ench. albidus	Rac50-50	7.9	23.0	2.0	EC$_{50}$	444	Measured	Fresh	Repr	45
RDX	Ench. crypticus	SSL	5.2	1.2	17.0	LOEC	>21,383	Measured	Fresh	Mort	48
RDX	Ench. crypticus	SSL	5.2	1.2	17.0	LOEC	>18,347	Measured	90-d W/A	Mort	48
RDX	Ench. crypticus	SSL	5.2	1.2	17.0	EC$_{20}$	3715	Measured	Fresh	Repr	48
RDX	Ench. crypticus	SSL	5.2	1.2	17.0	EC$_{50}$	51,413	Measured	Fresh	Repr	48
RDX	Ench. crypticus	SSL	5.2	1.2	17.0	EC$_{20}$	8797	Measured	90-d W/A	Repr	48
RDX	Ench. crypticus	SSL	5.2	1.2	17.0	EC$_{50}$	142,356	Measured	90-d W/A	Repr	48
RDX	E. fetida	SSL	5.2	1.2	17.0	EC$_{20}$	1.2	Measured	Fresh	Cocoons	42

(continued on next page)

TABLE 3.3 (continued)

Selected Ecotoxicological Benchmarks for Energetic Materials Established in Standardized Single-Species Toxicity Tests with Soil Invertebrates and in a Microcosm Test with Indigenous Soil Microinvertebrate Community

EM	Species	Soil Type	Soil pH	OM (%)	Clay (%)	Toxicity Benchmark	Value (mg kg⁻¹)	Concentration	Exposure	Endpoint	Reference
RDX	*E. fetida*	SSL	5.2	1.2	17.0	EC_{50}	3.7	Measured	Fresh	Cocoons	42
RDX	*E. fetida*	SSL	5.2	1.2	17.0	EC_{20}	19.2	Measured	90-d W/A	Cocoons	42
RDX	*E. fetida*	SSL	5.2	1.2	17.0	EC_{50}	59.6	Measured	90-d W/A	Cocoons	42
RDX	*E. fetida*	SSL	5.2	1.2	17.0	EC_{20}	1.6	Measured	Fresh	Repr	42
RDX	*E. fetida*	SSL	5.2	1.2	17.0	EC_{50}	5.0	Measured	Fresh	Repr	42
RDX	*E. fetida*	SSL	5.2	1.2	17.0	EC_{20}	4.8	Measured	90-d W/A	Repr	42
RDX	*E. fetida*	SSL	5.2	1.2	17.0	EC_{50}	14.9	Measured	90-d W/A	Repr	42
RDX	*E. fetida*	USEPA	6.0	10.0	20.0	LOEC	>500	Nominal	Fresh	Mort	41
HMX	*E. fetida*	USEPA	6.0	10.0	20.0	LOEC	>500	Nominal	Fresh	Mort	41
HMX	*E. andrei*	SFS	7.6	3.8	8.0	EC_{50}	106.0	Measured	Fresh	Ad wt	47
HMX	*E. andrei*	SFS	7.6	3.8	8.0	EC_{50}	15.8	Measured	Fresh	Repr	47
HMX	*E. andrei*	SFS	7.6	3.8	8.0	LOEC	29.5	Measured	Fresh	Repr	47
HMX	*E. andrei*	SFS	7.6	3.8	8.0	LOEC	15.6	Measured	Fresh	Cocoons	47
HMX	*E. andrei*	SFS	7.6	3.8	8.0	NOEC	711	Measured	Fresh	Mort	47
HMX	*F. candida*	Lufa 2.2	5.6	4.3	6.7	LOEC	>1000	Measured	30-d aged	Repr	44
HMX	*Ench. crypticus*	Lufa 2.2	5.6	4.3	6.7	LOEC	>1000	Measured	30-d aged	Repr	44
HMX	*Ench. albidus*	Rac50-50	7.9	23.0	2.0	LOEC	>918	Measured	Fresh	Mort	45
HMX	*Ench. albidus*	Rac50-50	7.9	23.0	2.0	LOEC	>918	Measured	Fresh	Repr	45
HMX	*Ench. crypticus*	SSL	5.2	1.2	17.0	LOEC	>21,750	Measured	Fresh	Mort/Repr	48
HMX	*Ench. crypticus*	SSL	5.2	1.2	17.0	LOEC	>17,498	Measured	90-d W/A	Mort/Repr	48
HMX	*E. fetida*	SSL	5.2	1.2	17.0	EC_{20}	2.7	Measured	Fresh	Cocoons	42
HMX	*E. fetida*	SSL	5.2	1.2	17.0	EC_{50}	8.5	Measured	Fresh	Cocoons	42

HMX	E. fetida	SSL	5.2	1.2	17.0	EC$_{20}$	0.4	Measured	Fresh	Repr	42
HMX	E. fetida	SSL	5.2	1.2	17.0	EC$_{50}$	1.2	Measured	Fresh	Repr	42
MNX	E. fetida	SanL	8.3	1.3	16.0	14-d LC$_{20}$	114.3	Measured	Fresh	Mort	49
MNX	E. fetida	SanL	8.3	1.3	16.0	14-d LC$_{50}$	262.1	Measured	Fresh	Mort	49
MNX	E. fetida	SiL	7.0	2.5	12.0	14-d LC$_{20}$	112.7	Measured	Fresh	Mort	49
MNX	E. fetida	SiL	7.0	2.5	12.0	14-d LC$_{50}$	244.6	Measured	Fresh	Mort	49
MNX	E. fetida	SanL	8.3	1.3	16.0	35-d EC$_{20}$	150.8	Measured	Fresh	Ad wt	49
MNX	E. fetida	SanL	8.3	1.3	16.0	35-d EC$_{50}$	525.8	Measured	Fresh	Ad wt	49
MNX	E. fetida	SiL	7.0	2.5	12.0	35-d EC$_{20}$	82.5	Measured	Fresh	Ad wt	49
MNX	E. fetida	SiL	7.0	2.5	12.0	35-d EC$_{50}$	228.9	Measured	Fresh	Ad wt	49
TNX	E. fetida	SanL	8.3	1.3	16.0	14-d LC$_{20}$	111.5	Measured	Fresh	Mort	49
TNX	E. fetida	SanL	8.3	1.3	16.0	14-d LC$_{50}$	251.3	Measured	Fresh	Mort	49
TNX	E. fetida	SiL	7.0	2.5	12.0	14-d LC$_{20}$	96.7	Measured	Fresh	Mort	49
TNX	E. fetida	SiL	7.0	2.5	12.0	14-d LC$_{50}$	216.3	Measured	Fresh	Mort	49
TNX	E. fetida	SanL	8.3	1.3	16.0	35-d EC$_{20}$	92.4	Measured	Fresh	Ad wt	49
TNX	E. fetida	SanL	8.3	1.3	16.0	35-d EC$_{50}$	237.7	Measured	Fresh	Ad wt	49
TNX	E. fetida	SiL	7.0	2.5	12.0	35-d EC$_{20}$	150.1	Measured	Fresh	Ad wt	49
TNX	E. fetida	SiL	7.0	2.5	12.0	35-d EC$_{50}$	364.4	Measured	Fresh	Ad wt	49
CL-20	Ench. crypticus	SSL	5.2	1.2	17.0	LC$_{20}$	6	Measured	Fresh	Mort	50
CL-20	Ench. crypticus	SSL	5.2	1.2	17.0	LC$_{50}$	18	Measured	Fresh	Mort	50
CL-20	Ench. crypticus	SSL	5.2	1.2	17.0	EC$_{20}$	0.1	Measured	Fresh	Repr	50
CL-20	Ench. crypticus	SSL	5.2	1.2	17.0	EC$_{50}$	0.3	Measured	Fresh	Repr	50
CL-20	Ench. crypticus	SSL	5.2	1.2	17.0	EC$_{20}$	0.035	Measured	83-d W/A	Repr	50
CL-20	Ench. crypticus	SSL	5.2	1.2	17.0	EC$_{50}$	0.1	Measured	83-d W/A	Repr	50
CL-20	Ench. crypticus	SSL	5.5	1.2	11.0	LC$_{20}$	0.2	Nominal	Fresh	Mort	45
CL-20	Ench. crypticus	SSL	5.5	1.2	11.0	LC$_{50}$	0.4	Nominal	Fresh	Mort	45
CL-20	Ench. crypticus	SSL	5.5	1.2	11.0	EC$_{20}$	0.04	Nominal	Fresh	Repr	45
CL-20	Ench. crypticus	SSL	5.5	1.2	11.0	EC$_{50}$	0.12	Nominal	Fresh	Repr	45

(continued on next page)

TABLE 3.3 (continued)

Selected Ecotoxicological Benchmarks for Energetic Materials Established in Standardized Single-Species Toxicity Tests with Soil Invertebrates and in a Microcosm Test with Indigenous Soil Microinvertebrate Community

EM	Species	Soil Type	Soil pH	OM (%)	Clay (%)	Toxicity Benchmark	Value (mg kg^{-1})	Concentration	Exposure	Endpoint	Reference
CL-20	Ench. albidus	SSL	5.5	1.2	11.0	LC$_{20}$	0.006	Nominal	Fresh	Mort	45
CL-20	Ench. crypticus	SSL	5.5	1.2	11.0	LC$_{50}$	0.2	Nominal	Fresh	Mort	45
CL-20	Ench. crypticus	RacAg	8.2	42.0	1.0	LC$_{20}$	0.003	Nominal	Fresh	Mort	45
CL-20	Ench. crypticus	RacAg	8.2	42.0	1.0	LC$_{50}$	0.1	Nominal	Fresh	Mort	45
CL-20	Ench. crypticus	RacAg	8.2	42.0	1.0	EC$_{20}$	0.001	Nominal	Fresh	Repr	45
CL-20	Ench. crypticus	RacAg	8.2	42.0	1.0	EC$_{50}$	0.08	Nominal	Fresh	Repr	45
CL-20	Ench. crypticus	Rac50-50	7.9	23.0	2.0	LC$_{20}$	0.3	Nominal	Fresh	Mort	45
CL-20	Ench. crypticus	Rac50-50	7.9	23.0	2.0	LC$_{50}$	0.7	Nominal	Fresh	Mort	45
CL-20	Ench. crypticus	Rac50-50	7.9	23.0	2.0	EC$_{20}$	0.03	Nominal	Fresh	Repr	45
CL-20	Ench. crypticus	Rac50-50	7.9	23.0	2.0	EC$_{50}$	0.62	Nominal	Fresh	Repr	45
CL-20	Ench. albidus	Rac50-50	7.9	23.0	2.0	EC$_{20}$	0.05	Nominal	Fresh	Repr	45
CL-20	Ench. albidus	Rac50-50	7.9	23.0	2.0	EC$_{50}$	0.19	Nominal	Fresh	Repr	45
TNT	E. fetida	USEPA	6.0	10.0	20.0	LOEC	>200	Nominal	Fresh	Mort	41
TNT	E. fetida	USEPA	6.0	10.0	20.0	NOEC	110	Nominal	Fresh	Ad wt	41
TNT	E. fetida	USEPA	6.0	10.0	20.0	LOEC	140	Nominal	Fresh	Ad wt	41
TNT	E. fetida	FS	3.8	5.9	10.5	LOEC	150	Nominal	Fresh	Ad wt	41
TNT	E. fetida	FS	3.8	5.9	10.5	LC$_{50}$	325	Nominal	Fresh	Mort	41
TNT	E. andrei	SFS	5.8	4.2	3.0	LC$_{50}$	143	Measured	Fresh	Mort	54
TNT	E. andrei	OECD	6.0	10.0	20.0	LOEC	420	Nominal	Fresh	Mort	55
TNT	E. andrei	OECD	6.0	10.0	20.0	LC$_{25}$	331	Nominal	Fresh	Mort	55
TNT	E. andrei	OECD	6.0	10.0	20.0	LC$_{50}$	365	Nominal	Fresh	Mort	55
TNT	E. andrei	SFS	5.9	4.2	3.0	LOEC	260	Nominal	Fresh	Mort	55

TNT	*E. andrei*	SFS	5.9	4.2	3.0	LC_{25}	Nominal	192	Fresh	Mort	55
TNT	*E. andrei*	SFS	5.9	4.2	3.0	LC_{50}	Nominal	222	Fresh	Mort	55
TNT	*E. andrei*	OECD	6.0	10.0	20.0	LOEC	Nominal	110	Fresh	Juv wt	43
TNT	*E. andrei*	OECD	6.0	10.0	20.0	EC_{25}	Nominal	102	Fresh	Juv wt	43
TNT	*E. andrei*	OECD	6.0	10.0	20.0	EC_{50}	Nominal	529	Fresh	Juv wt	43
TNT	*E. andrei*	OECD	6.0	10.0	20.0	LOEC	Nominal	220	Fresh	Repr	43
TNT	*E. andrei*	OECD	6.0	10.0	20.0	LOEC	Nominal	881	Fresh	Ad wt	43
TNT	*E. andrei*	OECD	6.0	10.0	20.0	EC_{25}	Nominal	495	Fresh	Ad wt	43
TNT	*E. andrei*	OECD	6.0	10.0	20.0	EC_{50}	Nominal	660	Fresh	Ad wt	43
TNT	*E. andrei*	OECD	6.0	10.0	20.0	EC_{20}	Measured	51.9	Fresh	Repr	47
TNT	*E. andrei*	SFS	7.6	3.8	8.0	EC_{50}	Measured	125.2	Fresh	Repr	47
TNT	*E. andrei*	SFS	7.6	3.8	8.0	EC_{20}	Measured	38.8	Fresh	Ad wt	47
TNT	*E. andrei*	SFS	7.6	3.8	8.0	EC_{50}	Measured	55.7	Fresh	Ad wt	47
TNT	*E. andrei*	SFS	7.6	3.8	8.0	LOEC	Measured	136	Fresh	Ad wt	47
TNT	*E. andrei*	SFS	7.6	3.8	8.0	LOEC	Measured	136	Fresh	Juv wt	47
TNT	*E. andrei*	SFS	7.6	3.8	8.0	LOEC	Measured	58.8	Fresh	Repr	47
TNT	*Ench. crypticus*	Lufa 2.2	5.6	4.3	6.7	7-d LC_{50}	Measured	1290	30-d aged	Mort	44
TNT	*Ench. crypticus*	Lufa 2.2	5.6	4.3	6.7	28-d EC_{50}	Measured	480	30-d aged	Repr	44
TNT	*F. candida*	Lufa 2.2	5.6	4.3	6.7	7-d LC_{50}	Measured	420	30-d aged	Mort	44
TNT	*F. candida*	Lufa 2.2	5.6	4.3	6.7	28-d EC_{50}	Measured	315	30-d aged	Repr	44
TNT	*Ench. crypticus*	Lufa 2.2	5.8	3.8	6.7	7-d LC_{50}	Measured	950	30-d aged	Mort	44
TNT	*Ench. crypticus*	Lufa 2.2	5.8	3.8	6.7	28-d EC_{50}	Measured	501	Fresh	Repr	29
TNT	*Ench. crypticus*	Lufa 2.2	5.8	3.8	6.7	7-d LC_{50}	Measured	140	Fresh	Mort	29
TNT	*F. candida*	Lufa 2.2	5.8	3.8	6.7	28-d EC_{50}	Measured	64.3	Fresh	Repr	29
TNT	*F. candida*	Lufa 2.2	5.8	3.8	6.7	28-d EC_{50}	Measured	119.6	7-d aged	Repr	29
TNT	*F. candida*	Lufa 2.2	5.8	3.8	6.7	28-d EC_{50}	Measured	155.3	15-d aged	Repr	29
TNT	*F. candida*	Lufa 2.2	5.8	3.8	6.7	28-d EC_{50}	Measured	186.1	30-d aged	Repr	29
TNT	*F. candida*	Lufa 2.2	5.8	3.8	6.7	28-d EC_{50}	Measured	230.2	60-d aged	Repr	29

(continued on next page)

TABLE 3.3 (continued)

Selected Ecotoxicological Benchmarks for Energetic Materials Established in Standardized Single-Species Toxicity Tests with Soil Invertebrates and in a Microcosm Test with Indigenous Soil Microinvertebrate Community

EM	Species	Soil Type	Soil pH	OM (%)	Clay (%)	Toxicity Benchmark	Value (mg kg^{-1})	Concentration	Exposure	Endpoint	Reference
TNT	*Ench. crypticus*	Lufa 2.3	6.5	1.2	7.4	7-d LC$_{50}$	640	Measured	Fresh	Mort	29
TNT	*Ench. crypticus*	Lufa 2.3	6.5	1.2	7.4	28-d EC$_{50}$	277	Measured	Fresh	Repr	29
TNT	*F. candida*	Lufa 2.3	6.5	1.2	7.4	7-d LC$_{50}$	104.5	Measured	Fresh	Mort	29
TNT	*F. candida*	Lufa 2.3	6.5	1.2	7.4	28-d EC$_{50}$	23.5	Measured	Fresh	Repr	29
TNT	*Ench. crypticus*	Lufa 3	7.1	1.9	9.6	7-d LC$_{50}$	1375	Measured	Fresh	Mort	29
TNT	*Ench. crypticus*	Lufa 3	7.1	1.9	9.6	28-d EC$_{50}$	624	Measured	Fresh	Repr	29
TNT	*F. candida*	Lufa 3	7.1	1.9	9.6	7-d LC$_{50}$	454	Measured	Fresh	Mort	29
TNT	*F. candida*	Lufa 3	7.1	1.9	9.6	28-d EC$_{50}$	40.6	Measured	Fresh	Repr	29
TNT	*Ench. crypticus*	Lufa 4	7.4	2.8	24.4	7-d LC$_{50}$	1099	Measured	Fresh	Mort	29
TNT	*Ench. crypticus*	Lufa 4	7.4	2.8	24.4	28-d EC$_{50}$	919	Measured	Fresh	Repr	29
TNT	*F. candida*	Lufa 4	7.4	2.8	24.4	7-d LC$_{50}$	416	Measured	Fresh	Mort	29
TNT	*F. candida*	Lufa 4	7.4	2.8	24.4	28-d EC$_{50}$	171	Measured	Fresh	Repr	29
TNT	*Ench. albidus*	OECD	6.0	10.0	20.0	LC$_{20}$-adult	317	Measured	Fresh	Mort	59
TNT	*Ench. albidus*	OECD	6.0	10.0	20.0	LC$_{50}$-adult	422	Measured	Fresh	Mort	59
TNT	*Ench. albidus*	OECD	6.0	10.0	20.0	EC$_{20}$	59	Measured	Fresh	Repr	59
TNT	*Ench. albidus*	OECD	6.0	10.0	20.0	EC$_{50}$	111	Measured	Fresh	Repr	59
TNT	*Ench. albidus*	OECD	6.0	10.0	20.0	LC$_{50}$-juvenile	44	Nominal	Fresh	Mort	59
TNT	*Ench. albidus*	OECD	6.0	10.0	20.0	LC$_{50}$-juvenile	89	Nominal	21-d aged	Mort	59
TNT	*Ench. crypticus*	SSL	5.2	1.2	17.0	LC$_{20}$	180	Measured	Fresh	Mort	52
TNT	*Ench. crypticus*	SSL	5.2	1.2	17.0	LC$_{50}$	360	Measured	Fresh	Mort	52
TNT	*Ench. crypticus*	SSL	5.2	1.2	17.0	EC$_{20}$	77	Measured	Fresh	Repr	52

Compound	Species	Soil				Endpoint	Value				Page
TNT	*Ench. crypticus*	SSL	5.2	1.2	17.0	EC$_{50}$	98	Measured	Fresh	Repr	52
TNT	*Ench. crypticus*	SSL	5.2	1.2	17.0	LC$_{20}$	100	Measured	83-d W/A	Mort	52
TNT	*Ench. crypticus*	SSL	5.2	1.2	17.0	LC$_{50}$	140	Measured	83-d W/A	Mort	52
TNT	*Ench. crypticus*	SSL	5.2	1.2	17.0	EC$_{20}$	37	Measured	83-d W/A	Repr	52
TNT	*Ench. crypticus*	SSL	5.2	1.2	17.0	EC$_{50}$	48	Measured	83-d W/A	Repr	52
TNT	Nematodes	FS	3.8	5.9	11.0	LOEC	>30	7-d Measured	Fresh	Abund	63
TNT	Microarthropods	FS	3.8	5.9	11.0	LOEC	>30	7-d Measured	Fresh	Abund	63
TNT	Oribatida	FS	3.8	5.9	11.0	LOEC	30	7-d Measured	Fresh	Abund	63
TNT	*E. andrei*	SFS	7.8	3.8	8.0	14-d LC$_{50}$	132	Measured	Fresh	Mort	57
2-ADNT	*E. andrei*	SFS	7.8	3.8	8.0	14-d LC$_{50}$	215	Measured	Fresh	Mort	57
4-ADNT	*E. andrei*	SFS	7.8	3.8	8.0	14-d LC$_{50}$	105	Measured	Fresh	Mort	57
2,4-DANT	*E. andrei*	SFS	7.8	3.8	8.0	LOEC	>100	Measured	Fresh	Mort	57
2,6-DANT	*E. andrei*	SFS	7.8	3.8	8.0	LOEC	>100	Measured	Fresh	Mort	57
TNB	*Ench. crypticus*	SSL	5.2	1.2	17.0	NOEC	45	Measured	Fresh	Mort	61
TNB	*Ench. crypticus*	SSL	5.2	1.2	17.0	LOEC	107	Measured	Fresh	Mort	61
TNB	*Ench. crypticus*	SSL	5.2	1.2	17.0	EC$_{20}$	5	Measured	Fresh	Repr	61
TNB	*Ench. crypticus*	SSL	5.2	1.2	17.0	EC$_{50}$	11	Measured	Fresh	Repr	61
TNB	*Ench. crypticus*	SSL	5.2	1.2	17.0	NOEC	76	Measured	90-d W/A	Mort	61
TNB	*Ench. crypticus*	SSL	5.2	1.2	17.0	LOEC	176	Measured	90-d W/A	Mort	61
TNB	*Ench. crypticus*	SSL	5.2	1.2	17.0	EC$_{20}$	9	Measured	90-d W/A	Repr	61
TNB	*Ench. crypticus*	SSL	5.2	1.2	17.0	EC$_{50}$	22	Measured	90-d W/A	Repr	61
2,4-DNT	*Ench. crypticus*	SSL	5.2	1.2	17.0	NOEC	41	Measured	Fresh	Mort	61
2,4-DNT	*Ench. crypticus*	SSL	5.2	1.2	17.0	LOEC	55	Measured	Fresh	Mort	61
2,4-DNT	*Ench. crypticus*	SSL	5.2	1.2	17.0	EC$_{20}$	19	Measured	Fresh	Repr	61
2,4-DNT	*Ench. crypticus*	SSL	5.2	1.2	17.0	EC$_{50}$	36	Measured	Fresh	Repr	61
2,4-DNT	*Ench. crypticus*	SSL	5.2	1.2	17.0	NOEC	37	Measured	90-d W/A	Mort	61
2,4-DNT	*Ench. crypticus*	SSL	5.2	1.2	17.0	LOEC	72	Measured	90-d W/A	Mort	61
2,4-DNT	*Ench. crypticus*	SSL	5.2	1.2	17.0	EC$_{20}$	14	Measured	90-d W/A	Repr	61

(continued on next page)

TABLE 3.3 (continued)

Selected Ecotoxicological Benchmarks for Energetic Materials Established in Standardized Single-Species Toxicity Tests with Soil Invertebrates and in a Microcosm Test with Indigenous Soil Microinvertebrate Community

EM	Species	Soil Type	Soil pH	OM (%)	Clay (%)	Toxicity Benchmark	Value (mg kg^{-1})	Concentration	Exposure	Endpoint	Reference
2,4-DNT	Ench. crypticus	SSL	5.2	1.2	17.0	EC$_{50}$	27	Measured	90-d W/A	Repr	61
2,6-DNT	Ench. crypticus	SSL	5.2	1.2	17.0	LOEC	>64	Measured	Fresh	Mort	61
2,6-DNT	Ench. crypticus	SSL	5.2	1.2	17.0	EC$_{20}$	37	Measured	Fresh	Repr	61
2,6-DNT	Ench. crypticus	SSL	5.2	1.2	17.0	EC$_{50}$	57	Measured	Fresh	Repr	61
2,6-DNT	Ench. crypticus	SSL	5.2	1.2	17.0	NOEC	37	Measured	90-d W/A	Mort	61
2,6-DNT	Ench. crypticus	SSL	5.2	1.2	17.0	LOEC	108	Measured	90-d W/A	Mort	61
2,6-DNT	Ench. crypticus	SSL	5.2	1.2	17.0	EC$_{20}$	18	Measured	90-d W/A	Repr	61
2,6-DNT	Ench. crypticus	SSL	5.2	1.2	17.0	EC$_{50}$	29	Measured	90-d W/A	Repr	61
Perchlorate	E. fetida	S/M	NA	17	NA	LC$_{50}$	2550	Nominal	Fresh	Mort	66
Perchlorate	E. fetida	S/M	NA	17	NA	21-d EC$_{50}$	1.3	Nominal	Fresh	Cocoons	66
Perchlorate	E. fetida	USEPA	6.0	10.0	20.0	28-d EC$_{50}$	350	Nominal	Fresh	Cocoons	66

Note: Ad wt = weight of adults, Abund = abundance of a microinvertebrate group (number per gram of soil), Cocoons = number of cocoons produced, *E.* = Eisenia, EM = energetic material, *F.* = Folsomia, *Ench.* = Enchytraeus, FS = forest soil, Juv wt = weight of juveniles (growth), Mort = adult mortality, NA = data not available OECD = Organization for Economic Co-operation and Development standard artificial soil, OM = organic matter, Repr = juvenile production, SanL = sandy loam soil, SFS = sandy forest soil, SilL = silt loam soil, S/M = sand/manure mixture, SSL = Sassafras sandy loam soil, USEPA = U.S. Environmental Protection Agency standard artificial soil, W/A = weathered and aged EM soil treatment.

in the standard Lufa 2.2 soil amended with up to 1000 mg kg^{-1} nominal RDX or HMX concentrations. These results comport with findings of Dodard et al. [45], who reported that survival of adult potworms *Ench. crypticus* Westheide & Graefe or *Ench. albidus* was not affected up to and including RDX concentration 658 mg kg^{-1} or HMX concentration 918 mg kg^{-1} (the respective greatest concentrations tested) in a formulated Rac50-50 soil. Juvenile production by either species was unaffected by HMX in all exposure concentrations ranging from 200 to 1000 mg kg^{-1}. In contrast, exposure of *Ench. albidus* to RDX significantly decreased juvenile production at 209 mg kg^{-1} (LOEC) and elicited a concentration-dependent response producing the EC_{20} and EC_{50} values of 161 and 444 mg kg^{-1}, respectively [45]. RDX did not significantly affect juvenile production by *Ench. crypticus* at 658 mg kg^{-1} [45].

Ecotoxicological data established on the basis of exposures in artificial soil (6.2% OM), standard Lufa 2.2 soil (4.6% OM), or formulated Rac50-50 soil (23% OM) can have limited application for ERA. Organisms inhabiting soils with lower OM contents can be exposed to a different chemical environment because bioavailability of nonpolar organic chemicals in soil is hypothesized to be affected by the soil OM content [46]. Greater toxicity of nitramine explosives was determined in studies that used natural soils with lower OM content compared with artificial soil or Lufa 2.2 soil. Simini et al. [42] determined the EC_{20} and EC_{50} values for juvenile production by *E. fetida* in freshly amended SSL soil (1.2% OM) of 1.6 and 5.0 mg kg^{-1}, respectively for RDX, and 0.4 and 1.2 mg kg^{-1}, respectively, for HMX. The LOEC values for RDX based on juvenile production by *E. andrei* and on the number of juveniles per cocoon were 91 and 47 mg kg^{-1}, respectively, in a sandy forest soil having 3.8% OM [47]. In the same study, Robidoux et al. [47] determined the EC_{50} values for HMX of approximately 106 and 16 mg kg^{-1} based on growth and reproduction of adult *E. andrei*, respectively, while adult survival was not affected up to approximately 700 mg kg^{-1}. However, considerably lower toxicities of RDX or HMX in SSL soil were reported for potworm *Ench. crypticus* [48]. The EC_{20} value of 3700 mg kg^{-1} for RDX and an unbounded (not followed by a greater concentration that caused statistically significant decrease from control) NOEC of 21,750 mg kg^{-1} for HMX based on juvenile production by *Ench. crypticus* were established by Kuperman et al. [48]. Neither RDX nor HMX affected survival of *Ench. crypticus* adults even at concentrations as high as 21,383 and 21,750 mg kg^{-1}, respectively. Furthermore, exposure of *Ench. crypticus* to HMX concentration 21,750 mg kg^{-1} in freshly amended SSL soil significantly stimulated juvenile production (11%–56% increase compared with control) [48].

The effects of reduced RDX metabolites hexahydro-1-nitroso-3,5-dinitro-1,3,5-triazine (MNX) and hexahydro-1,3,5-trinitroso-1,3,5-triazine (TNX) on survival and growth of adult *E. fetida* were investigated by Zhang et al. [49]. Using exposures in a sandy loam (1.3% OM, pH 8.3) and in a silt loam (2.5% OM, pH 7.0) soils, the authors [49] established the 14-d LC_{50} values of 262.1 and 244.6 mg kg^{-1}, respectively, for MNX, and 251.3 and 216.3 mg kg^{-1}, respectively, for TNX. The 35-d EC_{50} values for *E. fetida* growth in these natural soils were 228.9 and 525.8 mg kg^{-1}, respectively, for MNX, and 237.7 and 364.4 mg kg^{-1}, respectively, for TNX. These results suggest that properties of soils tested in this study affected sublethal toxicity of MNX and TNX.

In contrast with monocyclic RDX and HMX, the polycyclic nitramine CL-20 was highly toxic to potworms and earthworms. The EC_{50} values for reproduction in different soil types ranged from 0.08 to 0.62 mg kg^{-1} for *Ench. crypticus* and *Ench. albidus* [45,50], and from 0.05 to 0.09 mg kg^{-1} for *E. andrei* [51]. The LC_{50} value for adult *E. andrei* survival was estimated at 53 mg kg^{-1} SSL soil [51]. Weathering and aging of CL-20 in SSL increased the toxicity of test soil for *Ench. crypticus* by 147% based on the LOEC values for adult survival and by approximately 300% based on the EC_{20} or EC_{50} values for juvenile production [50]. These contrasting effects of cyclic nitramines clearly demonstrate that attempts to predict the potential ecotoxicity of explosives solely on the basis of similarities in molecular structure and functional groups, as was hypothesized for CL-20, RDX, and HMX, can lead to incorrect conclusions regarding ecological impacts of nitramines in the environment [50].

Soil type affected the toxicity of CL-20 to *Ench. crypticus* [45] and to *E. andrei* [51]. Reproduction toxicity benchmarks for *Ench. crypticus* were an order of magnitude lower (greater toxicity) in RacAg2002 soil formulation (41% OM, pH 8.2) compared with toxicity in SSL [45]. In contrast, all toxicity benchmarks for CL-20 were more than one order of magnitude greater (lower toxicity) for *E. andrei* in similarly formulated RacFor2002 soil used in the study by Robidoux et al. [51] compared with toxicity in SSL soil. The differential toxicities of CL-20 in these soil types suggest that future studies should include multiple soil types representing a wide range of soil properties (e.g., OM, clay content, pH) in order to assess the relationships among CL-20 toxicity endpoints, bioavailability, and soil properties.

Toxicological benchmarks determined in studies with *Ench. crypticus* showed that the toxicity of CL-20 was two orders of magnitude greater compared with TNT, which will be discussed later in this chapter, and more than five orders of magnitude greater compared with RDX (Table 3.3), based on the EC_{50} values for reproduction in the similarly designed studies with SSL soil [45,48,50,52]. The difference in toxicity to *Ench. crypticus* between CL-20 and HMX was even greater (Table 3.3), because potworms were not affected by exposure to HMX up to the greatest tested HMX concentration of 21,750 mg kg^{-1} in SSL soil [48]. Based on the results of these studies [45,48,50], the order of toxicity of cyclic nitramines to *Ench. crypticus* in SSL soil is (from greatest to least toxic) CL-20 > RDX > HMX. This order of toxicity for the three explosives parallels closely the order of their respective log K_{ow} values (1.92 > 0.90 > 0.17) [53], suggesting that greater toxicity of CL-20 can be related, at least partially, to its greater hydrophobicity and affinity toward OM, which increases its potential to partition into soil biota, and greater bioavailability and uptake potentials compared with either RDX or HMX. Issues related to the bioaccumulation of explosives in soil organisms are discussed in Chapter 10.

3.4.2   EFFECTS OF NITROAROMATIC COMPOUNDS

3.4.2.1   Acute Toxicity of Nitroaromatic Compounds

Assessment of the toxicity of nitroaromatic EMs to soil invertebrates has received considerably greater attention compared with cyclic nitramines. The effects of exposure to TNT and its metabolites in contaminated soils, composted explosives-contaminated soils, and experimentally amended soils have been investigated for earthworms [1,26,43,47,54–58], enchytraeids [44,52,59–61], collembola [44,60,62], and the nematode and microarthropod communities [63]. The majority of these studies focused on the effects of TNT. Fewer studies investigated the effects of TNT metabolites in soil [26,41,57,61].

The majority of reported toxicological data was established in studies with experimentally amended soils. Phillips et al. [41] reported no mortality for the earthworm *E. fetida* in the artificial soil up to the greatest tested TNT concentration of 200 mg kg^{-1} and a LOEC of 140 mg kg^{-1} for loss of body mass. A similar study with a forest silt loam soil (5.9% OM, pH 3.8) established LC$_{50}$ for adult survival and LOEC for loss of body mass of 325 and 150 mg kg^{-1}, respectively [41]. These data concur with findings by Robidoux et al. [55] for *E. andrei* exposed in artificial soil or forest sandy loam soil (4.2% OM, pH 5.9), which established the LC$_{50}$ values of 365 and 222 mg kg^{-1}, respectively. In a later study [43], Robidoux et al. found no mortality of *E. andrei* in artificial soil amended with up to 881 mg TNT kg^{-1}. Toxicity data reported by Phillips et al. [41] and Robidoux et al. [43,55] were based on nominal TNT concentrations and likely have overestimated the exposure (underestimated the toxicity) of earthworms to TNT, which undergoes rapid degradation in soil [52,54,56,59,64]. This conclusion is supported by findings of several studies that established lower LC$_{50}$ values for *E. andrei* based on the analytically determined TNT concentrations of 143 mg kg^{-1} [54] and 132 mg kg^{-1} [57] in different forest sandy loam soils (4.2% OM, pH 5.8). These discrepancies in reported toxicity data for TNT underscore the importance of determining ecotoxicological data for nitroaromatic EM based on the measured chemical concentration.

The acute toxicity of TNT was also investigated in studies with potworms and collembola. In a study with TNT aged for one month in Lufa 2.2 soil, Schäfer and Achazi [44] determined the 7-day LC$_{50}$ values of 1290 and 420 mg kg^{-1} for survival of adult *Ench. crypticus* and *F. candida*, respectively. Considerably greater acute toxicity was observed by Dodard et al. [59] for a different potworm species, *Ench. albidus*, with the pooled 21-day LC$_{50}$ value of 422 mg kg^{-1} based on the initial analytically determined TNT concentrations in freshly amended artificial soil. However, a time-course study examining the disappearance of TNT in soil showed that recovery of TNT was approximately 36% and 21% of the initial concentration after 7 and 21 days, respectively, suggesting a potentially greater toxicity of TNT to potworms than was established on the basis of the initial concentration. An indirect support for this hypothesis comes from studies by Kuperman et al. [52] who reported the LC$_{50}$ value of 360 mg kg^{-1} for *Ench. crypticus* in freshly amended SSL soil, which was allowed to equilibrate for 24 h after hydrating TNT-amended soil prior to exposing test organisms.

Several studies have shown that TNT is rapidly transformed in aerobic soils leading to formation of reduced amino-nitrotoluene metabolites (also see Chapter 2). Presence of these intermediates as well as their common co-occurrence with TNT, DNTs, and TNB in soil of contaminated sites makes it difficult to partition the effects of individual compounds on soil invertebrates. However, establishing toxicity data for individual compounds is a necessary step for understanding the mechanisms of toxicity of nitroaromatic EMs in soil. Acute toxicity data for these EMs is sparse. Lachance et al. [57] investigated the toxicity of TNT and its reduction products 2-ADNT, 4-ADNT, 2,4-DANT, and 2,6-DANT to adult *E. andrei* in a freshly amended sandy loam forest soil, and established the following order of toxicity: 4-ADNT > TNT > 2-ADNT based on the 14-day LC_{50} values of 105, 132, and 215 mg kg^{-1}, respectively. Exposure of *E. andrei* to 100 mg kg^{-1} of either 2,4-DANT or 2,6-DANT did not affect survival of adult earthworms. The LOEC values of 72, 108, and 176 mg kg^{-1} for 2,4-DNT, 2,6-DNT, and TNB, respectively, weathered and aged for three months in SSL soil were reported for adult *Ench. crypticus* by Kuperman et al. [52]. Simini et al. [26] assessed the in situ impacts of a mixture of EMs on earthworm *E. fetida* growth and survival endpoints using contaminated site soils. The authors reported that TNT and TNB had the strongest correlation with toxicity endpoints in all bioassays with r^2 values of 0.759 and 0.911 for TNT, and 0.773 and 0.814 for TNB established for soils from different locations investigated at that site. These values for 2,4-DNT were 0.613 and 0.358, whereas 2,6-DNT concentration had the weakest relationship with assessed endpoints based on r^2 values of 0.082 and 0.293.

3.4.2.2 Chronic Toxicity of Nitroaromatic Compounds

Toxicity data reported in the literature for earthworms [43,47,55,58,65], potworms [48,50,52,59–61], and collembola [44,60,62] showed that reproduction endpoints were generally more sensitive measures of EM toxicity to soil invertebrates compared with adult survival. Reproduction toxicity data for 2,4-DNT, 2,6-DNT, and TNB are limited to one published study with potworm *Ench. crypticus* exposed in SSL soil [61]. Based on the results of that study and those reported in the literature for reproduction endpoints, the order of toxicity of nitroaromatic EMs to *Ench. crypticus* in SSL soil is (from most to least toxic) TNB > 2,4-DNT > 2,6-DNT > TNT (Table 3.3). The effects of TNT on reproduction of soil invertebrates were investigated in several studies and are summarized in Table 3.3. The EC_{50} values for TNT ranged from 23 to 919 mg kg^{-1} for different soil types, test species, degree of weathering and aging of TNT in soil, and were derived using either nominal or analytically determined concentrations. This wide range of toxicity values for TNT clearly demonstrates that testing procedures for establishing ecotoxicological data for EMs require further standardization to produce more consistent and reliable benchmarks for use in ERA.

3.4.3 EFFECTS OF PERCHLORATE

Perchlorate (ClO_4^-) was used in explosives, propellants, and pyrotechnics in the forms of ammonium, sodium, and potassium perchlorate. Although perchlorate can persist as a contaminant in ground and surface water, assessment of the perchlorate toxicity to soil invertebrates received little attention due to its high aqueous solubility and mobility, which diminish potential exposure of soil organisms. Limited soil invertebrate toxicity data for sodium perchlorate was determined in studies with *E. fetida* exposed in sand/manure mixture (20:4 weight/weight) or artificial soil (filter paper contact tests were also conducted but results are not discussed in this chapter) during the acute or reproduction tests [66]. Based on nominal perchlorate ion concentrations in test substrate, these authors established the 21-d LC_{50} of 2550 mg kg^{-1} for adult survival in sand/manure mixture, and the EC_{50} values of 350 (28 d) and 1.3 (21 d) mg kg^{-1} for cocoon production in artificial soil or sand/manure mixture, respectively. Landrum et al. [66] attributed high mortality (>60% within 24 h) of *E. fetida* observed above 2000 mg kg^{-1} to the osmotic stress, rather than the toxicity of perchlorate itself and concluded that, based on calculated EC_{50} values, the perchlorate concentrations that can affect *E. fetida* are likely to occur only under extreme conditions.

3.5   EFFECTS OF WEATHERING AND AGING ENERGETIC MATERIALS IN SOIL ON TOXICITY TO SOIL ORGANISMS

Assessment of the effects of weathering and aging of contaminant explosives in soil on the exposed soil organisms is critical for developing toxicity benchmarks that adequately reflect potential ecological risks. Weathering and aging of EMs in soil may reduce exposure of soil organisms to the parent compound due to photodecomposition, hydrolysis, reaction with OM, sorption/fixation, precipitation, immobilization, occlusion, microbial transformation, and other fate processes that commonly occur at contaminated sites and are discussed in Chapter 2. Certain fate processes, including microbial transformation of EMs, can also produce transformation products that are more bioavailable or more toxic to soil organisms compared with parent EM compounds freshly introduced into soil [57].

Few studies have investigated the effects of weathering and aging of explosives in soil on the exposure of terrestrial organisms and consequent soil toxicity. Weathering and aging of some EMs in soil have been reported to alter their toxicities to plant [40] and soil invertebrate species [50,52,60,61]. Weathering and aging in soil significantly decreased the toxicity of TNT, TNB, and 2,6-DNT to Japanese millet and ryegrass based on seedling emergence, but significantly increased the toxicities to the three plant species based on shoot growth [40]. The authors [40] hypothesized that the formation of certain metabolites of the parent EMs, such as 2-ADNT, 4-ADNT, and 3,5-dinitroaniline (3,5-DNA) detected in that study, could have contributed to increased toxicity after weathering and aging EMs in soil. Kuperman et al. [50,52,61] reported that weathering and aging in SSL soil significantly increased the toxicities of TNT, 2,6-DNT, and CL-20 to *Ench. crypticus*, while the toxicities of 2,4-DNT or TNB were unaffected. In contrast, a decreased toxicity of TNT after aging in soil was reported for *Ench. albidus* in artificial soil [59] and for *F. candida* in Lufa 2.2 soil [60] or SSL soil [62]. No effects of weathering and aging of RDX or HMX in SSL soil on toxicity were reported for *Ench. crypticus* [48] and *E. fetida* [42]. Direct comparison of these results is difficult due to several factors, including differences in properties of soils used in the studies, the weathering and aging procedures employed, and resulting effects on bioavailability of EMs and their transformation/degradation products to the different organisms tested.

Specific mechanisms of changes in the toxicity following weathering and aging of EMs in soil are not well understood. Compounds produced due to TNT degradation or transformation during the weathering and aging process, such as ADNTs, which were detected in few studies, can be more toxic to soil organisms compared with the parent material, and can be one of the factors contributing to the increased toxicity in weathered-and-aged treatments [50,52,61]. Supporting this hypothesis are the results of studies by Lachance et al. [57] discussed earlier, which demonstrated that toxicity of the TNT reduction product 4-ADNT to adult earthworm *E. andrei* was greater compared with the toxicity of parent material. However, additional studies would be required to resolve the current uncertainties in our understanding of the mechanisms contributing to the increased or decreased toxicity of EMs following their weathering and aging in soil. These studies should be conducted with different soil types having ranges of properties that affect the fate and bioavailability of EMs to better understand the complex interactions among physical, chemical, and biological components that jointly contribute to the outcome of ecotoxicity testing.

3.6 STIMULATING EFFECTS OF ENERGETIC MATERIALS

Hormetic responses (significant stimulatory effects caused by low levels of potentially toxic chemicals, followed by inhibitory effects at higher concentrations) to nitroaromatic EM exposure have been documented for a variety of ecological receptors. Juvenile production by *Ench. crypticus* was hormetically stimulated in either freshly amended or weathered-and-aged TNT treatments in SSL soil [52]. Hormesis was similar in magnitude and occurred at 40 or 3 mg kg^{-1} in freshly amended or weathered-and-aged treatments, respectively. A similar hormetic response by *Ench. crypticus* was observed in a study with TNB freshly amended into SSL soil [61]. Stimulation of juvenile production was reported at 2.6 mg kg^{-1} TNB and resulted in a 19% increase in the average number of juveniles compared with control. Hormetic effects of nitroaromatic EMs were also observed in studies with terrestrial plants [40]. The authors reported hormetic effects for alfalfa, Japanese millet, and ryegrass exposed to TNT; for ryegrass exposed to TNB or 2,4-DNT; and for Japanese millet or ryegrass exposed to 2,6-DNT.

Hormesis is not unique to EMs. It has been reported for several plant and animal species exposed to heavy metals, pesticides, polycyclic aromatic hydrocarbons, and other organic chemicals [67–69]. Stebbing [68] suggested that hormesis is the cumulative consequence of transient and sustained overcorrections of biosynthesis, that is, a rate-controlled process controlled by the end-product inhibition. To date, however, no studies have identified the mechanisms responsible for hormetic effects of explosives at specific concentrations.

Nonhormetic stimulation within tested concentration ranges has been reported for several ecological receptors exposed to EMs. Such stimulation was observed for offspring production by *Daphnia magna* exposed to TNT concentrations of 0.08 mg L^{-1} [67]. Other explosives were also reported to elicit a stimulating effect on the measurement endpoints used in toxicity tests. Juvenile production by *Ench. crypticus* was stimulated by exposure to HMX in freshly amended SSL soil at concentrations ranging from 2210 to 21,750 mg kg^{-1} [48]. Bentley et al. [70] observed stimulation in egg production by fathead minnow exposed to 6.3 mg L^{-1} RDX. The density of green algae *Selenastrum capricornutum* Printz cells was increased following exposures to

HMX starting at 10 mg L^{-1} [71]. Although the mechanism of stimulation can vary, Steevens et al. [72] suggested that for hormetic responses, these mechanisms can include direct effects through the release of metabolic products of explosives having a specific effect on growth and reproduction of test organisms, and indirect effects through increased supply of nitrogen for bacteria, fungi, or algae (the important food sources for higher trophic levels) from mineralization of explosives. Similar mechanisms may be also involved in the nonhormetic stimulation.

3.7 MECHANISMS OF TOXICITY

Toxicological mechanisms or modes of action describe biological, chemical, and physical processes or events that can lead to lethal or sublethal effects on an exposed organism. Such effects can include mortality, inhibition of growth and/or reproduction, behavioral changes, or mutation (also see Chapter 9). Toxicants can be divided generally into seven categories according to the type of target molecules with which they react. These categories are: (1) enzyme inhibitors; (2) toxicants that disturb the chemical signal systems; (3) toxicants that generate very reactive molecules that destroy cellular components; (4) weak organic bases or acids that degrade the pH gradients across membranes; (5) toxicants that dissolve in lipophilic membranes and disturb their physical structure; (6) toxicants that disturb the electrolytic or osmotic balance or the pH; and (7) strong electrophiles, alkalis, acids, oxidants, or reducers that destroy tissue, DNA, or proteins [73]. Mechanisms of toxicity can be studied at different stages following the sequence of events that occur during exposure to a potentially toxic compound. For instance, one can study mechanisms of uptake, distribution, metabolism, elimination, and interactions with target molecules, as well as mechanisms of defense, tolerance, repair, or other biological responses [74]. A better understanding of EM toxicity mechanisms can improve environmental risk assessment by reducing uncertainty, better defining predictive models (such as a quantitative structure–activity relationship model), helping to design genetically engineered organisms for use in bioremediation, and identifying novel biomarkers for detection, monitoring, and discrimination of exposure to an energetic compound.

A traditional approach to studying toxicity mechanisms is hypothesis driven, where one or a few genes are selected for investigation at any given time. However, this approach is severely hampered by multiplicity of possible pathways and modes of action for a specific compound [74,75]. Advances in molecular biology and biotechnology have paved the way to developing a genomics or ecotoxicogenomics (genomics is a collective term for genetics, transcriptomics, proteomics, and metabonomics) approach in ecotoxicology. In contrast to the traditional approaches, the genomics approach simultaneously evaluates the expression of thousands of genes, proteins, or metabolites (endpoints). The genomics approach can facilitate identification or prediction of known and unknown mechanisms/modes of action, thus helping to define toxicity pathways, especially for poorly characterized or emerging contaminants [76,77]. This approach can enable extrapolation of chemical effects across species and in chemical mixtures, and will provide means for identification of new biomarkers of exposure and effects for use in both laboratory and field (monitoring) studies [76,77]. Although genomics will not fundamentally alter the environmental risk assessment process, it is expected to serve as a powerful tool for evaluating the exposure effects of environmental stressors.

Nitroaromatic compounds such as TNT and DNTs can be cytotoxic and geno-toxic (Chapters 8 and 9). The molecular structures of these EMs are similar to the structure of dinitroaniline herbicides, which can contribute to similarity in the plant responses to nitroaromatic EMs and dinitroaniline exposures, that is, inhibition of cell division and development. Using exposures on agar plates, Peterson et al. [38] investigated the mode of action for the effects of TNT or 4-ADNT on tall fescue root and shoot growth and metabolism, which decreased linearly as concentrations of TNT or 4-ADNT increased. Microscopic examination revealed that radicles (pri-mary roots) exposed to 15 mg L^{-1} TNT were swollen, stunted, and had no root hairs. Tissues of radicles were undifferentiated compared to roots that were untreated or exposed to 1.9 and 3.75 mg L^{-1} TNT, which had defined meristematic and elongation regions. Secondary roots exposed to 3.75 mg L^{-1} TNT or below were well developed with root hairs, whereas those exposed to 7.5 mg L^{-1} or above lacked root hairs and were underdeveloped. Similar symptoms were observed on radicles and secondary roots exposed to 4-ADNT concentrations ranging from 1.9 to 15 mg L^{-1}. The respi-ration rate was significantly reduced ($p < 0.0001$) in plants exposed to TNT but not to 4-ADNT. The symptoms of exposure to TNT or 4-ADNT described earlier were similar to those caused by dinitroaniline herbicides dithiopyr and pendamethalin [32]. This symptomatic similarity suggests that TNT and 4-ADNT can be disruptive for mitotic division; however, direct evidence is needed to confirm this hypothesis.

Gong et al. [78] demonstrated phytogenotoxicity of 2,4-DNT and 2,6-DNT using the *Tradescantia* micronucleus bioassay (Trad-MCN) [79] in a hydroponic culture. Exposure to 2,4-DNT concentrations ranging from 0 to 30 mg L^{-1} caused a linear concentration-dependent change in MCN frequency, whereas exposure to 2,6-DNT significantly increased MCN frequency at 135 mg L^{-1} compared to response in con-trol. Based on these results, the authors [78] established the minimum effective dose (MED) for 2,4-DNT and 2,6-DNT of 30 and 135 mg/l, respectively.

Cenas et al. [80] put forward multiple lines of evidence relating the cytotoxic-ity of explosives, including TNT and 2,4,6-trinitrophenyl-N-methylnitramine (tetryl) to oxidative stress (see Chapter 9). Symptoms were manifested in an increase in enzymatic single-electron reduction by NADPH:cytochrome P-450 reductase and ferredoxin:NADP(+) reductase, the protective effects of desferrioxamine and the antioxidant N,N'-diphenyl-p-phenylene diamine, and an increase in lipid peroxida-tion. In addition, DT-diaphorase may play a minor and equivocal role in the cytotox-icity of these explosives, indicating that exposure to explosives may trigger multiple toxicological pathways. Kumagai et al. [81] further determined the mechanism of TNT-induced oxidative stress and found that neuronal nitric oxide synthase (nNOS) catalyzed the single-electron reduction of TNT, resulting in decreased nitric oxide production and increased nNOS gene expression. The upregulation of nNOS repre-sents an acute adaptation to an increase in oxidative stress during exposure to TNT. Evidence from serial analysis of gene expression (SAGE) also supported an oxidative stress response to TNT exposure [82].

Cyclic nitramines RDX and CL-20 share neurobiological pathways in their toxic effects on earthworms. Gong et al. [83] observed that toxicity of CL-20 on the giant nerve fibers in earthworm *E. fetida* was reversible at low doses (e.g., ≤0.94 µg cm^{-2} of filter paper for up to 6 days of exposure) but irreversible at high doses (e.g., ≥1.90 µg cm^{-2} for as short as 3 days of exposure). RDX is a less potent neurotoxin compared with CL-20. To discern the molecular mechanisms underlying neurotoxicity, gene expression change in the central nervous system of CL-20 exposed earthworms is currently being investigated using microarray technology.

Using a model plant species *Arabidopsis thaliana* (L.) Heynh., Ekman et al. [84] identified genes known to respond to a variety of general stresses in RDX-exposed root tissues. Several genes encoding molecular chaperones and transcription factors as well as vacuolar proteins and peroxidases were highly induced, whereas those encoding ribosomal proteins, a cyclophilin, a katanin, and a peroxidase were strongly repressed. Comparison of transcriptional profiles for RDX or TNT exposures revealed significant differences in gene expression pattern, suggesting that *Arabidopsis* employs drastically different mechanisms for coping with these two compounds. Using a reverse-transcription PCR technique, Mezzari et al. [85] further demonstrated that the expression of three glutathione S-transferases (GSTs) and two nitroreductase enzymes in *Arabidopsis* increased substantially (up to 40-fold) after exposure to TNT but only increased slightly in response to RDX. Furthermore, Kutty and Bennett [86] characterized two gene products (NitA and NitB) encoding TNT transforming oxygen-insensitive nitroreductase from *Clostridium acetobutylicum* ATCC 824, and Cheong et al. [87] observed an increase in the expression of a fungal laccase gene for *Trametes versicolor* (L.: Fr.) Pilat during degradation of TNT and its catabolic intermediates. Seth-Smith et al. [88] determined that a gene responsible for the degradation of RDX in *Rhodococcus rhodochrous* strain 11Y is a constitutively expressed cytochrome P450-like gene, xplA, which is found in a gene cluster with an adrenodoxin reductase homologue, xplB. These findings demonstrated that mechanisms of EM toxicity to soil organisms differ between cyclic nitramine and nitroaromatic compounds, and can explain, at least in part, their differential effects on exposed soil receptors.

Using the subtractive suppression hybridization technique, Gong et al. [89] developed a cDNA library enriched for genes in earthworms affected by five ordnance related compounds (ORCs), including TNT, 2,6-DNT, HMX, and RDX. Among the 2208 cloned and sequenced cDNA inserts, 1262 sequences were identified as nonredundant expressed sequence tags (ESTs). Nearly half of the ESTs did not match with any sequences in the gene bank, suggesting that these ESTs are likely novel genes that may reveal unknown functions and biological pathways involved in response to EM exposure by the earthworms. Interestingly, genes encoding Cd-metallothionein, NADH dehydrogenase, and cytochrome oxidase were repeatedly found in the subtracted earthworm libraries, suggesting that exposure to ORCs may have triggered expression of these biomarker genes. Gong et al. [90] further constructed a high-density microarray containing the 2208 amplified cDNA clones and used it to assess the impact of different durations (i.e., 4 d, 14 d, or 28 d) of TNT exposure (100 mg kg^{-1}) on earthworm gene expression profiles. Twenty-nine genes were identified as significantly different from controls. A putative mitogen-activated protein kinase involved in signal transduction was induced after 4-d exposure. After the 4- and 14-d exposures, cellular protein degradation (ubiquitin activating enzyme), oxygen transport (hemoglobin linker chain, hemoglobin subunit B2, hemoglobin chain d1), and several unknown genes were downregulated. Exposure to TNT for 28 d led to induced genes involved in cellular protein degradation (ubiquitin activating enzyme and lumbrokinase-3), lysosomal degradation (beta-hexosaminidase), and induction of several unknown genes. The same exposure also downregulated several genes, including a putative transcription factor, a non-heme-iron oxygen transport protein, and a trichohyalin-similar gene. Lasting effects on hemoglobin genes were not observed. These results suggest that short-term exposure to TNT inhibits oxygen transport systems, while longer-term exposure can cause significant damage to cellular proteins and macromolecules. These results also demonstrated that, in spite of the technical challenges, toxicogenomics approaches can provide important tools for discerning molecular mechanisms of toxicity, even when applied to ecologically relevant organisms like earthworms for which there is little genomics information available [76,77].

3.8   CONCLUSIONS AND FUTURE OUTLOOK

Available ecotoxicological data showed that cyclic nitramine explosives RDX, HMX, and CL-20 have small or no adverse effects on soil microbial endpoints. In contrast, the nitroaromatic explosive TNT has been shown to selectively and adversely affect certain components of the soil microbial community, particularly Gram-positive organisms, which can result in long-term or permanent changes in composition and diversity of the microbial community. The biogeochemical cycles of carbon and nitrogen can also be disrupted by soil contamination with TNT. Exposure to TNT was shown to disrupt nitrous oxide reductase activity at very low concentrations leading to an increase in nitrate concentration in the soil ecosystem, thus, potentially short-circuiting the nitrogen cycle in soil at the impacted sites. The persistence of such impairment is unknown and requires further investigation. Furthermore, TNT can drastically decrease microbial metabolic efficiency, which in turn increases soil nitrate concentrations and diminishes carbon storage in the soil ecosystem. As soil C_{org} is an effective sorbent of TNT, such decreases could result in an increased bioavailability and toxicity of TNT in a contaminated soil ecosystem. The soil microbial toxicity data for TNT transformation and degradation products, as well as for products of TNT manufacturing, including the amino-nitro intermediates, DNTs and TNB have not been established and require additional investigations.

Terrestrial plants are relatively insensitive to the cyclic nitramine explosives RDX, HMX, and CL-20. Among these nitramines, RDX and HMX are highly mobile within the plant and concentrate in leaf and flower tissue. This effect can pose human health and ecological risks due to potential food-chain transfer to higher trophic levels. Plant species were adversely affected by exposure to nitroaromatic energetic compounds. Soil concentrations of nitroaromatic compounds that cause phytotoxicity varied with the soil type and plant species. Overall, the dinitrotoluenes were more toxic to plant species commonly used in toxicity tests compared with TNB or TNT. Shoot growth was generally a more sensitive measurement endpoint compared with seedling emergence. Weathering and aging of TNT, TNB, 2,4-DNT, and 2,6-DNT in soil significantly increased their phytotoxicities suggesting that transformation/

degradation products of the parent EMs can be more toxic or more bioavailable to plants. This hypothesis is being tested in the ongoing investigations with amino-nitro intermediates. Phytotoxicity of nitroglycerin has not been investigated but several studies that are currently underway should provide needed information. Although plants have expressed genotoxicity in response to nitroaromatic EMs, the available data are sparse and require confirmation with future research.

Available ecotoxicological data for soil invertebrates revealed that, in contrast to the effects on soil microorganisms and terrestrial plants, cyclic nitramines exhibited species-specific effects on the exposed organisms. Although RDX and HMX were highly toxic to reproduction of earthworms, low or no toxicity of these EMs was established for potworms. Toxicity data from studies with earthworms, potworms, and collembola showed that CL-20 was one of the most toxic compounds among all EMs tested. Studies with TNT established variable toxicological benchmark values depending on the soil type, test species, and exposure type. Weathering and aging of TNT, 2,6-DNT, and CL-20 in soil have altered their respective toxicity to soil invertebrates by either increasing or decreasing the exposure effects, depending on the species and soil type tested. The type of soil used for exposing soil invertebrates affected the toxicity of EMs tested in several investigations. However, the physical and chemical properties of soils used in these studies were limited to a relatively narrow range and provided insufficient data for establishing relationships between specific soil constituents and EM toxicity.

Overall, in spite of the substantial progress made over the last decade in EM ecotoxicology [1,91], considerable knowledge gaps still exist. Insufficient or no ecotoxicological data for soil organisms are available for the nitroso products of RDX degradation (MNX; hexahydro-5-nitro-1,3-dinitroso-1,3,5-triazine, DNX; and TNX); the amino-metabolites of TNT, TNB, and DNTs; tetryl (2,4,6-trinitrophe-nylmethylnitramine); nitroglycerin; and picric acid (2,4,6-trinitrophenol). Findings of altered toxicity of EMs in weathered-and-aged soil treatments reported in the literature clearly show that procedures for weathering and aging EMs in soils should be standardized and applied in the future studies to more completely investigate and resolve the toxicity of transformation and degradation products, and to determine mechanisms of such toxicity. Analogously, further investigation of the more toxic transformation compounds that arise within soils amended with EMs should also have a weathering-and-aging component, so that the level of persistence and long-term impact of the ecotoxicity of these products may also be assessed. Such studies should also be designed to generate toxicity benchmark data for EM products to provide more complete information on the ecotoxicological effects of energetic con-taminants in soil. Finally, the relationships among the physicochemical properties of soil and the toxicity of EMs to soil organisms should be determined by investigating the exposure effects in multiple natural soil types having wide ranges of constituents that affect the bioavailability of energetic compounds. Filling these data gaps will provide risk assessors and site managers with more reliable and accurate information necessary for distinguishing those sites that pose significant environmental risks from those that do not, for prioritizing contaminated sites by the level of risk posed, for quantifying the risks at each site, and for developing appropriate remedial actions and cleanup goals.

REFERENCES

1. Sunahara GI et al., Laboratory and field approaches to characterize the soil ecotoxicology of polynitro explosives, in *Environmental Toxicology and Risk Assessment: Science, Policy and Standardization—Implications for Environmental Decisions,* Vol. 10 STP 1403, Greenberg BM, Hull RN, Roberts MH Jr., Gensemer RW, Eds., American Society for Testing and Materials, West Conshohocken, PA, 2001, 293.

2. Gong P et al., Ecotoxicological effects of hexahydro-1,3,5-trinitro-1,3,5-triazine on soil microbial activities, *Environ. Toxicol. Chem.,* 20, 947, 2001.

3. Gong P et al., Toxicity of octahydro-1,3,5,7-tetranitro-1,3,5,7-tetrazocine (HMX) to soil microbes, *Bull. Environ. Contam. Toxicol.,* 69, 97, 2002.

4. Robidoux PY et al., Toxicity assessment of contaminated soils from an antitank range, *Ecotoxicol. Environ. Saf.,* 58, 300, 2004.

5. Gong P et al., Preliminary ecotoxicological characterization of a new energetic substance, CL-20, *Chemosphere,* 56, 653, 2004.

6. Fuller ME and Manning JF, Aerobic Gram-positive and Gram-negative bacteria exhibit differential sensitivity to and transformation of 2,4,6-trinitrotoluene (TNT), *Curr. Microbiol.,* 35, 77, 1997.

7. Fuller ME and Manning JF, Evidence for differential effects of 2,4,6-trinitrotoluene and other munitions compounds on specific subpopulations of soil microbial communities, *Environ. Toxicol. Chem.,* 17, 2185, 1998.

8. Fuller ME and Manning JF, Microbiological changes during bioremediation of explosives-contaminated soils in laboratory and pilot-scale bioslurry reactors, *Biores. Technol.,* 91, 123, 2004.

9. Siciliano SD and Greer CW, Plant-bacterial combinations to phytoremediate soil contaminated with high concentrations of 2,4,6-trinitrotoluene, *J. Environ. Qual.,* 29, 311, 2000.

10. Siciliano SD et al., Reduction in denitrification activity in field soils exposed to long term contamination by 2,4,6-trinitrotoluene (TNT), *FEMS Microbiol. Ecol.,* 32, 61, 2000.

11. Siciliano SD et al., Assessment of 2,4,6-trinitrotoluene toxicity in field soils by pollution induced community tolerance (PICT), denaturing gradient gel electrophoresis (DGGE) and seed germination assay, *Environ. Toxicol. Chem.,* 19, 2154, 1999.

12. Gong P et al., An in situ respirometric technique to measure pollution-induced microbial community tolerance in soils contaminated with 2,4,6-trinitrotoluene, *Ecotoxicol. Environ. Saf.,* 47, 96, 2000.

13. Comfort SD et al., TNT transport and fate in contaminated soil, *J. Environ. Qual.,* 24, 1174, 1995.

14. Richards SR and Knowles R, Inhibition of nitrous oxide reduction by a component of Hamilton Harbor sediment, *FEMS Microbiol. Ecol.,* 17, 39, 1995.

15. Bollag J-M and Kurek EJ, Nitrite and nitrous oxide accumulation during denitrification in the presence of pesticide derivatives, *Appl. Environ. Microbiol.,* 39, 845, 1980.

16. Zumft WG, Cell biology and molecular basis of denitrification, *Microbiol. Molec. Biol. Rev.,* 61, 533, 1997.

17. Gong P et al., Effects and bioavailability of 2,4,6-trinitrotoluene in spiked and field-contaminated soils to indigenous microorganisms, *Environ. Toxicol. Chem.,* 18, 2681, 1999.

18. Kowalchuk GA and Stephen JR, Ammonia-oxidizing bacteria: A model for molecular microbial ecology, *Annu. Rev. Microbiol.,* 55, 485, 2001.

19. Schmidt I et al., Aerobic and anaerobic ammonia oxidizing bacteria—competitors or natural partners? *FEMS Microbiol. Ecol.,* 39, 175, 2002.

20. Frische T and Hoper H, Soil microbial parameters and luminescent bacteria assays as indicators for in situ bioremediation of TNT-contaminated soils, *Chemosphere,* 50, 415, 2003.

21. Lee IS et al., Heavy metal concentrations and enzyme activities in soil from a contaminated Korean shooting range, *J. Biosci. Bioeng.,* 94, 406, 2002.

22. Rocheleau S et al., Toxicity of Nitro-Heterocyclic and Nitroaromatic Energetic Materials to Terrestrial Plants in a Natural Sandy Loam Soil, Technical Report No. ECBC-TR-351, U.S. Army Edgewood Chemical Biological Center, Aberdeen Proving Ground, MD, 2005.

23. Robidoux PY et al., Phytotoxicity of 2,4,6-trinitrotoluene (TNT) and octahydro-1,3,5,7-tetrazocine (HMX) in spiked artificial and natural forest soils, *Arch. Environ. Contam. Toxicol.*, 44, 198, 2003.

24. Rocheleau S et al., Toxicity and uptake of cyclic nitramine explosives in ryegrass *Lolium perenne. Environ. Pollut.*, 156, 199, 2008.

25. Harvey SD et al., Fate of the explosive heaxahydro-1,3,5-trinitro-1,3,5-triazine (RDX) in soil and bioaccumulation in bush bean hydroponic plants, *Environ. Toxicol. Chem.*, 10, 845, 1991.

26. Simini M et al., Evaluation of soil toxicity at Joliet army ammunition plant, *Environ. Toxicol. Chem.*, 14, 623, 1995.

27. Winfield LE, Rodgers JHJ, and D'Surney SJ, The responses of selected plants to short (<12 days) and long term (2, 4 and 6 weeks) hexahydro-1,3,5-trinitro-1,3,5-triazine (RDX) exposure. Part I: growth and developmental effects, *Ecotoxicol.*, 13, 335, 2004.

28. Cataldo DA, Harvey SD, and Fellows RJ, An Evaluation of the Environmental Fate and Behavior of Munitions Material (TNT, RDX) in Soil and Plant Systems, Environmental Fate and Behavior of RDX, U.S. Army Biomedical Research and Development Laboratory, Frederick, MD, 1990.

29. Schneider JF et al., Uptake of Explosives from Contaminated Soil by Existing Vegetation at the Iowa Army Ammunition Plant, Final/Technical Report, Report No. SFIM-AEC-ET-CR-95013, U.S. Army Environmental Center, Aberdeen Proving Ground, MD, 1995.

30. Major MA et al., Bioconcentration, Bioaccumulation and Biomagnification of Common Nitroaromatic and Nitramine Explosives and Their Metabolites and Environmental Breakdown Products, Final Technical Report, Toxicology Study No. 87-MA-4677-01, Health Effects Research Program, Directorate of Toxicology, U.S. Army Center for Health Promotion and Preventative Medicine, Aberdeen Proving Ground, MD, 2002.

31. Vila M et al., Phytotoxity to and uptake of RDX by rice, *Environ. Pollut.*, 145, 813, 2007.

32. Peterson DE et al., Herbicide Mode of Action, C-715, Kansas State University Agricultural Experiment Station and Cooperative Extension Service, Kansas State University, Manhattan, KS, January, 2001.

33. Lawton HC and Weatherill WT, Low Toxicity Extrudable Plastic Explosive, Final/Technical Report No. AD-0761087, Aerojet-General Corp., Downey, CA, 1963.

34. Schott CD and Worthley EG, The Toxicity of TNT and Related Wastes to an Aquatic Flowering Plant, *Lemna perpusilla* Torr., AD-0778158, Final/Technical Report, Edgewood Arsenal, Aberdeen Proving Ground, MD, 1974.

35. Palazzo AJ and Leggett DC, Toxicity, Uptake, Translocation, and Metabolism of TNT in Plants: A Literature Review, Final/Technical Report, U.S. Army Medical Research and Development Command, Fort Detrick, MD, 1983.

36. Palazzo AJ and Leggett DC, Effect and disposition of TNT in a terrestrial plant, *J. Environ. Qual.*, 15, 49, 1986.

37. Palazzo AJ and Leggett DC, Effect and Disposition of TNT in a Terrestrial Plant and Validation of Analytical Methods, Final/Technical Report, CRREL Report no. 86-15, U.S. Army Cold Regions Research and Engineering Laboratory, Hanover, NH, 1986.

38. Peterson MM et al., TNT and 4-amino-2,6-dinitrotoluene influence on germination and early seedling development of tall fescue, *Environ. Pollut.*, 93, 57, 1996.

39. Gong P, Wilke B-M, and Fleischmann S, Soil-based phytotoxicity of 2,4,6-trinitrotoluene (TNT) to terrestrial higher plants, *Arch. Environ. Contam. Toxicol.*, 36, 152, 1999.

40. Rocheleau S et al., Phytotoxicity of nitroaromatic energetic compounds freshly amended or weathered and aged in sandy loam soil, *Chemosphere*, 62, 545, 2006.

41. Phillips CT, Checkai RT, and Wentsel RS, Toxicity of Selected Munitions and munition-Contaminated Soil on the Earthworm (*Eisenia foetida*), Technical Report ERDEC-TR-037, U.S. Army Chemical and Biological Defense Agency, Aberdeen Proving Ground, MD, 1993.

42. Simini M et al., Reproduction and survival of *Eisenia fetida* in a sandy loam soil amended with the nitro-heterocyclic explosives RDX and HMX, *Pedobiologia*, 47, 657, 2003.

43. Robidoux PY et al., Chronic toxicity of energetic compounds in soil determined using the earthworm (*Eisenia andrei*) reproduction test, *Environ. Toxicol. Chem.*, 19, 1764, 2000.

44. Schäfer R and Achazi RK, The toxicity of soil samples containing TNT and other ammunition derived compounds in the Enchytraeid and Collembola biotest, *Environ. Sci. Pollut. Res. Int.*, 6, 213, 1999.

45. Dodard S et al., Survival and reproduction of enchytraeid worms (Oligochaeta) in different soil types amended with cyclic nitramine explosives, *Environ. Toxicol. Chem.*, 24, 2579, 2005.

46. Belfroid AC, Sijm DTHM, and Van Gestel CAM, Bioavailability and toxicokinetics of hydrophobic aromatic compounds in benthic and terrestrial invertebrates, *Environ. Rev.*, 4, 276, 1996.

47. Robidoux PY et al., TNT, RDX and HMX decrease earthworm (*Eisenia andrei*) life-cycle responses in a spiked natural forest soil, *Arch. Environ. Contam. Toxicol.*, 43, 379, 2002.

48. Kuperman RG et al., Survival and reproduction of *Enchytraeus crypticus* (Oligochaeta, Enchytraeidae) in a natural sandy loam soil amended with the nitro-heterocyclic explosives RDX and HMX, *Pedobiologia*, 47, 651, 2003.

49. Zhang B, Kendall RJ, and Anderson TA, Toxicity of the explosive metabolites hexahydro-1,3,5-trinitroso-1,3,5-triazine (TNX) and hexahydro-1-nitroso-3,5-dinitro-1,3,5-triazine (MNX) to the earthworm *Eisenia fetida*, *Chemosphere* 64, 86, 2006.

50. Kuperman RG et al., Toxicity of emerging energetic soil contaminant CL-20 to potworm *Enchytraeus crypticus* in freshly amended or weathered and aged treatments, *Chemosphere*, 62, 1282, 2006.

51. Robidoux PY et al., Acute and chronic toxicity of the new explosive CL-20 to the earthworm (*Eisenia andrei*) exposed to amended natural soils, *Environ. Toxicol. Chem.*, 23, 1026, 2004.

52. Kuperman RG et al., Weathering and aging of TNT in soil increases toxicity to potworm *Enchytraeus crypticus*, *Environ. Toxicol. Chem.*, 24, 2509, 2005.

53. Monteil-Rivera F et al., Physico-chemical measurements of CL-20 towards environmental applications: Comparison with RDX and HMX, *J. Chromatogr. A*, 1025, 125, 2004.

54. Renoux AY et al., Transformation of 2,4,6-trinitrotoluene in soil in the presence of the earthworm *Eisenia andrei*, *Environ. Toxicol. Chem.*, 19, 1473, 2000.

55. Robidoux PY et al., Acute toxicity of 2,4,6-trinitrotoluene (TNT) in the earthworm (*Eisenia andrei*), *Ecotoxicol. Environ. Saf.*, 44, 311, 1999.

56. Robidoux PY et al., Evaluation of tissue and cellular biomarkers to assess 2,4,6-trinitroluene (TNT) exposure in earthworms: Effects-based assessment in laboratory studies using *Eisenia andrei*, *Biomarkers* 7, 306, 2002.

57. Lachance B et al., Toxicity and bioaccumulation of reduced TNT metabolites in the earthworm *Eisenia andrei* exposed to amended forest soil, *Chemosphere*, 55, 1339, 2004.

58. Jarvis SA, McFarland VA, and Honeycutt ME, Assessment of the effectiveness of composting for the reduction of toxicity and mutagenicity of explosive-contaminated soil, *Ecotoxicol. Environ. Saf.*, 39, 131, 1998.

59. Dodard SG et al., Lethal and subchronic effects of 2,4,6-trinitrotoluene (TNT) on *Enchytraeus albidus* in spiked artificial soil, *Ecotoxicol. Environ. Saf.*, 54, 131, 2003.

60. Schäfer RK, Evaluation of the Ecotoxicological Threat of Ammunition Derived Compounds to the Habitat Function of Soil, Ph.D. thesis, Freien Universität, Berlin, Germany, 2002.

61. Kuperman RG et al., Toxicities of dinitrotoluenes and trinitrobenzene freshly amended or weathered and aged in a sandy loam soil to *Enchytraeus crypticus, Environ. Toxicol. Chem.,* 25, 1368, 2006.

62. Kuperman RG et al., Ecological soil screening levels for invertebrates at explosives-contaminated sites: Supporting sustainability of army testing and training, in *Proceedings of 25th Army Science Conference*, Orlando, FL, November 27–30, 2006.

63. Parmelee RW et al., Soil microcosm for testing the effects of chemical pollutants on soil fauna communities and trophic structure, *Environ. Toxicol. Chem.,* 12, 1477, 1993.

64. Sunahara GI et al., Ecotoxicological characterization of energetic substances using a soil extraction procedure, *Ecotoxicol. Environ. Saf.,* 43, 138, 1999.

65. Robidoux PY et al., Chronic toxicity of octahydro-1,3,5,7-tetranitro-1,3,5,7-tetrazocine (HMX) in soil using the earthworm (*Eisenia andrei*) reproduction test, *Environ. Pollut.,* 111, 283, 2001.

66. Landrum M et al., Effects of perchlorate on earthworm (*Eisenia fetida*) survival and reproductive success, *Sci. Total Environ.,* 363, 237, 2006.

67. Bailey HC et al., Toxicity of TNT Wastewaters to Aquatic Organisms, Vol 3—Chronic Toxicity of LAP Wastewater and 2,4,6-Trinitrotoluene, AD-A164 282, Final Report, SRI International, Menlo Park, CA, 1985.

68. Stebbing ARD, Hormesis: The stimulation of growth by low levels of inhibitors, *Sci. Total Environ.,* 22, 213, 1982.

69. Calabrese EJ, McCarthy ME, and Kenyon E, The occurrence of chemically induced hormesis, *Health Physics,* 52, 531, 1987.

70. Bentley RE et al., Acute Toxicity of HMX to Aquatic Organisms, U.S. Army Medical Research and Development Command, Final Report AD-A061 730, EG & G Bionomics, Wareham, MA, 1977.

71. Bentley RE, Petrocelli, SR, and Suprenant DC, Determination of the Toxicity to Aquatic Organisms of HMX and Related Wastewater Constituents. Part III. Toxicity of HMX, TAX and SEX to aquatic organisms, AD-A172 385, Final Report, Springborn Bionomics, Wareham, MA, 1984.

72. Steevens JA et al., Toxicity of the explosives 2,4,6-trinitrotoluene, hexahydro-1,3,5-trinitro-1,3,5-triazine, and octahydro-1,3,5,7-tetranitro-1,3,5,7-tetrazocine in sediments to *Chironomus tentans* and *Hyalella azteca*: Low-dose hormesis and high-dose mortality, *Environ. Toxicol. Chem.,* 21, 1475, 2002.

73. Stenersen J, *Chemical Pesticides: Mode of Action and Toxicology*, CRC Press, Boca Raton, FL, 2004.

74. Boelsterli UA, *Mechanistic Toxicology: The Molecular Basis of How Chemicals Disrupt Biological Targets*, Taylor & Francis Group, London, UK, 2003.

75. Escher BI and Hermens JL, Modes of action in ecotoxicology: Their role in body burdens, species sensitivity, QSARs, and mixture effects, *Environ. Sci. Technol.,* 36, 4201, 2002.

76. Ankley GT et al., Toxicogenomics in regulatory ecotoxicology, *Environ. Sci. Tech.,* 40, 4055, 2006.

77. USEPA (U.S. Environmental Protection Agency), Potential Implications of Genomics for Regulatory and Risk Assessment Applications at EPA, EPA 100/B-04/002, December 2004.

78. Gong P et al., Genotoxicity of 2,4- and 2,6-dinitrotoluene as measured by the *Tradescantia* micronucleus (Trad-MCN) bioassay, *Mutat. Res.,* 538, 13, 2003.

79. Ma TH et al., *Tradescantia* micronucleus (Trad-MCN) tests on 140 health-related agents, *Mutat. Res.,* 138, 157, 1984.

80. Cenas N et al., Quantitative structure-activity relationships in enzymatic single-electron reduction of nitroaromatic explosives: Implications for their cytotoxicity, *Biochim. Biophys. Acta.,* 1528, 31, 2001.

81. Kumagai Y, Neuronal nitric oxide synthase (nNOS) catalyzes one-electron reduction of 2,4,6-trinitrotoluene, resulting in decreased nitric oxide production and increased nNOS gene expression: Implication for oxidative stress, *Free Radic. Biol. Med.*, 37, 350, 2004.

82. Ekman DR et al., SAGE analysis of transcriptome responses in *Arabidopsis* roots exposed to 2,4,6-trinitrotoluene, *Plant Physiol.*, 133, 1397, 2003.

83. Gong P, Inouye LS, and Perkins EJ, Comparative neurotoxicity of two energetic compounds, hexanitrohexaazaisowurtzitane and hexahydro-1,3,5-trinitro-1,3,5-triazine, in the earthworm *Eisenia fetida*, *Environ. Toxicol. Chem.*, 26, 954, 2007.

84. Ekman DR, Wolfe NL, and Dean JF, Gene expression changes in *Arabidopsis thaliana* seedling roots exposed to the munition hexahydro-1,3,5-trinitro-1,3,5-triazine, *Environ. Sci. Technol.*, 39, 6313, 2005.

85. Mezzari MP et al., Gene expression and microscopic analysis of *Arabidopsis* exposed to chloroacetanilide herbicides and explosive compounds. A phytoremediation approach, *Plant Physiol.*, 138, 858, 2005.

86. Kutty R and Bennett GN, Biochemical characterization of trinitrotoluene transforming oxygen-insensitive nitroreductases from *Clostridium acetobutylicum* ATCC 824, *Arch. Microbiol.*, 184, 158, 2005.

87. Cheong S et al., Determination of laccase gene expression during degradation of 2,4,6-trinitrotoluene and its catabolic intermediates in *Trametes versicolor*, *Microbiol. Res.*, 161, 316, 2006.

88. Seth-Smith HM et al., Cloning, sequencing, and characterization of the hexahydro-1,3,5-trinitro-1,3,5-triazine degradation gene cluster from *Rhodococcus rhodochrous*, *Appl. Environ. Microbiol.*, 68, 4764, 2002.

89. Gong P et al., Construction of an earthworm cDNA microarray for diagnosing toxicological interactions of mixtures of ordnance related compounds, in *Abstract Book of the 4th SETAC World Congress and 25th Annual Meeting in North America*, Portland, OR, November 14–18, 2004, 449.

90. Gong P et al., Microarray analysis of 2,4,6-trinitrotoluene exposure effects in the earthworm *Eisenia fetida*, in *Abstract Book of SETAC North America 26th Annual Meeting*, Baltimore, MD, November 13–17, 2005, 193.

91. Talmage SS et al., Nitroaromatic munition compounds: Environmental effects and screening values, *Rev. Environ. Contam. Toxicol.*, 161, 1, 1999.

4 Aquatic Toxicology of Explosives

Marion Nipper, R. Scott Carr, and
Guilherme R. Lotufo

CONTENTS

4.1 Introduction 77
4.2 Aquatic Toxicity of Explosives, Propellants, and Related Chemicals 79
    4.2.1 Nitroaromatic Compounds 79
        4.2.1.1 Nitrobenzenes 79
        4.2.1.2 Nitrotoluenes 83
        4.2.1.3 Nitrophenols.. 93
        4.2.1.4 2,4,6-Trinitrophenylmethylnitramine (Tetryl) 98
        4.2.1.5 Photoactivation.. 99
    4.2.2 Cyclic Nitramines 105
        4.2.2.1 Hexahydro-1,3,5-trinitro-1,3,5-triazine (RDX) 105
        4.2.2.2 Octahydro-1,3,5,7-tetranitro-1,3,5,7-tetrazocine (HMX)... 105
        4.2.2.3 Hexanitrohexaazaisowurtzitane (CL-20).......................... 106
    4.2.3 Other Compounds.. 106
        4.2.3.1 Nitrocellulose.. 106
        4.2.3.2 Nitroguanidine 106
        4.2.3.3 Nitroglycerin 106
        4.2.3.4 Pentaerythritol Tetranitrate (PETN)............................. 107
        4.2.3.5 Diethylene Glycol Dinitrate (DEGDN) 107
4.3 Criteria and Screening Benchmarks.. 107
4.4 Conclusions and Recommendations 109
References.. 110

4.1 INTRODUCTION

The release of explosives into the environment can occur during several stages of munitions production, storage, transport, and usage, including manufacturing, handling, loading, and final dispersal or disposal. Disposal of munitions in the ocean has been practiced for many decades, particularly after the Second World War [1,2], mostly into depressions of the continental shelf or near continental slopes, but also into the deep sea [3]. The environmental implications of such contamination became the subject of scientific research, which intensified with the decommissioning of military installations in the United States and overseas in the 1990s and potential use of such grounds for nonmilitary activities. Explosive compounds, such as 2,4,6-trinitrotoluene (TNT), 1,3,5-trinitrobenzene (TNB), 1,3-dinitrobenzene (1,3-DNB), hexahydro-1,3,5-trinitro-1,3,5-triazine (RDX), octahydro-1,3,5,7-tetranitro-1,3,5,7-tetrazocine (HMX), and 2,4,6-trinitrophenylmethylnitramine (tetryl), have been detected in surface water of numerous military installations, sometimes at exceedingly elevated concentrations, particularly of TNT, which reached up to 14.9 μmoles L^{-1} (3.38 mg L^{-1}) [4].

Explosive compounds may undergo extensive transformation in aquatic systems, by microbial attack or abiotic mechanisms such as hydrolysis, oxidation, phototransformation, and so forth. Therefore, aquatic receptors may be exposed not only to energetic compounds released to the environment but also to their numerous transformation products. The key biotic and abiotic transformations of major explosives are discussed in Chapter 2.

Aquatic toxicology studies carried out during the early to mid twentieth century have addressed the toxicity of 2,4,6-trinitrophenol (picric acid) and related products [5–10], and of TNT wastewater [11], also called red or pink water due to color change by exposure to sunlight [4]. The relatively recent awareness of environmental contamination problems with explosives resulted in intense scientific research on their biological effects after the 1970s. Most investigations of the aqueous toxicity of explosives and related compounds were performed using freshwater organisms. The aqueous toxicity of a variety of energetic compounds has been investigated using several marine species [12–15].

Comprehensive reviews of the aquatic toxicology of energetic chemicals can be found elsewhere [4,16,17]. The present chapter will review and summarize the reported adverse biological effects of energetic materials to aquatic organisms including bacteria, algae, invertebrates, fish, and amphibians. This review will include the energetic compounds for which aquatic toxicity data have been reported in the available literature. For completeness, toxicity data on transformation products of energetic chemicals as well as on compounds of similar or related molecular structures to energetic compounds will also be included. This chapter will include the presentation and discussion of available criteria and screening benchmarks for energetic compounds.

Toxicological issues related to the presence of energetic chemicals in groundwater were not addressed in this review, because groundwater contaminants pose potential risk to human and livestock health, rather than to populations of aquatic organisms. The fate and biological effects of energetic chemicals in sediments are reviewed in Chapter 5.

In aquatic toxicology, quantal responses (e.g., dead or alive, fertilized or unfertilized) are expressed as LC_p (lethal concentration to p percent of the test organisms) or EC_p (effective concentration to p percent of the test organisms), whereas for graduated responses (e.g., reproduction or growth inhibition), the use of IC_p (inhibitory concentration to p percent of the test organisms) is recommended [18]. However, several of the papers cited in this review do not make this distinction and presented EC_{50} values for graduated responses. The data will be presented here as they were calculated in the original papers. Because the molecular mass of the compounds discussed in this chapter varies widely, aqueous concentrations are expressed on a molar basis, therefore allowing for more accurate toxic potency comparison and use of toxicity data in mixture interaction models.

4.2 AQUATIC TOXICITY OF EXPLOSIVES, PROPELLANTS, AND RELATED CHEMICALS

4.2.1 NITROAROMATIC COMPOUNDS

Ample evidence suggests that the toxicity of nitroaromatic explosive compounds and their transformation products is related to the number and orientation of nitro groups [15,16,19–22]. The manner in which toxicity is affected by these parameters varies among different categories of nitroaromatics and will be discussed in the following sections. Although the only nitroaromatic compounds used as energetic materials are 1,3-DNB, TNB, 2,4- and 2,6-DNT, TNT, 2,4,6-TNP, and tetryl, other data on related nitro-substituted benzenes, toluenes, and phenols, as well as their aminated breakdown products, will be included in this review. Many of these nonexplosive compounds are byproducts of manufacturing, or transformation products of explosives. Many other products, however, have been released in the aquatic environment as a result of industrial activities, such as in the manufacture of dyes.

4.2.1.1 Nitrobenzenes

In the nitrobenzene group of compounds, only 1,3-DNB and TNB have been used as explosives. Although there were some exceptions, a tendency for increased toxicity with increased degree of nitration of nitrobenzenes was identified [15,19–22]. The toxicity of these compounds is similar to that of chemicals with a narcotic mode of action [22]. It was found that hydrophobicity, as determined by the octanol/water partition coefficient, and rate of reduction of the nitro group contributes to toxicity of nitrobenzenes [21,23]. The position of the nitro groups strongly influences the toxicity of DNBs, with a tendency for lower toxicity of the explosive compound, 1,3-DNB (meta-DNB), relative to its ortho and para isomers (1,2- and 1,4-DNB, respectively) [21–23]. This tendency, however, seems to be species-dependent and holds true for freshwater fish and cladocerans [21–24], but not for the unicellular alga *Chlorella pyrenoidosa*, which had slightly higher sensitivity to meta-DNB [22,23] (Table 4.1).

The toxicity of DNBs and TNB to several freshwater test endpoints and species, including bacteria, microalgae, invertebrates, and fish, under several experimental conditions and exposure times, ranged from 0.06 to 295 μmoles L^{-1} and 1.8 to 12.7 μmoles L^{-1}, respectively, whereas the toxicity of the mono-nitrated benzene (NB) ranged from 15 to 950 μmoles L^{-1} (Table 4.1). Direct comparisons under the same test system indicate that the toxicity of DNBs is one to three orders of magnitude higher than that of NB to freshwater fish (*Poecilia reticulata*) 14-day survival [21], cladoceran (*Daphnia magna*) survival and reproduction [22], and population growth of the microalgae *Scenedesmus quadricauda*, *Microcystis aeruginosa* [25], and *C. pyrenoidosa* [22]. The bacterium *Pseudomonas putida*, however, is somewhat more sensitive to NB than to DNB, suggesting a different mode of action in these prokaryotic organisms [26]. An additional nitration of DNB, resulting in TNB, in general does not cause a dramatic change in toxicity, with a slight increase or

TABLE 4.1
Toxicity of Nitrobenzenes and Derivatives

Test Organism[a]	Habitat[b]	Biological Endpoint[c]	Statistical Endpoint	Chemical (μmoles L^{-1})									
				NB	1,2-DNB	1,3-DNB	1,4-DNB	1,3,5-TNB	2-NA	3-NA	4-NA	3,5-DNA	2,4-DNA
Fish													
Pimephales promelas [19,20,24,61,82–85]	FW	Survival	96-h LC$_{50}$	129.2; 950.4	0.36; 4.4	7.6; 10.0; 12.7; 41.6; 99.9	0.36; 1.1	2.4; 4.8; 5.2	—	—	737.0	112.6; 115.8; 119.0	—
Ictalurus punctatus [61,84]	FW	Survival	96-h LC$_{50}$	—	—	4.8	0.40	1.8	—	—	—	75.9	—
Oncorhynchus mykiss [84]	FW	Survival	96-h LC$_{50}$	—	—	1.0	—	2.4	—	—	—	16.4	—
Lepomis macrochirus [45,84]	FW	Survival	96-h LC$_{50}$	349.3	—	0.83	—	4.0	—	—	—	38.2	—
Cyprinus carpio [23]	FW	Survival	96-h LC$_{50}$	—	0.49	8.5	0.68	—	—	—	331.1	—	56.2
Oryzias latipes [86]	FW	Survival	48-h LC$_{50}$	162.5	—	—	—	—	117.5	446.7	—	—	—
Poecilia reticulata [21]	FW	Survival	14-d LC$_{50}$	501.2	0.71	2.3	0.23	—	70.8	371.5	389.0	—	—
Cyprinodon variegatus [46]	M	Survival	96-h LC$_{50}$	479.2	—	—	—	—	—	—	—	—	—
Sciaenops ocellatus [15]	M	Embryo survival	48-h EC$_{50}$	—	—	273.6	—	6.6	—	—	—	—	—
Arthropods													
Daphnia magna [19,20,22,28,44,47, 50,59,84,85]	FW	Immobility	24-h EC$_{50}$	487.4; 503.6	—	11.3	—	—	—	—	24.0; 51.3; 60.2; 75.9	—	37.1; 42.6; 49.0; 65.5
			48-h EC$_{50}$	219.3; 268.9	1.9; 31.5	8.7; 42.6; 163.0; 295.0	0.77	12.7; 14.1	—	—	—	73.3; 84.1	52.4
		Reproduct.	21-d EC$_{50}$	195.0	0.96	1.2	0.06	—	—	—	—	—	—

Species	Medium	Endpoint	Test								
Daphnia carinata [23]	FW	Immobility	48-h IC$_{50}$	0.81	7.9	0.40	75.9	151.4	135.0	—	107.1
Ceriodaphnia dubia [34]	FW	Reproduct.	7-d EC$_{50}$	—	—	6.6	—	—	—	—	—
Hyalella azteca [35]	FW	Survival	96-h LC$_{50}$	—	—	6.7	—	—	—	—	—
Americamysis bahia [15]	E	Survival	96-h LC$_{50}$	—	42.2	6.1	—	—	—	—	—
Annelids											
Aplexa hypnorum [61]	FW	Survival	96-h LC$_{50}$	—	—	2.52	—	—	—	—	—
Dinophilus gyrociliatus [15]	M	Survival	7-d LC$_{50}$	—	89.2	9.9	—	—	—	—	—
		Reproduct.	7-d EC$_{50}$	—	22.0	2.8	—	—	—	—	—
Echinoderms											
Arbacia punctulata [15]	M	Embryo dev.	48-h EC$_{50}$	—	547.2	6.1	—	—	—	—	—
		Fertilization	60-min EC$_{50}$	—	1535	394.2	—	—	—	—	—
Protozoans											
Tetrahymena pyriformis [57]	FW	Pop. growth	24-h EC$_{50}$	—	—	—	—	796.0	—	—	—
Entosiphon sulcatum [26]	FW	Pop. growth	72-h TT[d]	15.4	4.5	—	—	—	—	—	—
Algae											
Chlorella pyrenoidosa [22]	FW	Pop. growth	96-h EC$_{50}$	144.6	0.14	0.47	—	—	—	—	—
Scenedesmus obliquus [23]	FW	Pop. growth	96-h EC$_{50}$	0.91	1.1	—	467.8	331.1	398.1	—	20.9
Scenedesmus quadricauda [25]	FW	Pop. growth	8-d TT	268.1	4.2	—	—	—	79.6	152.9	—
Microcystis (Diplocystis) aeruginosa [25]	FW	Pop. growth	8-d TT	15.4	1.0	—	—	—	2.5	—	—
Ulva fasciata [15]	M	Zoosp. germ.	96-h EC$_{50}$	—	5.1	0.38	—	—	—	—	—

(continued on next page)

TABLE 4.1 (continued)
Toxicity of Nitrobenzenes and Derivatives

Test Organism[a]	Habitat[b]	Biological Endpoint[c]	Statistical Endpoint	Chemical (μmoles L^{-1})									
				NB	1,2-DNB	1,3-DNB	1,4-DNB	1,3,5-TNB	2-NA	3-NA	4-NA	3,5-DNA	2,4-DNA
Bacteria													
Pseudomonas putida [26]	FW	Pop. growth	16-h TT	56.9	—	83.3	—	—	—	—	—	—	—
Vibrio fischeri (Microtox) [22,30]	M	Bioluminesc.	15-min EC$_{50}$	144.6	58.9	239.7	1.1	—	—	—	—	—	—
			30-min EC$_{50}$	—	—	—	—	0.94	—	—	—	—	—

[a] Reference numbers in brackets.
[b] FW = freshwater; M = marine; E = estuarine.
[c] Reproduct. = reproduction; embryo dev. = embryo development; pop. growth = population growth; zoosp. germ. = zoospore germination; bioluminesc. = bioluminescence.
[d] TT = toxicity threshold.

decrease depending on test method and species [4]. Differences in toxicity between tests using the same method, species, and toxicant were considered acceptable from the biological standpoint when coefficients of variation of LC or EC_{50} values reached up to 64% [27]. Differences of orders of magnitude in toxicity test results would truly reflect distinct sensitivities of the test species, endpoints, or methods, rather than normal test variability.

Toxicity assessments with marine organisms revealed that the sensitivity to the two explosives 1,3-DNB and TNB varies widely with test species, method, and endpoint. However, TNB was consistently one to two orders of magnitude more toxic than 1,3-DNB in experiments with fish embryos, mysids, polychaetes, and sea urchin and macroalgae early life stages [15]. Toxicity tests with early life stages of different species of marine organisms were among the most and the least sensitive to nitrobenzenes. The most sensitive was the 96-h zoospore germination test with the green macroalga *Ulva fasciata*, with an EC_{50} of 5.1 and 0.38 µmoles L^{-1} for 1,3-DNB and TNB, respectively. The least sensitive was the 60-min fertilization assay with the sea urchin *Arbacia punctulata*, with EC_{50} values of 1535 and 394 µmoles L^{-1} for 1,3-DNB and TNB, respectively [15]. Sensitivity also varies broadly in freshwater tests with unicellular organisms, with large differences in the NB and 1,3-DNB toxicity threshold among the freshwater protozoan, *Entosyphum sulcatum*; bacterium, *P. putida*; and microalgae, *S. quadricauda*, *C. pyrenoidosa*, and *M. aeruginosa* (Table 4.1).

Deneer et al. [22] suggested that the toxicity of DNBs to freshwater organisms is more dependent on test duration than that of NBs, due to the formation of reactive DNB metabolites that are readily bound to various macromolecules in vivo and would, therefore, enhance toxicity. This suggestion, however, is not corroborated by data from other authors, who observed decreases in toxicity when DNBs and TNB were compared to their main metabolites, nitro- and dinitroanilines, in tests with fish, cladocerans, and algae [4,16,21,23] (Table 4.1).

4.2.1.2   Nitrotoluenes

The nitrotoluenes 2,4- and 2,6-DNT, and TNT have been used as explosives. Similarly to nitro-substituted benzenes, the toxicity of nitrotoluenes to a variety of fish, invertebrates, and microalgae tends to increase with the degree of nitration, particularly between NTs and DNTs. The toxicity of DNTs is also isomer dependent [19,22,23,25,26,28,29].

The two DNT (2,4- and 2,6-) isomers used as energetic materials, as well as 3,5-DNT, tend to be an order of magnitude less toxic than the 2,3-, 3,4- and 2,5-DNT isomers to a variety of freshwater organisms [16,19,20,22,26] (Table 4.2). The marine bacterium *Vibrio fischeri* (Microtox®), however, was slightly less sensitive to 2,3- and 3,4-DNT than to the 2,6-DNT isomer [22,26].

The two explosives, 2,4- and 2,6-DNT, tend to differ in their degree of toxicity to several organisms. Several freshwater and marine fish, as well as the Microtox test, exhibit higher sensitivity to 2,6- than to 2,4-DNT [15,22,30,31]. The order of toxicity of these two isomers to invertebrates, including freshwater and marine crustaceans, the polychaete *Dinophilus gyrociliatus* and the sea urchin *A. punctulata* is variable [15,22,31] (Table 4.2). Freshwater microalgae and early life stages of the marine macroalga, *U. fasciata*, exhibit higher sensitivity to 2,4-DNT [15,22,26,31].

TABLE 4.2
Toxicity of Nitrotoluenes

Test Organism[a]	Habitat[b]	Biological Endpoint[c]	Statistical Endpoint	2-NT	3-NT	4-NT	2,3-DNT	2,4-DNT	2,5-DNT	2,6-DNT	3,4-DNT	3,5-DNT	2,3,6-TNT	2,4,6-TNT
								Chemical (µmoles L^{-1})						
Amphibians														
Xenopus laevis embryo [33]	FW	Teratogen.	96-h LC$_{50}$	—	—	—	—	—	—	—	—	—	—	16.7
		Malformat.	96-h EC$_{50}$	—	—	—	—	—	—	—	—	—	—	9.8
Fish														
Pimephales promelas [19,24,37,41, 76,85,87]	FW	Survival	96-h LC$_{50}$	270.5; 277.1	237.0	362.4; 363.9	9.0; 10.4	156.5; 178.4; 180.1	7.1	101.6; 108.7	8.2	120.8; 124.1	0.44; 0.52	10.6; 11.4; 12.8; 13.2; 13.7; 16.3
		Behavior	96-h EC$_{50}$	—	—	—	—	—	—	—	—	—	—	2.0
		Total length	9-mo NOEC[d]	—	—	—	—	—	—	—	—	—	—	0.02,
			LOEC[d]	—	—	—	—	—	—	—	—	—	—	0.06
Cyprinus carpio [23]	FW	Survival	96-h LC$_{50}$	—	—	295.1	—	147.9	—	109.6	—	—	—	—
Salmo gairdneri [37,85,87]	FW	Survival	96-h LC$_{50}$	—	—	—	—	74.7	—	—	—	—	—	3.5; 5.3; 6.2
Ictalurus punctatus [37,87]	FW	Survival	96-h LC$_{50}$	—	—	—	—	136.2	—	—	—	—	—	10.6
Lepomis macrochirus [37,45,85,87]	FW	Survival	96-h LC$_{50}$	—	—	—	1.8	74.1	—	—	—	—	—	11.4; 13.2; 15.0

Species														
Poecilia reticulata [21]	FW	Survival	14-d LC$_{50}$	239.9	218.8	269.2	—	69.2	—	97.7	—	—	—	—
Cyprinodon variegatus [46]	E	Survival	96-h LC$_{50}$	—	—	—	12.6	—	—	—	—	—	—	—
Sciaenops ocellatus [15]	M	Embryo surv.	48-h EC$_{50}$	—	—	—	—	263.5	—	186.7	—	—	—	36.1
Arthropods														
Daphnia magna [13,19,20,22,28, 44,47,85,87–89]	FW	Immobility or survival	24-h EC/LC$_{50}$	306.3	255.2	80.2	17.6	120.8; 208.6; 210.3	—	76.9; 109.8	—	—	—	66.1
Daphnia magna [13,19,20,22,28, 44,47,85,87–89]	FW	Immobility or survival	48-h EC/LC$_{50}$	79.5; >562.2	55.0; 204.9	86.0; 88.2; 137.8	3.6; 25.8; 30.9	142.7; 143.8; 186.1; 192.2; 210.3; 260.8	17.0; 18.7	119.1; 119.7; 186.1	17.0; 20.3; 30.9	247.6; 248.2	3.0; 3.4; 4.0	>19.4; 51.5; 52.4; 52.9
Daphnia carinata [23]	FW		14-d LC$_{50}$	—	—	—	—	22.5	—	—	—	—	—	0.9
		Reproduct.	21-d EC$_{50}$	53.7	60.3	51.3	9.8	3.3; 3.6	—	52.5	6.0	—	—	3.5
		Immobility or survival	48-h IC$_{50}$	—	123.0	104.7	—	100.0	—	123.0	—	—	—	—
Ceriodaphnia dubia [34,76]	FW	Survival	48-h LC$_{50}$	—	—	—	—	—	—	—	—	—	—	17.8
		Survival	7-d LC$_{50}$	—	—	—	—	—	—	—	—	—	—	>26.4
		Reproduct.	7-d EC$_{50}$	—	—	—	—	—	—	—	—	—	—	14.5
			7-d NOEC, LOEC	—	—	—	—	—	—	—	—	—	—	7.2; 11.9
Hyalella azteca [35,37,87]	FW	Survival	48-h LC$_{50}$	—	—	—	—	>456.8	—	—	—	—	—	16.0; 28.6

(continued on next page)

TABLE 4.2 (continued)
Toxicity of Nitrotoluenes

Test Organism[a]	Habitat[b]	Biological Endpoint[c]	Statistical Endpoint	Chemical (µmoles L⁻¹)										
				2-NT	3-NT	4-NT	2,3-DNT	2,4-DNT	2,5-DNT	2,6-DNT	3,4-DNT	3,5-DNT	2,3,6-TNT	2,4,6-TNT
Tanytarsus dissimilis [37,85,87]	FW	Survival	48-h LC$_{50}$	—	—	—	—	123.5	—	—	—	—	—	109.3; 118.9
Paratanytarsus parthenogeneticus [76]	FW	Survival	48-h LC$_{50}$	—	—	—	—	—	—	—	—	—	—	189.0
Artemia salina [90]	M	Survival	24-h LC$_{50}$	—	—	—	—	—	—	—	—	—	—	128.2
Nitocra spinipes [13]	E	Survival	96-h LC$_{50}$	—	—	—	—	93.3	—	—	—	—	—	33.5
Schizopera knabeni [29]	E	Survival	96-h LC$_{50}$	798.5	—	—	—	—	—	357.0	—	—	—	—
		Embryo surv.	96-h EC$_{50}$	357.3	—	—	—	—	—	242.2	—	—	—	—
Americamysis bahia [15]	E	Survival	96-h LC$_{50}$	—	—	—	—	29.6	—	30.7	—	—	—	4.3
Annelids														
Lumbriculus variegatus [81,85,87]	FW	Survival	48-h LC$_{50}$	—	—	—	—	>456.8	—	—	—	—	—	21.6; 22.9; 24.2; >127.8
			14-d LC$_{50}$	—	—	—	—	177.9	—	—	—	—	—	59.0
		Reproduct.	21-d EC$_{50}$	—	—	—	—	278.6	—	—	—	—	—	26.5
Dinophilus gyrociliatus [15]	M	Survival	7-d LC$_{50}$	—	—	—	—	115.3	—	71.4	—	—	—	33.9
		Reproduct.	7-d EC$_{50}$	—	—	—	—	31.3	—	11.5	—	—	—	7.9

Species	Media	Endpoint	Metric								
Echinoderms											
Arbacia punctulata [15]	M	Embryo dev.	48-h EC$_{50}$	—	—	—	—	280.0	36.8	—	52.9
		Fertilization	60-min EC$_{50}$	—	—	—	—	373.3	461.2	—	453.7
Rotifers											
Brachionus calyciflorus [90,91]	FW	Survival	24-h LC$_{50}$	—	—	—	—	—	—	—	24.4; 40.1
Protozoans											
Tetrahymena pyriformis [57]	FW	Pop. growth	24-h EC$_{50}$	729.2	364.6	597.9	—	—	549.0	—	—
Plants											
Lemna minor [76]	FW	Frond prod.	96-h EC$_{50}$	—	—	—	—	—	—	—	7.0
Algae											
Microcystis aeruginosa [25,92]	FW	Survival	24-h LC100	—	—	—	—	—	—	—	35.2
Scenedesmus quadricauda [26]	FW	Pop. growth	8-d TT[c]	22.6	7.3	24.1	12.1	0.1	2.7	—	—
		Pop. growth	7-d TT	32.1	109.4	—	4.6	14.8	65.9	—	7.0
Scenedesmus obliquus [23]	FW	Pop. growth	96-h EC$_{50}$	—	—	182.0	—	—	87.1	—	—
Chlorella pyrenoidosa [22]	FW	Pop. growth	96-h EC$_{50}$	347.1	102.1	161.9	5.0	5.0	37.2	4.1	—
Pseudokirchneriella subcapitata (formerly *Selenastrum capricornutum*) [31,32,81,92]	FW	Pop. growth	96-h EC$_{50}$	—	—	—	—	14.3	90.3	—	2.5; 3.3
			7-d EC$_{50}$	—	—	—	—	—	—	—	10.2
			14-d EC$_{50}$	—	—	—	—	18.1	—	—	6.6

(continued on next page)

TABLE 4.2 (continued)
Toxicity of Nitrotoluenes

Test Organism[a]	Habitat[b]	Biological Endpoint[c]	Statistical Endpoint	Chemical (μmoles L^{-1})										
				2-NT	3-NT	4-NT	2,3-DNT	2,4-DNT	2,5-DNT	2,6-DNT	3,4-DNT	3,5-DNT	2,3,6-TNT	2,4,6-TNT
Ulva fasciata [15,29]	M	Zoospore germination	96-h EC$_{50}$	369.0	—	—	—	13.7	—	36.8; 73.4	—	—	—	11.0
Bacteria														
Bacterial Mix–Polytox [90]	FW	Resp. inhib.	21-min EC$_{50}$	—	—	—	—	—	—	—	—	—	—	>440.5
Vibrio fischeri (Microtox) [22,30–32, 36,38,90]	M	Bioluminesc.	5-min EC$_{50}$	—	—	—	—	—	—	—	—	—	—	47.6
			15-min EC$_{50}$	13.5	28.8	79.5	33.1	281.7	—	15.8	38.0	—	—	3.3; 7.4; 10.9
			15-min IC$_{50}$	—	—	—	—	269.0	—	16.4	—	—	—	3.4; 4.2; 5.8
			30-min EC$_{50}$	—	—	—	—	247.4	—	27.9	49.9	—	—	12.4; 15.8

[a] References in brackets.

[b] FW = freshwater; M = marine; E = estuarine.

[c] Teratogen. = teratogenesis; malformat. = malformation; embryo surv. = embryo survival; reproduct. = reproduction; embryo dev. = embryo development; pop. growth = population growth; frond prod. = frond production; resp. inhib = respiratory inhibition; bioluminesc. = bioluminescence.

[d] NOEC = no observed effect concentration; LOEC = lowest observed effect concentration.

[e] TT = toxicity threshold.

The explosive TNT tended to be more toxic than 2,4- and 2,6-DNT to a variety of species used in comparative studies, except for gametes and embryos of the sea urchin *A. punctulata* (Table 4.2). The less common isomer, 2,3,6-TNT, is 15- to 30-fold more toxic than the energetic material TNT to freshwater fish and cladocerans [16,19,20,24].

Successive reduction of TNT to aminodinitrotoluenes (ADNTs) and diaminonitrotoluenes (DANTs) leads to a sequential decrease of toxicity to the microalga, *Pseudokirchneriella subcapitata* (formerly *Selenastrum capricornutum*) [32] and in the Microtox test [30]. Further toxicity decrease occurs in the latter with an additional reduction to triaminotoluene (TAT) [30] (Table 4.2 and Table 4.3). The ADNTs were also less toxic than the parent compound to the freshwater fish *Pimephales promelas* [19], and embryos of the frog *Xenopus laevis* [33]. The reduction of 2,3,6-TNT to ADNTs promoted decreased toxicity in studies with *P. promelas* and *D. magna* [16,19,20]. However, ADNTs were more toxic than TNT to the cladoceran *D. magna* [19,20], and a sequential increase in toxicity along the reductive pathway of TNT was observed with *Ceriodaphnia dubia* [34] (Table 4.2 and Table 4.3). In aqueous exposures with the amphipod *Hyalella azteca*, 2-A-4,6-DNT was similarly toxic but 2,4-DA-6-NT was more toxic than TNT [35]. Therefore, changes in toxicity as TNT is transformed to its reduced products indicate a species-specific trend.

Similar to TNT, toxicity comparisons along the transformation pathways of DNTs suggest that relative toxicities along the reductive pathway or among isomers are species and chemistry dependent. The reduction of 2,4- and 2,6-DNT to aminonitrotoluenes (ANTs) resulted in more toxic metabolites to *V. fischeri* [36] and the copepod *Schizopera knabeni* [29], respectively, whereas 2-A-6-NT had lower toxicity than the parent compound to *U. fasciata* zoospores [29] (Table 4.2 and Table 4.4). Similarly to the marine crustacean, the freshwater cladoceran *D. magna* also exhibited higher sensitivity to ANTs than 2,4- and 2,6-DNT [19], whereas the freshwater alga *P. subcapitata* was less sensitive to ANTs than to their parent compounds [19,31]. The opposite was observed with further reduction to 2,4- and 2,6-DAT, with a sharp increase in toxicity to the microalga but decreased toxicity to Microtox [31].

Toxicity of 2,4- and 2,6-DNT amino metabolites to the fish *P. promelas* was isomer dependent, as 2-A-4-NT and 2-A-6-NT were less toxic than their respective parent compounds, whereas 4-A-2-NT and 2,4-DNT had similar toxicities [19,37] (Table 4.2 and Table 4.4). Similarly, 2,4-DNT and TNT metabolites with the amino group in the 4 position tend to be more toxic to *V. fischeri* than those with the amino in the 2 position [30,31,37,38] (Table 4.3 and Table 4.4).

Water quality parameters, such as pH, temperature, hardness, and salinity, can influence the effects of contaminants on aquatic life. The toxicity of TNT decreases slightly with increasing pH and temperature, but is not significantly affected by hardness [39]. The persistence of TNT in the environment is limited, and several biological and physical processes (photolysis, hydrolysis) influence its environmental fate [39]. The photolysis half-life of TNT appears to be season and latitude dependent, as it ranged from 14 h at latitude 20 in summer to 84 h at latitude 50 in winter in the northern hemisphere [40]. The aquatic toxicity of several photo- and biodegradation products of TNT has been studied. Pink water obtained by constant illumination of a TNT solution was more toxic than the TNT solution to *P. promelas* using

TABLE 4.3

Toxicity of TNT Reduction Products and Their Isomers

Test Organism[a]	Habitat[b]	Biological Endpoint[c]	Statistical Endpoint	Chemical (μmoles L⁻¹)									
				2-A-3, 6-DNT	2-A-4, 6-DNT	3-A-2, 4-DNT	3-A-2, 6-DNT	4-A-2, 6-DNT	4-A-3, 5-DNT	5-A-2, 4-DNT	2,4-DA- 6-NT	2,6-DA- 4-NT	TAT
Amphibians													
Xenopus laevis embryo [33]	FW	Teratogen.	96-h LC$_{50}$	—	166.0	—	—	115.0	—	—	—	—	—
		Malformat.	96-h EC$_{50}$	—	16.9	—	—	85.8	—	—	—	—	—
Fish													
Pimephales promelas [19,20]	FW	Survival	96-h LC$_{50}$	4.0; 4.6	75.1; 76.6	61.9	55.3; 57.3	35.0	>65.93	12.2	—	—	—
Arthropods													
Daphnia magna [19,20,34]	FW	Immobility	48-h EC$_{50}$	11.2; 12.7	22.8; 23.3	41.1; 48.7	23.8; 24.3	26.4; 27.4	>65.9; >66.4	15.7; 16.7	—	—	—
Ceriodaphnia dubia [34]	FW	Survival	7-d LC$_{50}$	—	24.9	—	—	33.5	—	—	>1.2	>12.0	>1.4
		Reproduct.	7-d EC$_{50}$	—	15.7	—	—	25.9	—	—	0.30	2.0	>1.4
Hyalella azteca [35]	FW	Survival	96-h LC$_{50}$	—	19.4	—	—	46.5	—	—	11.3	—	—
Tigriopus californicus [93]	M	Survival	72-h LC$_{50}$	—	>253.6	—	—	>253.6	—	—	—	—	—

	Type[b]	Endpoint	Test									
Molluscs												
Crassostrea gigas [93]	M	Survival	96-h LC$_{50}$	—	—	—	>253.6	>253.6	—	—	—	—
Algae												
Pseudokirchneriella subcapitata (formerly *S. capricornutum*) [32]	FW	Population growth	96-h IC$_{50}$	—	—	—	59.4	12.9	—	293.2	>306.40	—
Bacteria												
Vibrio fischeri (Microtox) [30,32,36,38]	M	Bioluminesc.	15-min EC$_{50}$	—	—	148.6; 169.4	773.04	1520.3	—	—	297.3; 302.7	—
			15-min IC$_{50}$	—	>131.4	—	>131.2; 773.0	>112.6	—	354.7	>306.4	257.4
			30-min EC$_{50}$	—	—	—	107.4; 618.4	>380.4; 1171	—	287.1	>598.2	739.4

a References in brackets.

b FW = freshwater; M = marine.

c Teratogen. = teratogenesis; malformat. = malformation; reproduct. = reproduction; bioluminesc. = bioluminescence.

TABLE 4.4
Toxicity of Reduction Products of Several DNT Isomers

Test Organism[a]	Habitat[b]	Biological Endpoint[c]	Statistical Endpoint	Chemical (μmoles L⁻¹)								
				2-A-3-NT	2-A-4-NT	2-A-5-NT	2-A-6-NT	3-A-4-NT	4-A-2-NT	4-A-3-NT	2,4-DAT	2,6-DAT
Fish												
Pimephales promelas [19,24]	FW	Survival	96-h LC₅₀	159.7	452.2; 468.6	569.8	328.0; 328.6	159.1; 167.6	169.6; 171.5	163.0	—	—
Arthropods												
Daphnia magna [19,20]	FW	Immobility	48-h EC₅₀	—	147.2; 147.9	—	86.8	38.1; 43.4	92.0; 93.3	—	—	—
Algae												
Pseudokirchneriella subcapitata (formerly *S. capricornutum*) [31]	FW	Population growth	96-h EC₅₀	—	101.5	—	145.7	—	97.7	—	0.08	0.47
Arthropods												
Schizopera knabeni [29]	E	Survival Embryo surv.	96-h LC₅₀ 96-h EC₅₀	— —	— —	— —	>277.4 52.2	— —	— —	— —	— —	— —
Algae												
Ulva fasciata [29]	M	Zoosp. germ.	96-h EC₅₀	—	—	—	>159.8	—	—	—	—	—
Bacteria												
Vibrio fischeri (Microtox) [30,31]	M	Bioluminesc.	15-min IC₅₀ 30-min EC₅₀	— —	94.3 143.5	— —	22.8 23.9	— —	43.5 62.6	— —	405.7 825.9	>609 4003

[a] Reference numbers in brackets.
[b] FW = freshwater; M = marine; E = estuarine.
[c] Embryo surv. = embryo survival; zoosp. germ. = zoospore germination; bioluminesc. = bioluminescence.

survival and behavioral endpoints, but less toxic to reproduction of the microalgae *P. subcapitata* and *M. aeruginosa* [41]. However, Rosenblatt et al. [16] reported that photolysis of TNT reduced its toxicity to several freshwater fish and *D. magna*, but no change in toxicity occurred in tests with freshwater amphipods and insect larvae. Several factors may have affected the results of the two latter investigations: species-specific sensitivity to phototransformation products of TNT, and the presence of different types and amounts of phototransformation products in the exposure water. Further research is necessary to improve the current understanding of the photolysis of nitrotoluenes, including identification and toxicity assessment of phototransformation products.

It was suggested by Zhao et al. [23] that the toxicity of 2,4- and 2,6-DNT would be lower than expected from QSAR studies because of the expected photolysis during typical experimental conditions. Such photolysis may not necessarily occur because of insufficient ultraviolet (UV) radiation in standard laboratory lighting. Under field conditions, however, the effects of photolysis would be expected. Phototransformation of 2,6-DNT began rapidly upon exposure to simulated solar radiation and promoted disappearance of the parent compound within 24 h in clean seawater, generating an orange-colored solution, which contained high molecular weight chemicals [42]. This phototransformed 2,6-DNT solution, however, was significantly more toxic than the original stock solution to several endpoints of zoospore germination tests with the macroalga *U. fasciata* and of the life-cycle test with the polychaete *D. gyrociliatus*, whereas copepod *S. knabeni*, survival and embryo hatching success tests exhibited lower toxicity of the phototransformed solution [29].

In summary, the toxicity of DNT differs among the various isomers and also varies considerably among different species. Likewise, the degradation of TNTs and DNTs into a variety of breakdown products can promote an increase or decrease in toxicity, which is also species dependent. Therefore, a general rule on the environmental effects of the biotic or abiotic transformation of nitrotoluenes cannot be established based on the available toxicity data. The transformation of nitrotoluenes by biotic and abiotic processes generates a mixture of compounds. A single study [36] on the toxicological interactions among nitroaromatic compounds revealed the possibility for nonadditive toxicological interactions among those compounds.

4.2.1.3 Nitrophenols

Picric acid (2,4,6-trinitrophenol or 2,4,6-TNP) has been used as an energetic material. Unlike TNB and TNT, picric acid was substantially less toxic than nitrophenols (NPs) with lower levels of nitration (i.e., less nitro- groups), as observed with several marine and freshwater uni- and multicellular organisms [25,28,29,43–48]. However, similar to nitro-substituted benzenes and toluenes, the toxicity of nitrophenols tends to increase with increasing nitration from NPs to dinitrophenols (DNPs), as observed with fish and higher invertebrates [28,43–46,49–51] (Table 4.5), although some unicellular organisms such as the microalga *S. quadricauda*, the protozoan *E. sulcatum*, and the bacterium *P. putida*, exhibited higher sensitivity to NPs than to 2,4-DNP [26,52] (Table 4.5). Therefore, as a general rule there is a trend for toxicity to increase from NPs to DNPs, followed by a toxicity decrease with additional nitration, from DNPs to picric acid.

TABLE 4.5
Toxicity of Nitrophenols and Derivatives

Test Organism[a]	Habitat[b]	Biological Endpoint[c]	Statistical Endpoint	Chemical (µmoles L⁻¹)												
				2-NP	3-NP	4-NP	2,4-DNP	2,5-DNP	2,6-DNP	2,4,6-TNP	2-AP	3-AP	4-AP	4-A-2-NP	2,4-DAP	2-A-4,6-DNP
Fish																
Pimephales promelas [50,61]	FW	Survival	96-h LC$_{50}$	—	—	294.7; 438.5	92.3	—	—	—	—	—	—	222.5	—	—
Salmo gairdneri [12]	FW	Survival	96-h LC$_{50}$	—	—	—	—	—	—	478.4	—	—	—	—	—	232.0
Lepomis macrochirus [45]	FW	Survival	96-h LC$_{50}$	—	—	59.7	3.4	—	—	742.0	—	—	—	—	—	—
Notopterus notopterus [94]	FW	Survival	96-h LC$_{50}$	—	—	—	7.3	—	—	—	—	—	—	—	—	—
Cyprinus carpio [23]	FW	Survival	96-h LC$_{50}$	263.0	125.9	—	—	—	—	—	—	—	—	—	—	—
Clupea harengus [83]	M	Egg division	48-h LC$_{50}$	—	—	—	5.0	—	—	—	—	—	—	—	—	—
Salmo salar [54]	M	Survival	96-h LT^d	—	—	—	3.8	—	—	—	—	—	—	—	—	—
Sciaenops ocellatus [15]	M	Embryo surv.	48-h EC$_{50}$	—	—	—	—	—	—	554.3	—	—	—	—	—	—
Cyprinodon variegatus [46]	E	Survival	96-h LC$_{50}$	—	—	194.1	157.5	—	—	567.4	—	—	—	—	—	—
Arthropods																
Daphnia magna [13,28,43,44,47, 49,51,59,89]	FW	Immobility or Survival	24-h EC$_{50}$ (pH 6.0)	173.7	177.8	—	6.6	—	—	—	—	—	—	—	—	—
			24-h EC$_{50}$ (pH 7.8)	229.1	195.0	—	20.9	—	—	—	—	—	—	—	—	—

Species	Med.	Response	Endpoint											
Daphnia carinata [23]	FW	Immobility	24-h EC$_{50}$ (pH 9.0)	309.0	295.1	—	32.4	—	—	—	—	—	—	—
			24-h EC$_{50}$	194.1; 1510	280.4	57.5; 79.1; 251.6	38.0; 103.2	—	371.0; 536.9; 632.9	—	20.2	—	2.8	—
			48-h EC$_{50}$	122.2	—	33.8; 60.4; 143.8; 158.1	22.3; 25.5; 25.6	—	371.0; 375.4; 392.8	—	10.1	—	2.2	—
			96-h LOEC	431.3	172.5	100.6	32.6	—	384.1	—	—	—	—	—
			48-h IC$_{50}$	144.6	123.0	—	—	—	—	—	—	—	—	—
Artemia salina [60]	M	Larvae surv.	24-h LC$_{50}$	46.7	—	158.9	18.5	—	—	—	—	—	—	—
Crangon septemspinosa [58]	E	Survival	96-h LT	236.5	—	189.8	-	—	—	—	—	—	—	—
Schizopera knabeni [29]	E	Survival	96-h LC$_{50}$	—	—	—	77.4	—	>593	—	—	—	—	>227
		Embryo surv.	96-h EC$_{50}$	—	—	—	53.8	—	191.8	—	—	—	—	47.7
Annelids														
Dinophilus gyrociliatus [15]	M	Survival	7-d LC$_{50}$	—	—	—	—	—	1156	—	—	—	—	—
		Reproduct.	7-d EC$_{50}$	—	—	—	—	—	675.6	—	—	—	—	—
Molluscs														
Crassostrea virginica [12]	M	Survival	144-h LC$_{50}$	—	—	—	—	—	1113	—	—	—	—	350.5
		Shell depos.	144-h EC$_{50}$	—	—	—	—	—	121.8	—	—	—	—	28.1

(continued on next page)

TABLE 4.5 (continued)
Toxicity of Nitrophenols and Derivatives

Test Organism[a]	Habitat[b]	Biological Endpoint[c]	Statistical Endpoint	Chemical (μmoles L⁻¹)												
				2-NP	3-NP	4-NP	2,4-DNP	2,5-DNP	2,6-DNP	2,4,6-TNP	2-AP	3-AP	4-AP	4-A-2-NP	2,4-DAP	2-A-4,6-DNP
Protozoans																
Tetrahymena pyriformis [48,57]	FW	Pop. growth	24-h EC$_{50}$	251.6	201.3	39.5	—				—	—	—	—	—	—
			48-h EC$_{50}$	1.5	—	—	—	0.60	1.6	6.2	—	—	—	—	—	—
Entosiphon sulcatum [26]	FW		72-h TT[e]	2.8	7.0	6.0	143.8	—	—	—	—	—	—	—	—	—
Algae																
Scenedesmus quadricauda [25,43]	FW	Pop. growth	96-h LOEC	258.8	201.3	517.6	217.3	—	—	1048	—	—	—	—	—	—
			8-d TT	30.9	54.6	53.2	86.9	—	—	266.2	—	—	—	—	—	—
Scenedesmus obliquus [23]	FW		96-h EC$_{50}$	309.0	177.8	269.1	—	—	—	—	—	—	—	—	—	—
Microcystis (Diplocystis) aeruginosa [25,92]	FW	Survival	8-d TT	194.1	122.2	402.6	—	—	—	174.6	18.3	9.2	9.2	—	—	—
			LC100	—	—	—	—	—	—	—	—	—	—	—	10.2	—
Ulva fasciata [15,29]	M	Zoospore germination	96-h EC$_{50}$	—	—	—	165.0	—	—	1811; 2029	—	—	—	—	—	>571.0

Bacteria

Escherichia coli [43]	FW	Pop. growth	48-h LOEC	>7189	2157	718.9	>543.2	—	—	—	—	—	—	—	>4365
Pseudomonas putida [26]	FW		16-h TT	6.5	50.3	28.8	826.7	—	—	—	—	—	—	—	—
Vibrio fischeri (Microtox) [30]	M	Bioluminesc.	30-min EC$_{50}$	—	—	—	—	—	—	—	—	—	—	1910	—

[a] Reference numbers in brackets.
[b] FW = freshwater; M = marine; E = estuarine.
[c] Embryo surv. = embryo survival; larvae surv. = larvae survival; reproduct. = reproduction; shell depos. = shell deposition; pop. growth = population growth; bioluminesc. = bioluminescence.
[d] LT = lethal threshold.
[e] TT = toxicity threshold.

Chemicals of a phenolic nature, including nitrophenols, have been described as nonspecific metabolic inhibitors [53], and the extent of inhibition depends on their pKa values, which determine the extent of their dissociation in water. It has been stated that phenols with a pKa value below 6.3 are capable of acting as weak acid respiratory uncouplers, whereas those with pKa above 6.3 would act as polar narcotics [48]. 2,4-Dinitrophenol (2,4-DNP) with pKa 4.03, is a well-known uncoupler of oxidative phosphorylation [54,55], whereas 2- and 3-NP, with pKa values of 6.80 and 8.27, respectively, likely act as polar narcotics [51]. An increase in pH promotes the dissociation of these three chemicals, which is accompanied by a toxicity decrease to *D. magna* [51]. Experiments were performed at pH values ranging from 6 to 9, which did not affect the organisms' survival in the absence of toxicants (Table 4.5). Similarly, the cessation of *A. punctulata* egg division when exposed to 3-NP and 2,4-DNP was more effective with a pH decrease, leading to an increased concentration of the chemicals in their undissociated form [7]. It has been hypothesized that the nonionized form of phenolic compounds penetrates biological membranes more easily than the ionized form that occurs at higher pH. Therefore, nonionized forms are the strongest contributors to the toxic effect of a given phenolic compound, although some toxicity should also be expected from the ionized form [53].

The effects of mono- and dinitrophenols as well as aminophenols (APs) on photosynthesis, respiration, and chlorophyll production of the microalga *C. pyrenoidosa* were analyzed [52]. These compounds caused hormesis (dose–response phenomenon characterized by low dose stimulation and high dose inhibition, resulting in either a J-shaped or an inverted U-shaped dose response curve) on chlorophyll production at concentrations of NP, DNP, and AP ranging from 92 to 458 μmoles L^{-1}. The reader is directed to Chapter 3 of this book for a discussion of hormetic responses of terrestrial receptors to energetic materials.

The toxicity of mononitrophenols is isomer dependent, but no general trend could be established for a variety of test systems and species. The ortho isomer (2-NP; pKa = 6.8) is usually less toxic than the meta form (3-NP; pKa = 8.3) [23,28,43,52,56,57] (Table 4.5), possibly due to its lower polarity, higher volatility, lower solubility [53], and higher pKa. The para isomer (4-NP) also tended to be more toxic than the ortho to a wide variety of organisms [23,28,43,57–59], although there were some exceptions, such as the microalgae *S. quadricauda* and *M. aeruginosa* [25,43], the bacterium *P. putida* [26], and the brine shrimp *Artemia salina* [60].

The few direct toxicity comparisons using nitro-substituted phenols and their reduced metabolites indicate that the relationship between reduction of nitro to amino groups, and toxicity varies with the degree of nitration of the parent compound and the test species. The reduction of picric acid to picramic acid (2-A-4,6-DNP) and of 4-NP to 4-AP promoted an order of magnitude toxicity increase in lethal and sublethal assessments with several marine and freshwater invertebrates, fish, and algae [12,29,59] (Table 4.5). The reduction of 2,4-DNP to 4-A-2-NP, however, decreased toxicity to *P. promelas* by 2.5-fold [50,61].

4.2.1.4   2,4,6-Trinitrophenylmethylnitramine (Tetryl)

2,4,6-Trinitrophenylmethylnitramine (tetryl) is highly toxic to marine organisms, with EC_{50} values to marine fish, algae, bacteria, and a variety of invertebrates ranging from 0.07 to 11.1 µmoles L^{-1} [15,30] (Table 4.6). Tetryl has not been manufactured for several decades [4], and it is not known if the parent compound or its degradation products are a present environmental concern. High performance liquid chromatography (HPLC) of tetryl-spiked marine sediments and pore waters exhibited a number of unidentified peaks likely corresponding to currently unknown degradation products [62]. Toxicity assessments of these samples indicated strong adverse effects to marine species, but further studies are necessary to determine the nature, stability, and toxicity of the breakdown products of tetryl.

4.2.1.5 Photoactivation

Photoactivation, also known as photoenhanced or photoinduced toxicity, consists of the enhancement of adverse effects of bioaccumulated chemicals by exposure to UV radiation. Photoactivation is caused by changes in the energetic state of some kinds of bioaccumulated organic molecules. It is a property that depends on molecular structure [63] and is shared by a variety of organic chemicals, for example, several polycyclic aromatic hydrocarbons (PAHs). Although photoactivation of a variety of nitroaromatic compounds may occur, there are no data providing definitive confirmation of this phenomenon. Irradiation of test chambers containing 2,4-DNT with UV light (wavelength of 295–340 nm) for 2 h, at the end of a toxicity test with this compound, resulted in significantly increased acute toxicity to the freshwater cladoceran *D. magna*, but not to the estuarine copepod *Nitocra spinipes* [13]. Variation in toxicity may be a result of phototransformation of 2,4-DNT, rather than photoactivation of the bioaccumulated compound.

Coexposure of *D. magna* to nitroaromatic compounds (TNT, 2-A-4,6-DNT, or 4-A-2,6-DNT) and near-ultraviolet radiation (near-UV; 312–354 nm), after an initial 12-h exposure to the chemicals in the dark, promoted a significant toxicity increase (four- to fivefold) of the three chemicals [64]. Similar experiments with the planarian *Dugesia dorotocephala* only exhibited increased toxicity of 2-A-4,6-DNT (tenfold) with near-UV coexposure, but not of the other two chemicals [64]. In the absence of organisms, TNT was phototransformed, that is, transformed into breakdown products, by near-UV exposure (from 336 to 400 nm), whereas 2-A-4,6-DNT and 4-A-2,6-DNT remained stable. Therefore, photoactivation would be the likely cause of the enhanced effect of the ADNTs, whereas phototransformation products may have contributed to the increase in toxicity exhibited by simultaneous exposure of the organisms to TNT and near-UV. Alternatively, it is possible that the phototransformation products were photoactivated. Coexposure to near-UV for 30 min, but not for 15 min, also enhanced toxicity of these chemicals to the bacterium *V. fischeri* (Microtox test) [38]. It is not clear if this is a result of the formation of a new, more toxic TNT metabolite in 30 min, but not 15 min, or if 30 min would be the minimum time for a sufficient amount of the chemicals to be bioaccumulated to enable the occurrence of the photoactivation process within the cells.

Coexposure of sea urchin *Lytechinus variegatus* eggs or embryos to near-UV and different nitroaromatic chemicals (TNT; 2,3-, 2,4-, 2,6-, and 3,4-DNT; 2,3-, 2,4-, 2,6-, or 3,4-DAT) enhanced their adverse effects, relative to exposure to the same compounds in the dark [65]. Coexposure of *Daphnia magna* to these chemicals and near-UV, after an initial 12-h exposure to the chemicals in the dark, caused

TABLE 4.6
Toxicity of Various Ordnance Compounds

Test Organism[a]	Habitat[b]	Biological Endpoint	Statistical Endpoint	Chemical (μmoles L^{-1})							
				RDX	HMX	Tetryl	NQ	NG	NC	DEGDN	PETN
Fish											
Pimephales promelas [66,72,75–78,95]	FW	Survival	96-h LC$_{50}$	26.1; 29.7; 57.2	—	—	>26079; >31902	11.0; 13.2; 13.6; 15.8	>957.0	2506	—
		Early life stage dry weight	28-d EC$_{50}$	25.5	—	—	—	—	—	—	—
		Early life stage survival		40.8	—	—	—	—	—	—	—
		Early life stage length	28-d NOEC^c, LOEC^c	—	—	—	10090; 19506	—	—	—	—
		Early life stage dry weight	28-d NOEC, LOEC	6.1, 10.6	—	—	—	—	—	—	—
		Egg survival	96-h LC$_{50}$	>450.2	—	—	—	>79.3	—	—	—
		1-h posthatch survival		193.6	—	—	—	24.2	—	—	—
		7-d posthatch survival		17.1	50.6	—	—	9.2	—	—	—
			7-d LOEC	—	—	—	—	1.0	—	—	—
		30-d posthatch survival	96-h LC$_{50}$	72.0	—	—	—	9.2	—	—	—

Species [refs]		Effect	Test								
Ictalurus punctatus [66,72,75,77–79]	FW	30-d posthatch survival	30-d LOEC					0.53			
		60-d posthatch survival	96-h LC50	49.5				15.0			
		Survival	48-h LC50	27.0							
		Survival	96-h LC50	18.5; 58.5	>108.0		>25330	14.1	>957	1419	
		30-d post-hatch survival	30-d LOEC					1.4			
Salmo gairdneri [66,75,77,78]	FW	Survival	96-h LC50	28.8			>15740	12.3	>957	1449	
		Fry growth	42-d LOEC				16364				
Oncorhynchus mykiss [72,76]	FW	Survival	96-h LC50		>108.0		>14894	8.4			
		60-d post-hatch dry weight	60-d NOEC, LOEC					0.53, 0.88			
Lepomis macrochirus [66,72,75,77–79]	FW	Survival at various pH and hardnesses	96-h LC50	16.2–38.3	>108.0		>25310	6.1–11.9	>957	1316	
Sciaenops ocellatus [5]	M	Embr. hatch. and survival	48-h EC50	>299.0		6.3					
Arthropods											
Daphnia magna [13,66,72,75,78,79]	FW	Immobility or survival	48-h EC50	67.5; >450.2	>108.0	0.0	>9840; >27,271	140.9; 202.6	>957	459.4	>155.0
Ceriodaphnia dubia [67,76]	FW	Survival	48-h LC50	>76.5			25925	78.5			
		Reproduction	7-d NOEC, LOEC	13.4, 27.1			2498, 4228	14.2, 24.1			
Hyalella azteca [75]	FW	Survival	48-h LC50				>26,233				
Gammarus minus [75]	FW	Survival	48-h LC50				>26,134				

(continued on next page)

TABLE 4.6 (continued)
Toxicity of Various Ordnance Compounds

Test Organism[a]	Habitat[b]	Biological Endpoint	Statistical Endpoint	Chemical (µmoles L⁻¹)							
				RDX	HMX	Tetryl	NQ	NG	NC	DEGDN	PETN
Gammarus fasciatus [66,72,78]	FW	Immobility	48-h EC$_{50}$	>450.2	>108.0	—	—	220.2	>957	—	—
Gammarus pseudolimnaeus [79]	FW	Survival	48-h LC$_{50}$	—	—	—	—	—	—	1812	—
Asellus militaris [66,72,78]	FW	Immobility	48-h EC$_{50}$	>450.2	>108.0	—	—	220.2	>957	—	—
Chironomus tentans [66,72,78]	FW	Immobility	48-h EC$_{50}$	>450.2	>108.0	—	—	88.1; 242.2	>957	—	—
Hexagenia bilineata [79]	FW	Survival	48-h LC$_{50}$	>67.5	—	—	—	—	—	—	—
	FW	Survival	48-h LC$_{50}$	—	—	—	—	—	—	1747	—
Paratanytarsus parthenogeneticus [67,76,79]	FW	Survival	48-h LC$_{50}$	>131.46	—	—	—	153.8	—	814.8	—
		Growth and egg product.	10-d LOEC	>93.64	—	—	—	—	—	—	—
Paratanytarsus dissimilis [75]	FW	Survival	48-h LC$_{50}$	—	—	—	>32623	—	—	—	—
Americamysis bahia [15]	E	Survival	96-h LC$_{50}$	>211.6	—	4.5	—	—	—	—	—
Nitocra spinipes [13]	E	Survival	96-h LC$_{50}$	—	>121.6	—	>816.8	—	—	—	>101.2
Annelids											
Lumbriculus variegatus [75]	FW	Survival	48-h LC$_{50}$	—	—	—	>27559	—	—	—	—
Dinophilus gyrociliatus [15]	M	Survival	7-d LC$_{50}$	>220.6	—	0.21	—	—	—	—	—
		Reproduction	7-d EC$_{50}$	117.0	—	0.07	—	—	—	—	—

Echinoderms

Species		Effect	Endpoint								
Arbacia punctulata [15]	M	Embryo devel.	48-h EC$_{50}$	>330.0	—	0.28	—	—	—	—	—
		Fertilization	60-min EC$_{50}$	>330.0	—	11.1	—	—	—	—	—

Cnidarians

Species		Effect	Endpoint								
Hydra littoralis [67,76]	FW	Survival	48-h LC$_{50}$	>145.4	—	—	19,804	76.8	—	—	—

Algae

Species		Effect	Endpoint								
Microcystis aeruginosa [66,72,77]	FW	Population growth	96-h EC$_{50}$	>144.1	>108.0	—	—	>44.0	—	—	—
		Chlorophyll a reduction		—	—	—	—	>44.0	—	—	—
Anabaena flos-aquae [72,77]	FW	Population growth	96-h EC$_{50}$	—	>108.0	—	—	>44.0	—	—	—
		Chlorophyll a reduction		—	—	—	—	>44.0	—	—	—
Navicula pelliculosa [66,72,77]	FW	Population growth	96-h EC$_{50}$	>144.1	>108.0	—	—	14.5	>957	—	—
			96-h LOEC	—	—	—	—	1.41	—	—	—
		Chlorophyll a reduction		—	—	—	—	4.4	—	—	—
Pseudokirchneriella subcapitata (formerly Selenastrum capricornutum) [32,68,72,74–76,78]	FW	Population growth	96-h EC$_{50}$	>144.1; >165.2	>108.0	—	—	1.8; 5.1	>957	—	—
			96-h IC$_{50}$	>181.0	>22.0	—	—	—	—	—	—
			120-h LOEC	—	—	—	>36,063	—	—	297.8	—
			120-h EC$_{50}$	—	—	—	18,613	4.4	—	—	—
		Chlorophyll a reduction	96-h EC$_{50}$	—	—	—	—	—	554.3	—	—
		Dry weight reduction		—	—	—	20,621	—	—	—	—

(continued on next page)

TABLE 4.6 (continued)
Toxicity of Various Ordnance Compounds

Test Organism[a]	Habitat[b]	Biological Endpoint	Statistical Endpoint	Chemical (μmoles L^{-1})							
				RDX	HMX	Tetryl	NQ	NG	NC	DEGDN	PETN
M. aeruginosa [74]	FW	Population growth	96-h EC$_{50}$	—	—	—	—	—	>957	—	—
		Chlorophyll a reduction		—	—	—	—	—	>957	—	—
A. flos-aquae [74]	FW	Population growth	96-h EC$_{50}$	—	—	—	—	—	>957	—	—
		Chlorophyll a reduction		—	—	—	—	—	>957	—	—
Ulva fasciata [15]	M	Zoospore germination	96-h EC$_{50}$	54.0	—	2.3	—	—	—		
Bacteria											
Vibrio fischeri (Microtox) [30,32]	M	Bioluminescence	15-min IC$_{50}$	>181.0	>21.7	—	—	—	—	—	—
			30-min EC$_{50}$	328.3	>84.4	1.6	>19,218	—	—	—	46.0

a Reference numbers in brackets.
b FW = freshwater; M = marine; E = estuarine.
c NOEC = no observed effect concentration; LOEC = lowest observed effect concentraiton.

an increase in the toxicity of 2,3- and 3,4-substituted DNTs and DATs, but not of 2,4- and 2,6-substituted chemicals [65]. In the latter study, exposure to phototransformation products and photoactivation of accumulated compounds may both have contributed to the observed effects.

4.2.2 Cyclic Nitramines

4.2.2.1 Hexahydro-1,3,5-trinitro-1,3,5-triazine (RDX)

Hexahydro-1,3,5-trinitro-1,3,5-triazine, commonly known as royal demolition explosive or RDX, appears to be more acutely toxic to freshwater fish (*P. promelas, Salmo gairdneri, Lepomis macrochirus, Ictalurus punctatus*) than to invertebrates and algae, including a variety of crustaceans; the hydrozoan *Hydra littoralis*; and the microalgae *P. subcapitata, M. aeruginosa, Anabaena flos-aquae,* and *Navicula pelliculosa* [66–68] (Table 4.6). Adverse effects of RDX to *L. macrochirus* were not affected by a pH between 6.0 and 8.0, or by hardness in the range of 35 to 250 mg L^{-1} $CaCO_3$, but increased with temperature [66]. The sensitivity of *P. promelas* was age dependent, being higher one week after hatching than as an earlier posthatch or after 30 days of age [66].

Embryo hatching success and posthatch larval survival of the marine fish *Sciaenops ocellatus* were not affected by RDX at the highest concentration achieved without the use of a solvent carrier (299 µmoles L^{-1}) [15] (Table 4.6). At its solubility limit, this chemical was also neither acutely toxic to marine crustaceans and polychaetes, nor exhibited sublethal effects in tests with echinoids [15], whereas adverse effects were observed with the macroalga *U. fasciata* zoospore germination, the polychaete *D. gyrociliatus* reproduction, and the 30-min Microtox biolumines cence endpoints [15,30] (Table 4.6).

Chemical analyses of RDX in aqueous solution indicate that RDX is resistant to degradation under aerobic conditions, since no loss occurred after 90 d of incubation in river water [69]. Unchanged toxicity to fish of RDX solutions aged up to 96 h also suggested a lack of aerobic degradation [69]. Photolysis has been described as the primary physical mechanism degrading RDX in aqueous solution and as a potential method of reducing its toxicity [16]. Photolysis of a 45 µmoles L^{-1} aqueous stock solution of RDX in sunlight for 28 h caused the parent compound to reach nondetectable concentrations in HPLC analyses, and resulted in a nontoxic solution to *C. dubia* survival and reproduction [70], whereas the LOEC (lowest observed effect concentration) of the parent compound to reproduction of this species is 27 µmoles L^{-1} [67].

4.2.2.2 Octahydro-1,3,5,7-tetranitro-1,3,5,7-tetrazocine (HMX)

Octahydro-1,3,5,7-tetranitro-1,3,5,7-tetrazocine, most commonly known as HMX, has low solubility (~22.3 µmoles L^{-1} at 20°C) [71] and, based on little available information, low toxicity to aquatic organisms. Its reported 48-h and 96-h LC_{50} values to several freshwater fish, microcrustaceans, and algae were above solubility [72] (Table 4.6). Similarly, saturated solutions of HMX were not toxic in 15- and 30-min Microtox tests [30,32], and in experiments with the estuarine copepod *N. spinipes* [13].

4.2.2.3 Hexanitrohexaazaisowurtzitane (CL-20)

The polycyclic polynitramine, hexanitrohexaazaisowurtzitane (CL-20) has recently been developed as an explosive compound [73]. The toxicity of this chemical to two marine organisms has been assessed. No toxicity to *V. fischeri* (Microtox) or to the alga *P. subcapitata* was exhibited at concentrations up to its solubility limit (8.2 μmoles L^{-1} at 25°C) [73].

4.2.3 OTHER COMPOUNDS

4.2.3.1 Nitrocellulose

Studies with nitrocellulose (NC) indicated no toxicity at concentrations up to 957 μmoles L^{-1} when tested with several species of freshwater fish (*P. promelas*, *L. macrochirus*, *S. gairdneri*, and *I. punctatus*), invertebrates (*D. magna*, *Gammarus fasciatus*, *Asellus militaris*, and *Chironomus tentans*), and microalgae (*M. aeruginosa*, *A. flos-aquae*, and *N. pelliculosa*) [74] (Table 4.6). The freshwater microalga *P. subcapitata* exhibited an EC_{50} of 554.3 μmoles L^{-1}. These values, however, were derived from nominal concentrations of active ingredient in static toxicity tests using "a slurry of poacher pit fines" collected at an army ammunition plant [74], with nitrocellulose at 11.8% active ingredient. The overall lack of toxicity of nitrocellulose is likely a result of its insolubility in water. Toxicity to microalga was likely caused by decreased photosynthesis due to cloudiness of the test solution, rather than chemical toxicity. Further studies to verify this speculation are recommended.

4.2.3.2 Nitroguanidine

Nitroguanidine (NQ) was not acutely toxic to four species of freshwater fish, one cladoceran, two amphipods, midge larvae, and oligochaetes at its solubility limit, which ranged from 15.7 mmoles L^{-1} at 12°C to 32.6 mmoles L^{-1} at 22°C [75] (Table 4.6). Chronic toxicity to rainbow trout early life stages was also low, with a LOEC of 16.3 mmoles L^{-1} for fry growth but no significant effect on survival [75], and it also produced no toxicity to estuarine copepods at very high concentrations [13] (Table 4.6). NQ exhibits a linear chain molecule and, therefore, is unlikely to be photoactivated. The mixture of chemicals produced by its photolysis, on the other hand, promoted a two orders of magnitude toxicity enhancement in tests with *D. magna*, *P. promelas,* and *P. subcapitata* [75,76]. Therefore, adverse environmental effects should not be expected from the use of NQ in aquatic environments, unless residues are accumulated in highly sunlit areas, where photolysis would be expected.

4.2.3.3 Nitroglycerin

Freshwater fish tended to exhibit higher sensitivity than invertebrates to nitroglycerin (NG), with LC_{50} values to fish ranging from 8.4 to 15.8 μmoles L^{-1}, and 48-h EC_{50} values to several species of invertebrates in the range of 14.2 to 242.2 μmoles L^{-1}, depending on test conditions [77,78] (Table 4.6). Eggs and 1-h posthatch fry of *P. promelas* were less sensitive than later life stages. The sensitivity of phytoplankton to NG was variable among species, with 96-h EC_{50} values in the range of 1.8 to

>44 µmoles L^{-1} (Table 4.6). LOEC values for weight and survival with 7-, 30- and 60-d fish fry ranged from 0.53 to 1.4 µmoles L^{-1} [77] (Table 4.6).

4.2.3.4   Pentaerythritol Tetranitrate (PETN)

The toxicity of pentaerythritol tetranitrate (PETN) was higher to bacteria than crustaceans, with a 30-min EC_{50} of 46 µmoles L^{-1} in the Microtox test [30] (Table 4.6). Tests with the estuarine copepod *N. spinipes* and freshwater cladoceran *D. magna* resulted in 96- and 48-h LC_{50} and EC_{50} values >101 and >155 µmoles L^{-1}, respectively [13].

4.2.3.5   Diethylene Glycol Dinitrate (DEGDN)

Microalgae were the most sensitive of several freshwater organisms used in toxicity tests with diethylene glycol dinitrate (DEGDN), with 80% reduction in biomass of *P. subcapitata* after a 120-h exposure to 298 µmoles L^{-1} of this chemical [79] (Table 4.6). The freshwater crustaceans *D. magna* and *Gammarus pseudolimnaeus* exhibited 48-h LC_{50} values of 459 and 1812 µmoles L^{-1}, respectively, and 96-h LC_{50} values for fish ranged from 1316 to 2506 µmoles L^{-1} [79] (Table 4.6).

4.3   CRITERIA AND SCREENING BENCHMARKS

Although a variety of toxicity data are available for numerous explosives and related compounds, as discussed throughout this chapter, only a few of these chemicals have been intensely used and become relatively abundant in some aquatic ecosystems. Therefore, the development of water quality criteria has only been attempted with a few compounds, which are considered munitions of environmental concern. Talmage et al. [4] reviewed the biological effects of several energetic compounds and calculated the water quality criteria (WQC) for the protection of aquatic life when sufficient data were available. The derivation of WQC requires the use of results of at least eight acute toxicity tests with a wide variety of animals from specific taxonomic groups, results of at least three chronic toxicity tests using fish and invertebrates, and results of at least one test with an aquatic plant, or more if plants are among the most sensitive organisms tested with the substance [80].

Talmage et al. [4] analyzed an extensive TNT toxicity database and derived freshwater final acute and chronic values (FAV and FCV, respectively) of 4.99 and 0.410 µmoles L^{-1}, and the respective criterion maximum concentration (CMC) of 2.50 µmoles L^{-1}, that is, half the FAV. The lowest chronic effect value for fish, of 0.176 µmoles L^{-1}, was suggested as a better screening benchmark than the calculated FCV until a sufficient chronic toxicity database becomes available [4]. All these values are above the LOEC from nine-month life-cycle tests with fathead minnows, of 0.06 µmoles L^{-1} [76] (Table 4.2), suggesting that the proposed chronic values need to be revised for adequate long-term protection of aquatic life.

Proposed interim TNT WQC for the protection of marine life were 0.376 and 0.125 µmoles L^{-1} for acute and chronic values, respectively [15]. These criteria are expected to be protective of marine life, based on the sensitivity of marine and estuarine species to TNT given in the present review (Table 4.2).

TABLE 4.7

Final and Interim Acute and Chronic Values
Derived as Water Quality Criteria (µmoles L⁻¹)
for the Protection of Aquatic Life

Chemical[a]	Acute Value[b]	Chronic Value[b]	Habitat[c]
2,4,6-TNT [4]	4.99	0.176	FW
2,4,6-TNT [15]	0.376	0.125	M
1,3,5-TNB [4]	0.282	0.052	FW
1,3-DNB [4]	1.28	0.101	FW
3,5-DNA [4]	2.51	0.322	FW
2-A-4,6-DNT [4]	1.78	0.096	FW
HMX[4]	12.66	1.11	FW
HMX [72]	2.53	—	FW
RDX [4]	6.26	0.837	FW
RDX [66]	1.58	—	FW
RDX [81]	27.88	27.67	FW
RDX [81]	3.87	3.84	M
RDX [81]	12.24	12.15	Combined
NG [77]	1.8	0.03	FW
NC [74]	47.9	—	FW

a Reference in brackets.
b See text for details on acute and chronic values derivation.
c FW = freshwater; M = marine; Combined = criterion developed
   using combined data from marine and freshwater toxicity tests.

Data available in the scientific literature were not sufficient for the calculation of Tier I acute or chronic WQC with other energetic compounds and their transformation products. Tier II secondary freshwater acute and chronic values (SAV and SCV, respectively) were derived by Talmage et al. [4] for TNB, 1,3-DNB, 3,5-DNA, 2-A-4,6-DNT, HMX, and RDX, by dividing the lowest genus mean acute value (GMAV) by the secondary acute factor (SAF; Table 4.7).

An early freshwater quality criterion of 2.53 µmoles L⁻¹ had been derived for HMX [72] by using a 0.05 application factor to the 96-h LC_{50} to 7-d old fry of fathead minnows (50.6 µmoles L⁻¹), which was the only freshwater organism that was sensitive to this chemical below its solubility limit. Talmage et al. [4], however, used data by Bentley et al. [72] and applied the formulas recommended by USEPA [80] to calculate SAV and SCV values, resulting in a fivefold higher SAV (Table 4.7).

Recent FAV and FCV values calculated for RDX [81] (Table 4.7) proposed much higher freshwater quality criteria than the secondary values derived by Talmage et al. [4] (Table 4.7), or the interim criterion of 1.58 µmoles L⁻¹ suggested by Bentley et al. [66]. The recently derived freshwater values [81] (>27 µmoles L⁻¹) are high, and likely insufficiently protective, considering some of the acute toxicity data for freshwater fish reported in Table 4.6, for example, 96-h EC_{50} values of 16.2 and 17.1 µmoles L⁻¹

to different life stages of *L. macrochirus* and *P. promelas*. The recently suggested marine water quality criteria [81] seem protective of marine organisms, based on the results of tests conducted with several species [15] (Table 4.6). Although the suggested combined acute value is below all acute toxicity data reported on Table 4.6, it is higher than the 96-h LC_{50} values derived for freshwater fish. The combined acute value also does not consider the sensitivity of native species in the receiving environment, which may be higher than that of the laboratory test species. The combined chronic value is above the 28-d LOEC for *P. promelas* growth and, therefore, is not expected to be sufficiently protective of aquatic communities chronically exposed to RDX.

An application factor was used to derive an interim freshwater quality criterion for nitrocellulose (NC) [74]. This chemical was not toxic to several species of fish, invertebrates, and algae, but had an EC_{50} of 554.3 µmoles L^{-1} to the freshwater alga *P. subcapitata*. The value proposed for the protection of freshwater aquatic life was 47.9 µmoles L^{-1}, approximately 10-fold below the lowest measured toxicity value. However, due to NC's lack of solubility in water, it is speculated if the phytotoxicity is due to adverse effects of the chemical per se or to obscuration of the test solution by the presence of excessive NC. No criteria have been proposed for the protection of marine life.

Bentley et al. [78] proposed that a concentration of 0.044 µmoles L^{-1} of nitroglycerin would be protective of freshwater aquatic life, by applying the lower estimate of several application factors specific for this chemical. Sullivan et al. [77] were slightly more conservative regarding long-term effects, proposing NG final acute and chronic values of 1.8 and 0.03 µmoles L^{-1}, respectively. These authors suggested that the 24-h average NG concentration in freshwater should not exceed the FCV. All proposed values for NG seem protective of freshwater species with respect to the data presented on Table 4.6.

4.4 CONCLUSIONS AND RECOMMENDATIONS

A relatively broad variety of aquatic toxicity studies exists for nitro-substituted phenol, toluene, and benzene explosives and related compounds, but very little toxicological information is available for tetryl, cyclic nitramines, and the other energetic compounds discussed in this chapter. Several explosives, such as tetryl, are no longer manufactured and are, therefore, of diminishing environmental concern, although their persistence and the nature, stability, and toxicity of their breakdown products is not understood in sufficient detail and should be further investigated. A variety of other energetic compounds, for example, perchlorates, are used in military operations, and due to environmental concerns with their release, additional studies on their fate and effects in aquatic systems are recommended.

Aquatic sites contaminated with military-related materials are likely to contain mixtures of a variety of energetic compounds and their transformation products, as well as other classes of chemicals. Although individual compounds may be present below effects concentrations at contaminated sites, simultaneous exposure to a mixture of these compounds and their transformation products may result in adverse toxicological effects. However, the ecological risks associated with contaminants are often considered on an individual chemical basis without consideration of chemical interactions affecting bioaccumulation and toxicity. Interacting chemicals result in toxicological effects or responses that are difficult to predict based upon single chemical toxicological data. Limited information is available regarding the interaction in aqueous mixtures of explosives and related compounds [36], and studies addressing mixture interactions between explosives and other classes of contaminants in aqueous exposures were not found in the current literature. Future investigations on the toxicological interactions of aqueous mixtures containing explosives are recommended.

Although several energetic compounds are present in marine environments around the world, only a relatively small fraction of aquatic toxicity data is available for marine or estuarine species. Therefore, further toxicity studies including organisms from those environments are recommended.

Comparisons of lethal and sublethal endpoints with different categories of explosives suggest that the use of sublethal endpoints could be useful for the interpretation of potential long-term environmental effects of energetic materials. Concentrations associated with reproductive effects were substantially lower than those promoting mortality of the cladocerans *D. magna* [22] and *C. dubia* [34], and the marine polychaete *D. gyrociliatus* [15] for a variety of explosives and related compounds.

Fisher et al. [79] speculated that algae are sensitive to nitrogen-containing munitions, possibly due to nitrate esters, which are believed to alter membrane permeability. Higher sensitivity of unicellular organisms to energetic compounds, relative to invertebrates and fish, was evident from examining the data presented in this chapter. Freshwater microalgae and zoospores of the marine macroalga *U. fasciata* are the most sensitive organisms to a variety of the chemicals discussed in this chapter. Therefore, it is recommended that assessments aiming at understanding the environmental effects of energetic compounds should include microalgae or early life stages of macroalgae, in addition to vertebrates and invertebrates. Very little information is available on the toxic mode of action of energetic materials, and research on this subject should also be encouraged.

REFERENCES

1. Dave G, Field-test of ammunition (TNT) dumping in the ocean, in *Sediment Quality Assessment and Management: Insight and Progress*, Munawar M, Ed., Aquatic Ecosystem Health and Management Society, Burlington, Ontario, Canada, 2003, p. 213.
2. Darrach MR, Chutjian A, and Plett GA, Trace explosives signatures from World War II unexploded undersea ordnance, *Environ. Sci. Technol.*, 32, 1354, 1998.
3. Ahnert A and Borowski C, Environmental risk assessment of anthropogenic activity in the deep sea, *J. Aquat. Ecosyst. Stress Recovery*, 7, 99, 2000.
4. Talmage S et al., Nitroaromatic munition compounds: Environmental effects and screening values, *Rev. Environ. Contam. Toxicol.*, 161, 1, 1999.
5. Krahl M, Clowes G, and Taylor J, Action of metabolic stimulants and depressants on cell division at varying carbon dioxide tensions, *Biol. Bull.*, 71, 400, 1936.
6. Krahl M and Clowes G, Physiological effects of nitro- and halo-substituted phenols in relation to extracellular and intracellular hydrogen ion concentration. I. Dissociation constants and theory, *J. Cell. Comp. Physiol.*, 11, 1, 1938.
7. Krahl M and Clowes G, Physiological effects of nitro- and halo-substituted phenols in relation to extracellular and intracellular hydrogen ion concentration. II. Experiments with *Arbacia* eggs, *J. Cell. Comp. Physiol.*, 2, 21, 1938.

8. Lefevre P, Certain chemical factors influencing artificial activation of *Nereis* eggs, *Biol. Bull.*, 89, 144, 1945.

9. Lefevre P, Further chemical aspects of the sensitization and activation reactions of *Nereis* eggs, *Biol. Bull.*, 95, 333, 1948.

10. Grindley J, Toxicity to rainbow trout and minnows of some substances known to be present in waste water discharged to rivers, *Ann. Appl. Biol.*, 33, 103, 1946.

11. Degani JG, Studies of the toxicity of ammunition plants wastes to fishes, *Trans. Am. Fish. Soc.*, 73, 45, 1943.

12. Goodfellow WJ et al., Acute toxicity of picric acid and picramic acid to rainbow trout, *Salmo gairdneri*, and American oyster, *Crassostrea virginica*, *Water Resour. Bull.*, 19, 641, 1983.

13. Dave G, Nilsson E and Wernersson A-S, Sediment and water-phase toxicity and UV-activation of six chemicals used in military explosives, *Aquat. Ecosyst. Health Manage.*, 3, 291, 2000.

14. Dave G and Nilson S, Laboratory assay of TNT (2,4,6-trinitro-toluene) fate and toxicity in seawater and sediment, in *Sediment Quality Assessment and Management: Insight and Progress*, Munawar M, Ed., Aquatic Ecosystem Health and Management Society, Burlington, Ontario, Canada, 2003, p. 207.

15. Nipper M et al., Development of marine toxicity data for ordnance compounds, *Arch. Environ. Contam. Toxicol.*, 41, 308, 2001.

16. Rosenblatt D et al., Organic explosives and related compounds, in *The Handbook of Environmental Chemistry*, Hutzinger O, Ed., Springer-Verlag, Berlin, Germany, 1991, p. 195.

17. Hovatter P et al., Ecotoxicity of nitroaromatics to aquatic and terrestrial species at army superfund sites, *Environ. Toxicol. Risk Assess.: Model. Risk*, 6, 117, 1997.

18. Environment Canada, Guidance Document on Application and Interpretation of Single-species Tests in Environmental Toxicology, EPS 1/RM/34, Environment Canada, Ottawa, Ontario, Canada, 1999.

19. Pearson J et al., An approach to the toxicological evaluation of a complex industrial wastewater, in *Aquatic Toxicology*, ASTM STP 667, Marking L and Kimerle R, Eds., American Society for Testing and Materials, Philadelphia, PA, USA, 1979, p. 284.

20. Liu D, Bailey H and Pearson J, Toxicity of a complex munitions wastewater to aquatic organisms, in *Aquatic Toxicology and Hazard Assessment: Sixth Symposium,* ASTM STP 802, Bishop W, Cardwell R, and Heidolph B, Eds., American Society for Testing and Materials, Philadelphia, PA, USA, 1983, p. 135.

21. Deneer J et al., Quantitative structure-activity relationships for the toxicity and biocon-centration factor of nitrobenzene derivatives towards the guppy (*Poecilia reticulata*), *Aquat. Toxicol.*, 10, 115, 1987.

22. Deneer J et al., QSAR study of the toxicity of nitrobenzene derivatives towards *Daphnia magna*, *Chlorella pyrenoidosa* and *Photobacterium phosphoreum*, *Aquat. Toxicol.*, 15, 83, 1989.

23. Zhao Y-H et al., Quantitative structure-activity relationships of nitroaromatic compounds to four aquatic organisms, *Chemosphere*, 34, 1837, 1997.

24. Bailey H and Spanggord R, The relationship between the toxicity and structure of nitro-aromatic chemicals, in *Aquatic Toxicology and Hazard Assessment: Sixth Symposium, ASTM STP 802*, Bishop W, Cardwell R, and Heidolph B, Eds., American Society for Testing and Materials, Philadelphia, PA, USA, 1983, p. 98.

25. Bringmann G and Kuhn R, Testing of substances for their toxicity threshold: Model organisms *Microcystis* (*Diplocystis*) *aeruginosa* and *Scenedesmus quadricauda*, *Mitt. Internat. Verin. Limnol.*, 21, 275, 1978.

26. Bringmann G and Kuhn R, Comparison of the toxicity thresholds of water pollutants to bacteria, algae, and protozoa in the cell multiplication inhibition test, *Water Res.*, 14, 231, 1980.

27. Gentile JH et al., Marine ecotoxicological testing with crustaceans, in *Ecotoxicological Testing for the Marine Environment,* Persoone G, Jaspers E, and Claus C, Eds., State University of Ghent and Institute of Marine Scientific Research, Bredene, Belgium, 1984, p. 479.

28. Bringmann V and Kuhn R, Befunde des Schadwirkung wassergefahrender Stoffe gegen *Daphnia magna, Zeitschr. Wasser Abwass. Forsch.*, 10, 161, 1977.

29. Nipper M et al., Fate and effects of picric acid and 2,6-DNT in marine environments: Toxicity of degradation products, *Mar. Pollut. Bull.*, 51, 1205, 2005.

30. Drzyzga O et al., Toxicity of explosives and related compounds to the luminescent bacterium *Vibrio fischeri* NRRL-B-11177, *Arch. Environ. Contam. Toxicol.*, 28, 229, 1995.

31. Dodard S et al., Ecotoxicity characterization of dinitrotoluenes and some of their reduced metabolites, *Chemosphere*, 38, 2071, 1999.

32. Sunahara G et al., Development of a soil extraction procedure for ecotoxicity characterization of energetic compounds, *Ecotoxicol. Environ. Saf.*, 39, 185, 1998.

33. Saka M, Developmental toxicity of p,p′-dichlorodiphenyltrichloroethane, 2,4,6-trinitrotoluene, their metabolites, and benzo[a]pyrene in *Xenopus laevis* embryos, *Environ. Toxicol. Chem.*, 23, 1065, 2004.

34. Griest W et al., Chemical and toxicological testing of composted explosives-contaminated soil, *Environ. Toxicol. Chem.*, 12, 1105, 1993.

35. Sims J and Steevens J, The role of metabolism in the toxicity of 2,4,6-trinitrotoluene and its degradation products to the aquatic amphipod *Hyalella azteca, Ecotoxicol. Environ. Safety,* 70, 38, 2008.

36. Hankenson K and Schaeffer D, Microtox assay of trinitrotoluene, diaminonitrotoluene, and dinitromethylamine mixtures, *Bull. Environ. Contam. Toxicol.*, 46, 550, 1991.

37. Bailey H and Liu D, *Lumbriculus variegatus*, a benthic oligochaete, as a bioassay organism, in *Aquatic Toxicology, ASTM STP* 707, Eaton J, Parrish P, and Hendricks A, Eds., American Society for Testing and Materials, Philadelphia, PA, USA, 1980, p. 202.

38. Johnson L et al., Phototoxicology. 2. Near-ultraviolet light enhancement of Microtox assays of trinitrotoluene and aminodinitrotoluenes, *Ecotoxicol. Environ. Saf.*, 27, 23, 1994.

39. Ryon MG, Water quality criteria for 2,4,6-trinitrotoluene (TNT), Oak Ridge National Laboratory, Oak Ridge, TN, USA, 1987.

40. Spanggord RJ et al., Environmental Fate Studies on Certain Munition Wastewater Constituents, Phase II, Laboratory Studies, SRI International, Menlo Park, CA, USA, 1980.

41. Smock L, Stoneburner D, and Clark J, The toxic effects of trinitrotoluene (TNT) and its primary degradation products on two species of algae and the fathead minnow, *Water Res.*, 10, 537, 1976.

42. Nipper M et al., Degradation of picric acid and 2,6-DNT in marine sediments and waters: The role of microbial activity and ultra-violet exposure, *Chemosphere*, 56, 519, 2004.

43. Bringmann G and Kuhn R, Vergelichende wasser-toxikologische Untersuchungen and Bakterien, Algen und Kleinkrebsen, *Gesundh. Ingen.*, 80, 115, 1959.

44. LeBlanc G, Acute toxicity of priority pollutants to water flea (*Daphnia magna*), *Bull. Environ. Contam. Toxicol.*, 24, 684, 1980.

45. Buccafusco R, Ells S, and LeBlanc G, Acute toxicity of priority pollutants to bluegill, *Bull. Environ. Contam. Toxicol.*, 26, 446, 1981.

46. Heitmuller P, Hollister T, and Parrish P, Acute toxicity of 54 industrial chemicals to sheepshead minnows (*Cyprinodon* variegatus), *Bull. Environ. Contam. Toxicol.*, 27, 506, 1981.

47. Kuhn R et al., Results of the harmful effects of water pollutants to *Daphnia magna* in the 21 day reproduction test, *Water Res.*, 23, 501, 1989.

48. Schultz T, Bearden A, and Jaworska J, A novel QSAR approach for estimating toxicity of phenols, *SAR QSAR Environ. Res.*, 5, 99, 1996.

49. Kopperman H, Carlson R, and Caple R, Aqueous chlorination and ozonation studies I. Structure-activity correlations of phenolic compounds to *Daphnia magna, Chem.-Biol. Interact.*, 9, 245, 1974.

50. Phipps G, Holcombe G, and Fiandt J, Acute toxicity of phenol and substituted phenols to the fathead minnow, *Bull. Environ. Contam. Toxicol.*, 26, 585, 1981.

51. Cronin M, Zhao Y, and Yu R, pH-dependence and QSAR analysis of the toxicity of phenols and anilines to *Daphnia magna, Environ. Toxicol.*, 15, 140, 2000.

52. Huang J-C and Gloyna E, Effect of organic compounds on photosynthetic oxygenation— I. Chlorophyll destruction and suppression of photosynthetic oxygen production, *Water Res.*, 2, 347, 1968.

53. Buikema AJ, McGinniss M, and Cairns JJ, Phenolics in aquatic ecosystems: A selected review of recent literature, *Mar. Environ. Res.*, 2, 87, 1979.

54. Zitko V et al., Toxicity of alkyldinitrophenols to some aquatic organisms, *Bull. Environ. Contam. Toxicol.*, 16, 508, 1976.

55. Terada H, Uncouplers of oxidative phosphorylation, *Environ. Health Perspect.*, 87, 213, 1990.

56. Gersdorff W, Effect of the introduction of the nitro group into the phenol molecule on toxicity to goldfish, *J. Cell. Comp. Physiol.*, 14, 61, 1939.

57. Yoshioka Y, Testing for the toxicity of chemicals with *Tetrahymena pyriformis, Sci. Total Environ.*, 43, 149, 1985.

58. McLeese D, Zitko V, and Peterson M, Structure-lethality relationships for phenols, anilines and other aromatic compounds in shrimp and clams, *Chemosphere*, 2, 53, 1979.

59. Kuhn R et al., Results of the harmful effects of selected water pollutants (anilines, phenols, aliphatic compounds) to *Daphnia magna, Water Res.*, 23, 495, 1989.

60. Barahona M and Sanchez-Fortun S, Comparative sensitivity of three age classes of *Artemia salina* larvae to several phenolic compounds, *Bull. Environ. Contam. Toxicol.*, 56, 271, 1996.

61. Holcombe G et al., The acute toxicity of selected substituted phenols, benzenes and benzoic acid esters to fathead minnows *Pimephales promelas, Environ. Poll. (Ser. A)*, 35, 367, 1984.

62. Nipper M et al., Toxicological and chemical assessment of ordnance compounds in marine sediments and pore waters, *Mar. Pollut. Bull.*, 44, 789, 2002.

63. Mekenyan O et al., QSARs for photoinduced toxicity: I. Acute lethality of polycyclic aromatic hydrocarbons to *Daphnia magna, Chemosphere*, 28, 567, 1994.

64. Johnson L et al., Phototoxicology. 3. Comparative toxicity of trinitrotoluene and aminodinitrotoluenes to *Daphnia magna, Dugesia dorotocephala*, and sheep erythrocytes, *Ecotoxicol. Environ. Saf.*, 27, 34, 1994.

65. Davenport R, Phototoxicology. 1. Light-enhanced toxicity of TNT and some related compounds to *Daphnia magna* and *Lytechinus variagatus* embryos, *Ecotoxicol. Environ. Saf.*, 27, 14, 1994.

66. Bentley RE et al., Laboratory Evaluation of the Toxicity of Cyclotrimethylene Trinitramine (RDX) to Aquatic Organisms, US Army Medical Research and Development Command, Fort Detrick, MD, USA, 1977.

67. Peters G et al., The acute and chronic toxicity of hexahydro-1,3,5-trinitro-1,3,5-triazine (RDX) to three freshwater invertebrates, *Environ. Toxicol. Chem.*, 10, 1073, 1991.

68. Burton D, Turley S, and Peters G, The toxicity of hexahydro-1,3,5-trinitro-1,3,5-triazine (RDX) to the freshwater green alga *Selenastrum capricornutum, Water Air Soil Poll.*, 76, 449, 1994.

69. Etnier EL, Water Quality Criteria for Hexahydro-1,3,5-trinitro-1,3,5-triazine (RDX), Oak Ridge National Laboratory, Oak Ridge, TN, USA, 1986.

70. Burton D and Turley S, Reduction of hexahydro-1,3,5-trinitro-1,3,5-triazine (RDX) toxicity to the cladoceran *Ceriodaphnia dubia* following photolysis in sunlight, *Bull. Environ. Contam. Toxicol.*, 55, 89, 1995.

71. Bausum HT, Recommended Water Quality Criteria for Octahydro-1,3,5,7-tetranitro-1,3,5,7-tetrazocine (HMX), US Army Biomedical Research and Development Laboratory, Frederick, MD, USA, 1989.

72. Bentley RE et al., Acute Toxicity of 1,3,5,7-Tetranitro-octahydro-1,3,5,7-tetrazocine (HMX) to Aquatic Organisms, Final Report, EG & G Bionomics, Wareham, MA, USA, 1977.

73. Gong P et al., Preliminary ecotoxicological characterization of a new energetic substance, CL-20, *Chemosphere*, 56, 653, 2004.

74. Bentley RE et al., Laboratory Evaluation of the Toxicity of Nitrocellulose to Aquatic Organisms, Army Medical Research Development Command, Washington, DC, USA, 1977.

75. van der Schalie W, The Toxicity of Nitroguanidine and Photolyzed Nitroguanidine to Freshwater Aquatic Organisms, Report No. AD-A153045, Army Medical Bioengineering Research and Development Laboratory, Fort Detrick, MD, USA, 1985.

76. Burton DT, Turley SD, and Peters GT, Toxicity of Nitroguanidine, Nitroglycerin, Hexahydro-1,3,5-trinitro-1,3,5-triazine (RDX), and 2,4,6-Trinitrotoluene (TNT) to Selected Freshwater Aquatic Organisms, Report No. WREC 93-B3, The University of Maryland System, Agricultural Experiment Station, Wye Research and Education Center, Queenstown, MD, USA, 1993.

77. Sullivan JH Jr et al., A Summary and Evaluation of Aquatic Environmental Data in Relation to Establishing Water Quality Criteria for Munitions-Unique Compounds. Part 2: Nitroglycerin, Water and Air Research, Inc., Gainesville, FL, USA, 1979.

78. Bentley RE et al., Laboratory Evaluation of the Toxicity of Nitroglycerin to Aquatic Organisms, Defense Technical Information Center, Alexandria, VA, USA, 1978.

79. Fisher D, Burton D, and Paulson R, Comparative acute toxicity of diethyleneglycol dinitrate to freshwater aquatic organisms, *Environ. Toxicol. Chem.*, 8, 545, 1989.

80. U.S. EPA, Quality Criteria for Water, Report EPA 440/5-86-001, U.S. Environmental Protection Agency, Washington, DC, USA, 1986.

81. ENSR International Inc. and Parametrix, Derivation of Toxicity Reference Values for the Acute and Chronic Toxicity of Hexahydro-1,35-trinitro-1,3,5-triazine (RDX) to Marine Aquatic Organisms, Final Report, U.S. Army Public Works, ENSR Project No. 09000-279-400, Redmond, WA, USA, 2005.

82. Curtis M and Ward C, Aquatic toxicity of forty industrial chemicals: Testing in support of hazardous substances spill prevention regulation, *J. Hydrol.*, 51, 359, 1981.

83. Curtis M, Curran C, and Ward C, Aquatic toxicity testing as fundament for a spill prevention program, in *Control of Hazardous Material Spills: Proceedings of the 1980 National Conference*, Nashville, TN, USA, 1980, p. 284.

84. van der Schalie W, The Acute and Chronic Toxicity of 3,5-Dinitroaniline, 1,3-Dinitrobenzene, and 1,3,5-Trinitrobenzene to Freshwater Aquatic Organisms, Tech. Report 8305, AD-A138408, U.S. Army Medical Bioengineering Research and Development Laboratory, Fort Detrick, MD, USA, 1983.

85. Liu DHW et al., Toxicity of TNT Wastewaters to Aquatic Organisms, vol. I. Acute Toxicity of LAP Wastewater and 2,4,6-Trinitrotoluene, Report AD-A142144, SRI Int., Menlo Park, CA, USA, 1983.

86. Tonogai Y et al., Actual survey on TLm (median tolerance limit) values of environmental pollutants, especially on amines, nitriles, aromatic nitrogen compounds and artificial dyes, *J. Toxicol. Sci.*, 7, 193, 1982.

87. Liu DHW et al., Toxicity of TNT Wastewaters to Aquatic Organisms. Volume II—Acute Toxicity of Condensate Wastewater and 2,4-Dinitrotoluene, Report DSU-4262, SRI Int., Menlo Park, CA, USA, 1983.

88. Bailey H, Development and testing of a laboratory model ecosystem for use in evaluating biological effects and chemical fate of pollutants, in *Aquatic Toxicology and Hazard Assessment: Fifth Conference, ASTM STP* 766, Pearson J, Foster R, and Bishop W, Eds., American Society for Testing and Materials, Philadelphia, PA, USA, 1982, p. 221.

89. Randall T and Knopp P, Detoxification of specific organic substances by wet oxidation, *J. Water Pollut. Control Fed.*, 52, 2117, 1980.

90. Toussaint M et al., A comparison of standard acute toxicity tests with rapid-screening toxicity tests, *Environ. Toxicol. Chem.*, 14, 907, 1995.

91. Snell T and Moffat B, A 2-d life cycle test with the rotifer *Brachionus calyciflorus*, *Environ. Toxicol. Chem.*, 11, 1249, 1992.

92. Fitzgerald G, Gerloff G, and Skoog F, Studies on chemicals with selective toxicity to blue-green algae, *Sewage Ind. Wastes: J. Fed. Sewage Works Assoc.*, 24, 888, 1952.

93. Won W, DiSalvo L, and Ng J, Toxicity and mutagenicity of 2,4,6-trinitrotoluene and its microbial metabolites, *Appl. Environ. Microbiol.*, 31, 576, 1976.

94. Verma S, Rani S, and Dalela R, Synergism, antagonism, and additivity of phenol, pentachlorophenol, and dinitrophenol to a fish (*Notopterus notopterus*), *Arch. Environ. Contam. Toxicol.*, 10, 365, 1981.

95. Burton DT, Turley SD, and Peters GT, The acute and chronic toxicity of hexahydro-1,3,5-trinitro-1,3,5-triazine (RDX) to the fathead minnow (*Pimephales promelas*), *Chemosphere*, 29, 567, 1994.

5 Fate and Toxicity of Explosives in Sediments

Guilherme R. Lotufo, Marion Nipper,
R. Scott Carr, and Jason M. Conder

CONTENTS

5.1 Introduction 117
5.2 Methodology for Amending Explosives to Sediments 118
5.3 Fate and Lethal Toxicity of Explosives in Sediment Exposures 119
    5.3.1 Trinitrotoluene and Related Compounds 119
    5.3.2 2,6-Dinitrotoluene (2,6-DNT) 125
    5.3.3 2,4,6-Trinitrophenol (Picric Acid) 126
    5.3.4 2,4,6-Trinitrophenylmethylnitramine (Tetryl) 127
    5.3.5 Cyclic Nitramines 127
5.4 Effects of Explosives on Reproduction and Growth of Invertebrates 128
5.5 Toxicity of Porewater Extracted from Sediments Spiked with
    Explosives 128
5.6 The Use of Biomimetic Devices to Assess Explosives Bioavailability 129
5.7 Conclusions and Research Recommendations 131
References 133

5.1 INTRODUCTION

Contamination of soils, sediment, groundwater, and surface water with explosives is associated with military activities at ammunition production sites and military training facilities. At ammunition plants, this contamination occurs mainly as a result of contaminated runoff, effluent from the facilities, liquid waste lagoons, and spills [1]. Major explosive contaminants such as 2,4,6-trinitrotoluene (TNT), 1,3,5-trinitrobenzene (TNB), hexahydro-1,3,5-trinitro-1,3,5-triazine (RDX), octahydro-1,3,5,7-tetranitro-1,3,5,7-tetrazocine (HMX), and 2,4,6-trinitrophenylmethylnitramine (tetryl) and their transformation products have been detected in freshwater sediment samples collected from military installations at concentrations ranging from low (e.g., less than 6.7 mg kg^{-1}) to exceedingly high (up to 711,000 mg kg^{-1}) [1]. Those extremely high concentrations ("hot spots") are indicative of large amounts of nondissolved and nonsorbed explosives in some contaminated sites. Low concentrations (<3.5 mg kg^{-1}) of 2,6-dinitrotoluene (2,6-DNT), tetryl, and 2,4,6-trinitrophenol (picric acid) were detected in marine sediments near a naval facility in Ostrich Bay, Puget Sound, Washington [2].

As a result of wartime activities, dumping, and accidents, unexploded ordnance (UXO) is an additional source of explosive contaminants to sediments in marine [3] and freshwater environments [4] throughout the world. Field experiments using cleaved shells containing TNT demonstrated that leakage from breached munitions can serve as a source of sediment contamination [5].

The fate of explosive compounds in soils has been well characterized [6] and was reviewed in Chapter 2 of this book. In contrast, reports on the fate and biological effects of explosives and related compounds in sediments are scarce, despite the potential ecological risk associated with their presence in this compartment of aquatic environments. Recent studies characterizing the processes that occur following the initial contact of explosives and related compounds with sediments revealed rapid rates of transformation for a variety of nitroaromatic compounds [7–14]. Investigations of the toxicity of sediment-associated explosives were relatively recent and used sediment amended with contaminants in the laboratory [8–11,13,14].

This chapter reviews the fate and toxicity of sediment-associated explosives and related compounds, as well as the challenges associated with development of accurate reference and screening values for use in ecological risk assessment. In addition, this chapter provides a detailed review of the adverse biological effects of explosive compounds and their transformation products to benthic invertebrates exposed to spiked sediments and corresponding porewaters. The use of novel approaches and future research needs are discussed in this chapter to address the research challenges and highlight the knowledge gaps.

5.2   METHODOLOGY FOR AMENDING EXPLOSIVES TO SEDIMENTS

Investigations of the fate, toxicity, and bioavailability of sediment-associated explosives and their transformation products used exposure matrices prepared by mixing uncontaminated sediments with prescribed amounts of chemicals using a variety of methods:

- RDX, HMX, and TNT and their transformation products were dissolved in organic solvent carrier (methanol or acetone), spiked into wet sediment, and mixed vigorously by vortex [8,9,13,15].
- TNT was dissolved in acetonitrile, added to dry sediment and mixed by hand. Following evaporation of the solvent, the sediment was wetted and mixed vigorously [11].
- Tetryl, picric acid, or 2,6-DNT were dissolved in water, added to sediments dewatered by centrifugation, and then mixed vigorously by vortex [10].
- Picric acid was dissolved in water, added to sediment in glass jars, then placed on a rolling table for 6 h for homogenization [12].
- 2,6-DNT dissolved in methanol was added to glass jars, placed on a rolling table for evaporation of the solvent, followed by addition of wet sediment and rolled for 6 h [12].
- Quartz sand coated with TNT or crystalline RDX was vigorously mixed with wet sediment [14].

One of the main challenges associated with sediment spiking and mixing methods is obtaining adequately uniform compound distribution in the solid matrix [16]; therefore, vigorous and thorough postspiking mixing is recommended. To assess the homogeneity of the exposure substrate after mixing procedures, chemical concentrations should be analytically determined in subsamples of spiked sediment. Compound distribution was reported as adequately uniform using all the methods described earlier. Spiking methods that do not require organic solvent carriers are preferable as those solvents have potential effects on the partitioning behavior of spiked compounds [16].

As explosives degrade quickly in sediment [7–14], transformation is likely to occur during spiking and mixing for all the methods described. Therefore, concentrations of compounds and their transformation products in spiked sediments should be adequately characterized for use as metrics of toxicity. The concentration of spiked compounds in sediment decreases with increasing sediment–chemical contact time because of transformation processes [10–12]. Therefore, short mixing periods are expected to minimize the decrease of the parent compound concentration, whereas longer mixing and storage (or aging) periods are expected to produce sediments with lower but more slowly changing concentrations of parent compound [11].

5.3 FATE AND LETHAL TOXICITY OF EXPLOSIVES IN SEDIMENT EXPOSURES

5.3.1 TRINITROTOLUENE AND RELATED COMPOUNDS

Fate kinetics and mass balance studies conclusively demonstrated that the fate of TNT in wet sediment is dependent on a combination of reductive transformation and sorption processes [7]. TNT in sediment–water systems was rapidly reduced to aminonitrotoluenes in both aerobic and anaerobic conditions, and irreversible sorption of reduced transformation products to the sediment phase was near complete under aerobic conditions [7]. The reduction of TNT was strongly regioselective, with favored formation of 4-amino-2,4-dinitrotoluene (4-ADNT) and 2,4-diamino-4-nitrotoluene (2,4-DANT). The presence of intermediates (e.g., hydroxylamines) was demonstrated, and the presence of final products in the TNT reduction pathway (i.e., triaminotoluene) as well as azoxy derivatives was suspected in sediment spiked with TNT [7]. When [14C]TNT was spiked into sediment, the parent compound rapidly disappeared while increasing amounts of the activity, mostly corresponding to 2,4-DANT, became strongly associated with sediment particles and resisted organic solvent extraction under aerobic conditions as the reaction time (24 h) progressed [7]. The fraction of solvent-resistant transformation products of TNT, likely associated with the sediment through covalent binding, also increased with increasing sediment–contaminant contact time in sandy and fine-grained marine sediments spiked with [14C]TNT [14]. The extent of reductive transformation and sorption of [14C]TNT to sediment particles was higher in spiked fine-grained sediment compared to sandy sediment [14], corroborating the previously postulated [7] influence of redox conditions on the fate of TNT in sediments.

When TNT was spiked to freshwater or marine sediments for toxicity and bio-availability investigations, aminodinitrotoluenes (ADNTs) and diaminonitrotoluenes (DANTs) were detected following the short (0.5 to 4 h) postspiking mixing periods, with transformation continuing during storage or toxicological exposure [9,11,13,14]. An inverse relationship between TNT target concentration and transformation rate was reported for a diversity of sediments [8,11,13,14]. Decreasing TNT transformation at high spiking concentrations was also observed in soil studies [17,18] and likely resulted from increasing saturation of binding sites in the sediment or soil particles and inhibition of microbial activity. Transformation of TNT to aminated products was substantially lower in coarse-grained than in fine-grained sediment when both were spiked with the same initial organic carbon-normalized concentrations [14].

The measured sum concentration of TNT and its major transformation products, ADNTs and DANTs (hereafter referred to as ΣTNT [14]), in sediments spiked with TNT was typically lower than the target (or nominal) concentration following a short mixing period (0.5 to 4 h) [9,11,13,14] (Table 5.1), suggesting that rapid transformation of spiked TNT to compounds that sorb strongly to sediment and resisted organic solvent extraction likely occurs in a variety of sediment types. Differences between nominal and measured concentrations of ΣTNT were substantially greater in an earlier study [9] compared to a later study [13] that used the same sediment and spiking methods (Table 5.1). The spiked sediments in the earlier study [9] were air-dried prior to solvent extraction as per a standard method for chemical analysis of explosives in soils and sediments samples [19], whereas in the later study [13] sediment samples were extracted without air-drying or any other manipulation. Air-drying had greatly decreased the extraction efficiency of 2,6-DNT [12] and likely resulted in unrealistically lower measured concentrations of ΣTNT in air-dried samples, especially at low nominal concentrations. Overall, the percent difference between nominal and measured concentration of ΣTNT in spiked sediments was similar for different studies when sediments were extracted wet [13,14]. Therefore, the disappearance of spiked TNT by transformation processes was not influenced by nominal concentration or sediment organic carbon content (Table 5.1).

Measured concentrations of 2-ADNT and 2,4-DANT in spiked sediments at the end of a short mixing period (up to 4 h) were also substantially lower than nominal concentrations [8,9,13], as expected based on the description of the fate of those compounds in sediment [7]. Spiking of TNB to sediment resulted in the formation of 1-amino-3,5-dinitrobenzene, also called 3,5-dinitroaniline [13]. Similar to TNT, reduced transformation products of TNB can form covalent bonds with the sediment and resist solvent extraction, thus explaining the substantial difference between the analytically determined concentrations of extractable compounds and the expected concentrations. Similar to TNT, air-drying of sediment spiked with TNB or 2,4-DANT prior to solvent extraction appears to promote a decrease of the extraction efficiency of parent and transformation products, as lower extraction efficiencies were observed with air-drying [8] than without air-drying [13] for those compounds in separate studies. A direct comparison of the extraction efficiency of spiked sediment extracted wet and air-dried, conducted for 2,6-DNT [12], should be performed for other major explosives for validation of the recommendation of wet extraction [12] of sediments contaminated with explosive compounds.

TABLE 5.1

Nominal Concentration of TNT and Concentration of the Sum of TNT and its Transformation Products (ΣTNT) Measured after Mixing into Spiked Sediments of Different Types

Sediment Type	Sediment TOC (%)	Concentration (nmol g⁻¹)		Percent Difference	Mixing Period (h)	Moisture Condition	Reference
		Nominal	Measured				
Fine grained	0.65	55	<0.4[a]	100.0	0.5	Dry	9
Fine grained	0.65	110	<0.4[a]	100.0	0.5	Dry	9
Fine grained	0.65	220	<0.4[a]	99.8	0.5	Dry	9
Fine grained	0.65	441	19	95.7	0.5	Dry	9
Fine grained	0.65	881	293	66.8	0.5	Dry	9
Fine grained	0.65	1762	675	61.7	0.5	Dry	9
Fine grained	0.65	127	105	17.3	1	Wet	13
Fine grained	0.65	238	198	16.8	1	Wet	13
Fine grained	0.65	605	404	33.2	1	Wet	13
Fine grained	0.65	1365	886	35.1	1	Wet	13
Fine grained	0.36	110	103	6.4	4	Wet	14
Fine grained	0.36	220	227	−3.2	4	Wet	14
Fine grained	0.36	440	405	8.0	4	Wet	14
Fine grained	0.36	880	652	25.9	4	Wet	14
Fine grained	0.36	1760	1539	12.6	4	Wet	14
Sandy	0.09	28	13	53.6	4	Wet	14
Sandy	0.09	55	34	38.2	4	Wet	14
Sandy	0.09	110	79	28.2	4	Wet	14
Sandy	0.09	220	161	26.8	4	Wet	14
Sandy	0.09	440	278	36.8	4	Wet	14

[a] Numerical value represents detection.

Exposure chambers consisting of sediment amended with TNT, including [¹⁴C] TNT, and uncontaminated overlying water were used for toxicity testing [11,14]. A substantial portion (up to 50%) of the radioactivity or analytically quantified compounds originally associated with the sediment particles and pore water was present in the nonrenewed overlying water at the end of exposures (10 or 28 d). Such high partitioning was expected given the reported relatively high aqueous solubility of TNT and its transformation products [20]. Renewal of overlying waters is typically performed during chronic sediment exposures using benthic invertebrates to maintain adequate water quality, especially when feeding is required [8]. However, exchange of overlying water during invertebrate exposures to TNT-spiked sediment is expected to rapidly deplete contaminants from the exposure system. Therefore, it is recommended that toxicity evaluations of nitroaromatic compounds spiked to sediments should use exposure conditions (e.g., duration, water quality requirements) that do not require water changes during the experimental period [11].

Because TNT is rapidly transformed in sediment and the concentration of ΣTNT decreases during the toxicity exposures, establishing accurate concentration–response relationships for TNT and related compounds using data from sediment toxicity testing is challenging. In addition, partitioning of TNT and its transformation products to the overlying water renders the latter as a potentially significant source of contaminant uptake to test organisms, relative to the sediment compartment, confounding the extrapolation of sediment toxicity test results to realistic field scenarios.

Because nominal concentrations overestimate, sometimes by orders of magnitude, measured concentrations, they are not representative of effects concentrations. Analytically determined initial chemical concentrations more accurately represent the effects concentrations, especially when the toxic response occurs shortly after exposure initiation. Rapid manifestation of toxicity is expected for TNT and its major transformation products as these chemicals reach steady state in small organisms very quickly (see review in Chapter 6). Lethal effects of TNT were observed during the initial few days of exposure in sediment toxicity tests [11,13]. Therefore, the use of short exposure duration for the evaluation of the effects of sediment-associated explosives is desirable. For sublethal toxicity testing requiring longer exposure periods that allow greater transformation of parent compound with the concomitant decrease in toxicity, multiple analytical determinations of sediment concentrations should be conducted to determine the appropriate exposure concentration associated with the observed effects [11].

The results of toxicity investigations using sediments spiked with TNT and related compounds as well as other explosive compounds are summarized in Table 5.2. Toxicity exposures referenced in Table 5.2 were conducted following benthic invertebrate sediment toxicity text methods described in published guidelines [21,22] or using minor modifications of those methods.

For each study, the no observable effect concentrations (NOEC) and lowest observable effect concentrations (LOEC) are provided as initial concentrations expressed as ΣTNT concentrations. The earliest study of the toxicity of TNT in sediment [15] employed [^{14}C]TNT and used radioactivity measurements as a surrogate for the sediment concentrations. Most of the radioactivity detected in the sediments in that study [15] likely corresponded to mostly nonsolvent-extractable transformation products of TNT. This was later confirmed in similar spiked-sediment studies performed using sediments from the same site (Lotufo, unpublished). Therefore, the effects concentrations reported in that study are likely substantially higher than the concentration of ΣTNT, explaining the apparent high relative tolerance of the amphipod *Leptocheirus plumulosus* and polychaete *Neanthes arenaceodentata* to sediments spiked with this chemical. In subsequent studies [9,11,13,14], sediment concentrations of ΣTNT at initiation of the toxicity exposure were chemically measured, therefore providing more accurate determination of the exposure concentration suitable for expressing lethal toxicity. However, concentrations measured using solvent extraction of air-dried sediments [8,9] likely underestimated the actual concentration of extractable compounds in the sediments and therefore resulted in the determination of lower than actual effects concentrations.

TABLE 5.2

Toxicity of Explosives and Their Transformation Products Determined for Aquatic Invertebrates Using Spiked Sediment Exposures

Spiked Chemical	Species	TOC	NOEC (mg kg⁻¹)	NOEC (nmol g⁻¹)	LOEC (mg kg⁻¹)	LOEC (nmol g⁻¹)	Duration	Reference
TNT	*Leptocheirus plumulosus*	2.70	116	511	228	1004	28	15
TNT	*Neanthes arenaceodentata*	2.70	275	1211	508	2238	28	15
TNT	*Hyalella azteca*	0.65	<0.1[a]	<0.4[a]	0.1	0.5	10	9
TNT	*Chironomus tentans*	0.65	4	19	67	293	10	9
TNT	*Chironomus tentans*	0.65	0	0	37	188	10	13
TNT	*Tubifex tubifex*	0.96	25	112	69	304	28	11
TNT	*Eohaustorius estuarius*	0.09	24	107	44	194	10	14
TNT	*Eohaustorius estuarius*	0.36	20	116	38	209	10	14
2,4-DANT	*Leptocheirus plumulosus*	2.70	<0.1[a]	<0.4[a]	3.0	18	28	8
2,4-DANT	*Neanthes arenaceodentata*	2.70	187	1120	574	3437	28	8
2,4-DANT	*Hyalella azteca*	0.65	<0.1[a]	<0.4[a]	0.3	1.8	10	9
2,4-DANT	*Chironomus tentans*	0.65	110	661	ND	ND	10	9
2,4-DANT	*Chironomus tentans*	0.65	33	199	59	356	10	13
2-ADNT	*Chironomus tentans*	0.65	23	115	59	301	10	13
TNB	*Leptocheirus plumulosus*	2.70	7	33	13	61	28	8
TNB	*Neanthes arenaceodentata*	2.70	107	512	ND	ND	28	8
TNB	*Hyalella azteca*	0.65	<0.1[a]	<0.4[a]	1	5	10	9
TNB	*Chironomus tentans*	0.65	8	36	8	36	10	9
TNB	*Chironomus tentans*	0.65	59	353	131	784	10	13

(continued on next page)

TABLE 5.2 (continued)

Toxicity of Explosives and Their Transformation Products Determined for Aquatic Invertebrates Using Spiked Sediment Exposures

Spiked Chemical	Species	TOC	NOEC		LOEC		Duration	Reference
			(mg kg^{-1})	(nmol g^{-1})	(mg kg^{-1})	(nmol g^{-1})		
2,6-DNT	*Ampelisca abdita*	0.10	5	25	ND	ND	10	10
2,6-DNT	*Ampelisca abdita*	1.10	0.5	3	ND	ND	10	10
Picric acid	*Ampelisca abdita*	0.10	73	319	162	708	10	10
Picric acid	*Ampelisca abdita*	1.10	BLD[b]	BLD[b]	BLD[b]	BLD[b]	10	10
Tetryl	*Ampelisca abdita*	0.10	0.5	2	4	12	10	10
Tetryl	*Ampelisca abdita*	1.10	0.1	0.3	ND	ND	10	10
RDX	*Leptocheirus plumulosus*	2.70	891	4014	ND	ND	10	8
RDX	*Neanthes arenaceodentata*	2.70	891	4014	ND	ND	10	8
RDX	*Hyalella azteca*	0.65	102	460	ND	ND	10	9
RDX	*Chironomus tentans*	0.65	711	3203	ND	ND	10	9
RDX	*Eohaustorius estuarius*	0.09	2229	10040	ND	ND	10	14
HMX	*Leptocheirus plumulosus*	2.70	115	389	ND	ND	10	8
HMX	*Neanthes arenaceodentata*	2.70	353	1193	ND	ND	10	8
HMX	*Hyalella azteca*	0.65	126	426	ND	ND	10	9
HMX	*Chironomus tentans*	0.65	146	493	ND	ND	10	9

Note: TOC = total organic carbon (% of dry weight), NOEC = no observed effect concentration, LOEC = lowest observed effect concentration, ND = not determined due to insufficient mortality, BDL = below detection limit. NOEC and LOEC are reported as measured sum concentrations of parent and transformation products at experiment initiation. Effects were significant reductions in survival.

[a] Numerical value represents detection limit.

[b] Detection limit not reported.

Except for *H. azteca*, lethality was observed at remarkably similar ΣTNT concentrations for a variety of experimental species and sediment organic carbon content. However, accurate comparisons of the toxicity of TNT-spiked sediments using different sediments and species should take into account the variable relative concentration of parent and transformation products in the different exposure sediments because of potential toxicity differences among compounds and nonadditive mixture interactions.

The toxicity of TNT, 2-ADNT, 2,4-DANT spiked into sediments was compared using the midge *Chironomus tentans* [13] using the same sediment matrix and exposure conditions. Lethal toxicity was similar for all compounds, with differences in LOEC values never exceeding a factor of 4 (Table 5.2) [13]. Overall, reduced transformation products of TNT appear to be less lethal than TNT to marine [8] and freshwater invertebrates [9] (Table 5.2). Toxic interaction among TNT and its major transformation products in spiked sediment exposures was additive in the single study addressing the mixture toxicity of nitroaromatic compounds in sediment, suggesting the toxicity of a nitroaromatic mixture would be reasonably predicted using single-compound toxicity data [13].

Comparative studies revealed that the lethal concentrations of ΣTNT, 2,4-DANT, and TNB were lower for amphipods than other invertebrates [8,9,13] (Table 5.2), suggesting a potentially high relative sensitivity of amphipods to sediment-associated nitroaromatic compounds. Amphipods and other crustaceans were also relatively sensitive to the toxic effects of nitroaromatic compounds compared to other invertebrate taxa and fish in water-only exposures (see Chapter 4). Therefore, the use of amphipods to assess the toxicity investigations of sediments from explosives-contaminated sites is recommended to assure reasonable protectiveness.

The among-studies variability in the toxicity of sediments spiked with TNT and related compounds likely resulted from differences in species sensitivity, compound bioavailability, and the relative concentration of parent and transformation compounds among the exposure sediments. Inaccuracies in quantifying exposure concentrations also likely contributed to the observed differences among studies.

5.3.2   2,6-Dinitrotoluene (2,6-DNT)

The fate of 2,6-DNT in spiked marine sediments has been investigated using fine-grained and sandy substrates [12]. These studies showed that biotransformation of 2,6-DNT started soon after the initial contact of the chemical with the sediments and proceeded for several months. Transformation rates were greater in fine-grained than in sandy sediment and increased with temperature. The transformation product 2-amino-6-nitrotolune (2-A-6-NT) was formed by reduction and the compound 2-nitrotoluene (2-NT) was formed by loss of a nitro group. After prolonged incubation, these chemicals tended to be replaced by high molecular weight polymers, possibly derived from polymerization of 2,6-DNT or its transformation products. Complete disappearance of the parent compound and known transformation products from fine-grained and sandy sediments occurred during a 56-d incubation at 20°C [12]. The possibility of mineralization of at least part of the 2,6-DNT was speculated. Although the majority of 2,6-DNT in spiked sediments is biotransformed, abiotic transformation took place in sterilized sediments [12]. Several electron donors capable of reducing nitroaromatic compounds—for example, reduced iron and sulfur species, reduced substances of natural organic matter such as hydroquinone structures, and complexes of other transition metals such as copper and cobalt— are present in sediments [23], and hydrolysis or other breakdown processes might contribute to the transformation of ordnance compounds. Abiotic reduction rates between different environments can differ by orders of magnitude [23], which would explain a greater loss of 2,6-DNT in sterilized sandy sediment when compared to the fine-grained substrate [12]. The reduction of 2,6-DNT by abiotic reactions seems to have occurred at a greater rate than that of 2-A-6-NT, leading to an accumulation of the latter in sediments as 2,6-DNT was reduced.

Analytically determined concentrations of 2,6-DNT and 2-A-6-NT, as well as picric acid, in marine sediments were much lower when samples were dried at room temperature and extracted with acetonitrile [12] according to standard procedures for soils and sediments [19], than when sediments were not dried prior to extraction [12]. A portion of the parent compounds could be transformed during the drying procedure, and the authors suggested that sediments contaminated with explosives should not be air-dried prior to extraction for chemical measurements.

The toxicity of fine-grained and sandy sediments spiked with 2,6-DNT was examined using the amphipod *Ampelisca abdita* in 10-d exposures [10]. No significant lethal effects were observed in any treatment (Table 5.2). Because the highest 2,6-DNT measured concentration in sediments used in toxicity evaluations was relatively low (25 nmol g^{-1}), the effects of 2,6-DNT at concentrations representative of the lethal range for other nitroaromatic compounds are unknown. Reports of the toxicity of sediments spiked with 2,6-DNT transformation products were not found in the available literature.

5.3.3 2,4,6-TRINITROPHENOL (PICRIC ACID)

The fate of picric acid in marine sediments was investigated by Nipper et al. [12] using the same sediments and experimental conditions described for 2,6-DNT. Transformation rates were highest in fine-grained sediment incubated at 20°C. Major transformation products included 4-dinitrophenol, aminodinitrophenols (including picramic acid), 3,4-diaminophenol, amino nitrophenol, and nitrodiaminophenol. High performance liquid chromatography analysis of the sediment extracts revealed unknown compounds that were speculated to correspond to aminonitrotoluene isomers, diaminotoluenes, dinitrotoluenes, and nitrotoluenes. Complete degradation of the parent compound and known transformation products occurred during the 56-d storage period, but unidentified transformation products remained in the sandy and fine-grained sediments. Similar to the fate of 2,6-DNT, picric acid disappeared from sterilized spiked marine sediments, suggesting a certain degree of abiotic transformation [12].

The toxicity of both sediment types spiked with picric acid was examined using *A. abdita* in 10-d exposures [10]. Significant amphipod mortality was observed in the sandy sediment at picric acid concentrations relatively similar to lethal concentrations reported for TNT and its major breakdown products (Table 5.2). In the fine-grained sample, mortality was highest at an intermediate concentration in which picric acid was below the detection limit (nominal concentration 116 nmol g^{-1}), with greater survival in the two higher and lower concentrations, thus generating a U-shaped concentration–response relationship. The toxicity in treatments with picric acid below the detection limit was attributed to unknown transformation products. Inhibition of microbial transformation at the highest concentrations tested in that study could have contributed to the lower mortality at higher concentrations compared to mortality at the intermediate concentrations.

5.3.4 2,4,6-Trinitrophenylmethylnitramine (Tetryl)

Information on the fate of tetryl in sediments is limited to a study by Nipper et al. [10]. Unknown breakdown products were formed in sandy and fine-grained marine sediments spiked with tetryl. Picric acid, a primary hydrolysis product of tetryl in seawater, was detected in the sandy sediment only [10]. Tetryl-spiked sandy sediments produced significant amphipod mortality at relatively low concentrations in a 10-d exposure using *A. abdita* [10] (Table 5.2). Significant lethal effects were not observed in the tetryl-spiked fine-grained sediment (Table 5.2), in which the highest concentration (0.3 nmol g^{-1}) was much lower than lethal levels in the sandy sediment. Since tetryl has not been produced since the early 1970s [1], recent research efforts to understand its fate and effects in sediments have been minimal. However, given its elevated toxicity to a variety of species [10], additional studies of its fate and biological effects in different types of sediment are recommended.

5.3.5 Cyclic Nitramines

The cyclic nitramine explosives RDX and HMX have lower water solubility and significantly less binding affinity with soil organic matter than TNT [1,6]. Therefore, major differences in the fate of those compounds in aquatic systems are expected. Studies using [^{14}C]-compounds suggested, but did not conclusively demonstrate by chemical analysis of transformation products, that RDX and HMX spiked into fine-grained, organically rich sediments were likely degraded during the course of toxicity exposures [8]. A substantial portion of the radioactivity originally associated with [^{14}C]RDX or [^{14}C]HMX spiked sediments partitioned to the overlying water in exposure chambers during the course of toxicity tests using fine-grained sediment [8]. That radioactivity was likely associated with transformation products more soluble in water than the parent compounds. Radiolabel mass balance studies revealed substantial losses of radioactivity from exposure chambers in static exposures to [^{14}C] RDX, which may have resulted from volatilization of RDX transformation products or loss via mineralization [8]. In contrast to the fate of RDX in organically rich sediment, transformation was minimal in sandy sediment, and a much smaller fraction of the spiked radioactivity was associated with the overlying water at termination of a 10-d toxicity exposure [14]. In the latter study, RDX was added to sediments in excessive amounts, up to 0.24% of the sediment dry mass, to attain concentrations expected to promote toxicity to an amphipod. In that study, most of the spiked RDX remained in the sediment/porewater compartment of the exposure chamber, and likely corresponded to undissolved crystals rather than to compound bound to sediment particles. RDX has demonstrated low sorption affinity to soil particles (see Chapter 2).

The toxicities of RDX or HMX were investigated using spiked fine-grained marine sediments with *L. plumulosus* and *N. arenaceodentata* 10-d exposures [8]. The toxicity of RDX was also investigated using sandy sediment [14]. Significant mortality was not observed in any exposure, even at exceedingly high measured concentrations (Table 5.2). Reports of the toxicity of sediments spiked with RDX or HMX transformation products were not found in the available literature.

5.4   EFFECTS OF EXPLOSIVES ON REPRODUCTION AND GROWTH OF INVERTEBRATES

A variety of sublethal effects have been observed in aqueous exposures to explosives and related compounds with several invertebrates (see Chapter 5). However, few records of biological effects other than lethality are available for sediment exposures to explosives. In studies where effects other than mortality were investigated, nonlethal effects (i.e., significant decreases in growth [9] or reproduction [11,15]) have been reported either only in treatments where survival was also significantly decreased and therefore were not truly sublethal manifestations of sediment exposure to explosives. However, a sublethal decrease in offspring production was observed in *L. plumulosus* at the initial exposure concentration of 4014 nmol g^{-1} 2,4-DA-6-NT, and a sublethal decrease in biomass was exhibited by juvenile *N. arenoceodentata* at an initial concentration of 335 nmol g^{-1}, with lethal effects at the initial exposure concentration of 3437 nmol g^{-1} [8]. Lethal effects of RDX to *L. plumulosus* and *N. arenoceodentata* were not observed [8] (Table 5.2). More studies are necessary to provide adequate understanding of the sensitivity of sublethal endpoints relative to lethality to sediment-exposed benthic invertebrates.

Exposure to explosive-spiked sediments has promoted stimulatory effects on growth of a few species of invertebrates. Significantly increased biomass was observed for *C. tentans* exposed to RDX, HMX [9], TNT, and TNB [13] in sediment treatments that did not cause significant lethality. Similar effects were observed for *H. azteca* exposed to TNB or RDX [9] and for *N. arenaceodentata* exposed to TNT [15]. The mechanisms responsible for hormetic-like concentration–response relationships reported in these studies are unknown, but were speculated to be related to explosives-induced changes in the nutritional status of the sediment, and biochemical and physiological alterations in invertebrates [9].

5.5   TOXICITY OF POREWATER EXTRACTED FROM SEDIMENTS SPIKED WITH EXPLOSIVES

The toxicity of porewaters from fine-grained and sandy marine sediments spiked individually with 2,6-DNT, tetryl, or picric acid, and allowed to equilibrate at 4°C for 5–10 d, varied with both the compound and the nature of the spiked sediment [10]. Toxicity was assessed using sea urchin (*Arbacia punctulata*) embryological development, macroalgae (*Ulva fasciata*) zoospore germination and germling growth, and polychaete (*Dinophilus gyrociliatus*) survival and reproduction tests. Tetryl and picric acid were the most and least toxic of the three chemicals, respectively, in porewaters from both kinds of sediments [10]. Samples spiked with 2,6-DNT contained a transformation product identified as 2-A-6-NT. Unidentified peaks, possibly transformation products, were also seen in some of the picric acid- and tetryl-spiked samples. Porewater from the fine-grained sediment spiked with 2,6-DNT and tetryl was up to three orders of magnitude more toxic than porewater from the sandy sediment (Table 5.3), possibly due to the presence of relatively toxic nonidentified transformation products.

Further assessments of the toxicity of transformation products of 2,6-DNT and picric acid in porewaters from spiked sandy and fine-grained marine sediments were performed over a six-month sediment incubation period at 10°C and 20°C, during which incremental transformation was reported [24]. Toxicity of porewater from sediments spiked with 2,6-DNT decreased for the macroalgae *U. fasciata* zoospores with sediment incubation time, as transformation progressed, but increased for the copepod *Schizopera knabeni* nauplii. The primary transformation product of 2,6-DNT, 2-A-6-NT, was also more toxic than the parent compound to copepod nauplii, but not to algal zoospores, in exposures to spiked seawater [24]. Toxicity of porewater from either picric acid-spiked sediment to copepods and macroalgae zoospores increased at the beginning of the transformation process, but decreased with sediment incubation time, as transformation progressed. Picramic acid and 2,4-dinitrophenol (2,4-DNP), which are transformation products of picric acid, were more toxic than their parent compound in both macroalgae zoospore and copepod toxicity tests [24]. The porewater of fine-grained sediment spiked with either 2,6-DNT or picric acid had lower toxicity than its sandy counterpart after six months, suggesting faster microbial transformation in fine-grained sediments resulting in further transformation of picramic acid and 2,4-DNP into less toxic products [24].

In summary, the toxicity of porewaters from sediments spiked with 2,6-DNT, tetryl, and picric acid indicated the following toxicity gradient (from least to most toxic): picric acid, 2,6-DNT, and tetryl. Further investigations suggest that some transformation products of those chemicals are more toxic than their parent compounds, but that transformation, primarily of biotic nature, proceeds toward the formation of less toxic products given sufficient time.

5.6 THE USE OF BIOMIMETIC DEVICES TO ASSESS EXPLOSIVES BIOAVAILABILITY

Traditional chemical techniques used to estimate total sediment toxicant concentrations poorly predict bioavailability and toxicity. Typically, organic compounds must dissolve in sediment porewaters or digestive fluids in order to be accessible for uptake by an organism (bioavailability). As traditional chemical techniques aim to extract the whole mass of toxicants associated with sediments, sediment concentrations often provide inaccurate estimates of bioavailability and toxicity. Organismal concentrations represent a more accurate and theoretically sound measure of bioavailability.

TABLE 5.3

Toxicity of Porewater Extracted from Sandy or Fine-Grained Sediments Spiked with 2,6-DNT, Tetryl, or Picric Acid

Sediment Type	Chemical	EC50 (μmol L^{-1})							
		Urchin Embryo Development	Polychaete		Macroalgae Zoospore			
			Survival	Eggs/Adult	Germination	Germling Length	Germling Cell Number
Sandy	2,6-DNT	202.6	115.8	44.8	31.2	18.0	28.2
Fine grained	2,6-DNT	0.236	0.252	0.126	0.505	<0.478	<0.478
Sand	Tetryl	0.940	0.019	0.023	2.8	1.7	2.1
Fine grained	Tetryl	0.002[a]	0.007	0.007	<0.010[a]	<0.010[a]	<0.010
Sand	Picric acid	2587	556	364	2509	363	748
Fine grained	Picric acid	830	744	283	686	49	206

Source: Ref. 10.

However, determination and interpretation of explosive concentrations in organisms are faced with significant challenges, as explosives typically bioaccumulate at low whole body concentrations relative to environmental concentrations and are efficiently biotransformed and/or quickly eliminated following cessation of exposure (see Chapter 6).

Biomimetic techniques can be used to partially overcome the challenges associated with determination of true exposure concentrations of rapidly degrading or transforming explosives. These techniques are chemical methods, which attempt to measure concentrations of bioavailable toxicants rather than solvent-extractable concentrations of toxicants in environmental matrices [25]. The use of solid phase microextraction fibers (SPMEs) as biomimetic devices to predict toxicity and bioavailability of organic compounds in soil, sediment, and water has received considerable attention [26–28]. The SPME polymer coating sorbs dissolved or weakly dissociable molecules hypothesized to be bioavailable to organisms [25,29]. A disposable SPME technique was developed to measure bioavailable TNT and its transformation products in sediment during toxicity and bioaccumulation experiments [30]. Simple SPME deployment and extraction techniques were more rapid and less expensive than traditional chemical measurements of TNT or its transformation products in TNT-spiked sediments [30]. Additionally, these techniques enabled nondestructive, minimal-disturbance sampling during experiments with organisms [30]. The SPMEs provided accurate and precise measurements of toxic concentrations of TNT to *Tubifex tubifex*, even among dissimilar matrices such as water, sediment, and tissues [28,31]. Furthermore, the SPMEs predicted more accurately the whole-body concentrations of TNT and its transformation products in *T. tubifex* exposed to TNT-spiked water, sediment, and carbon-amended sediment compared with traditional measurement of sediment concentrations [28]. Therefore, biomimetic approaches show promise for measuring explosives and other organic compounds in water or sediment media, and should be viewed as supplementary tools to better characterize bioavailability of these chemicals in exposure matrices [26].

5.7 CONCLUSIONS AND RESEARCH RECOMMENDATIONS

The nitroaromatic explosives TNT, TNB, 2,6-DNT, picric acid, and tetryl undergo fast transformation when added to sediment, more rapidly in fine-grained sediments and at relatively low spiking concentrations. Reported major decreases in measured concentrations of explosives and their transformation products following a short contact period with contaminants and sediment resulted, at least in part, from irreversible binding of explosive transformation products to sediment. Such rapid changes in the exposure matrices have precluded determination of the accurate concentration–response relationships and toxicity values in those studies. Therefore, determination of temporal changes of whole sediment and porewater concentrations for parent compound and its transformation products during whole-sediment toxicity experiments is recommended to better understand concentration–response relationships. Proper interpretation of toxicity data from sediment exposures is also hampered by lack of sediment toxicity data for transformation products of explosive compounds, as only few studies reported comparative sediment toxicity data for parent compounds and their transformation products. Selections of appropriate sediment matrix and exposure conditions that ensure minimal temporal changes in exposure concentrations are recommended for reducing the uncertainty associated with assessing toxicity of explosives in spiked sediments. Transformation of a variety of explosive compounds was relatively slow in sandy, organic-carbon poor sediments. Therefore, the use of sandy sediments, coupled with the use of short (e.g., 4 d) exposures is recommended for deriving spiked-sediment toxicity data for explosives.

Sediments spiked with the explosives TNT, TNB, picric acid, tetryl, and the TNT breakdown products 2-ADNT and 2,4-DANT promoted significant invertebrate mortality in laboratory exposures. Exposure to 2,6-DNT in sandy and fine-grained sediments did not result in significant lethality in the only sediment toxicity study with that compound. Limited available toxicity data suggests that amphipods are more sensitive to TNT and its transformation products than the other invertebrates investigated, and that TNT and its major breakdown products 2-ADNT and 2,4-DANT are lethal at similar concentrations in sediment exposures.

Only a few studies have reported the fate of the cyclic nitramine compounds RDX or HMX in sediment matrices. Therefore, further research is needed to determine sorption–desorption dynamics and transformation rates of these compounds in different sediment types and exposure conditions. Unlike exposures to sediments spiked with nitroaromatic compounds, exposure of benthic invertebrates to sediments spiked with RDX and HMX did not elicit lethal toxicity, even at exceedingly high concentrations, and therefore are likely to pose minimal risk to benthic invertebrates at contaminated sites.

Sediment toxicity information for nine explosives and their transformation products exists for three freshwater and four marine invertebrate species, including bulk (e.g., *Neanthes arenaceodentata* and *Tubifex tubifex*) and selective (e.g., *Hyalella azteca* and *Chironomus tentans*) deposit feeders. In contrast, aqueous toxicity of energetic compounds and related products has been determined for a larger variety of compounds using numerous freshwater and marine species (see Chapter 4). Future studies should be designed to establish experimental data that enable comparisons of the bioaccumulation and toxicity of explosive compounds to benthic invertebrates belonging to different functional groups using the same sediment exposure conditions.

Reports on the presence of explosives in sediments from contaminated field sites exist. However, a lack of adequate information on spatial distribution of contamination, and the bioavailability and toxicity of sediment-associated explosives in historically contaminated military (e.g., former manufacturing plants) and active training sites (e.g., Navy firing ranges) preclude an accurate evaluation of the local and global environmental significances of the presence of explosives in aquatic systems. We encourage additional research to fill this substantial data gap.

Existing laboratory sediment toxicity data for explosives provides a low level of certainty for extrapolating ecological risks at contaminated sites from laboratory-based toxicity benchmark values. Therefore, the uncertainty associated with toxicity values reported in studies referenced in this chapter should be addressed when deriving toxicity reference values for assessing risks of adverse biological effects or deriving numerical guidelines for use in sediment quality evaluations. Effects-based evaluations using benthic toxicity tests of sediment collected from study sites are recommended for ecological risk assessment of aquatic sites contaminated with explosive compounds.

REFERENCES

1. Talmage SS et al., Nitroaromatic munition compounds: Environmental effects and screening values, *Rev. Environ. Contam. Toxicol.*, 161, 1, 1999.
2. EA Engineering, Science and Technology, Inc. Remedial investigation—Jackson Park Housing Complex/Naval Hospital Bremerton—Engineering Field Activity Northwest, Contract No. N44255-94-D-7309, 1996.
3. Darrach MR, Chutjian A, and Plett GA, Trace explosives signatures from World War II unexploded undersea ordnance, *Environ. Sci. Technol.*, 32, 1354, 1998.
4. Cusson B et al., Environmental Study on Sediment Quality in the Southern Portion of Lake Saint-Pierre Used by the Nicolet Munitions Experimental Test Centre (METC), Final Report, Environment Canada, Quebec, Canada, 2003.
5. Dave G, Field test of ammunition (TNT) dumping in the ocean, in *Quality Assessment and Management: Insight and Progress*, Munawar, M, Ed., Aquatic Ecosystem Health and Management Society, Washington, DC, 2003, 213.
6. Pennington JC and Brannon JM, Environmental fate of explosives, *Thermochim. Acta*, 384, 163, 2002.
7. Elovitz MS and Weber EJ, Sediment mediated reduction of 2,4,6-trinitrotoluene and fate of the resulting aromatic (poly)amines, *Environ. Sci. Technol.*, 33, 2617, 1999.
8. Lotufo GR et al., Toxicity of sediment-associated nitroaromatic and cyclonitramine compounds to benthic invertebrates, *Environ. Toxicol. Chem.*, 20, 1762, 2001.
9. Steevens JA et al., Toxicity of the explosives 2,4,6-trinitrotoluene, hexahydro-1,3,5-trinitro-1,3,5-triazine, and octahydro-1,3,5,7-tetranitro-1,3,5,7-tetrazocine in sediments to *Chironomus tentans* and *Hyalella azteca*: Low-dose hormesis and high-dose mortality, *Environ. Toxicol. Chem.*, 21, 1475, 2002.
10. Nipper M et al., Toxicological and chemical assessment of ordnance compounds in marine sediments and porewaters, *Mar. Pollut. Bull.*, 44, 789, 2002.
11. Conder JM et al., Recommendations for the assessment of TNT toxicity in sediment, *Environ. Toxicol. Chem.*, 23, 141, 2004.
12. Nipper M et al., Degradation of picric acid and 2,6-DNT in marine sediments and waters: The role of microbial activity and ultra-violet exposure, *Chemosphere*, 56, 519, 2004.
13. Lotufo GR and Farrar JD, Comparative and mixture sediment toxicity of trinitrotoluene and its major transformation products to the freshwater midge *Chironomus tentans*, *Arch. Environ. Contam. Toxicol.*, 49, 333, 2005.
14. Rosen G and Lotufo GR, Toxicity of two munitions constituents to the estuarine amphipod *Eohaustorius estuarius* in spiked sediment exposures, *Environ. Toxicol. Chem.*, 24, 2887, 2005,
15. Green AS, Moore DW, and Farrar D, Chronic toxicity of 2,4,6-trinitrotoluene to a marine polychaete and an estuarine amphipod, *Environ. Toxicol. Chem.*, 18, 1783, 1999.
16. Northcott GL and Jones KC, Spiking hydrophobic organic compounds into soil and sediment: A review and critique of adopted procedures, *Environ. Toxicol. Chem.*, 19, 2418, 2000.
17. Hundal LS et al., Long-term TNT sorption and bound residue formation in soil, *J. Environ. Qual.*, 26, 896, 1997.
18. Gong P, Wilke BM, and Fleischmann S, Soil-based phytotoxicity of 2,4,6-trinitrotoluene (TNT) to terrestrial higher plants, *Arch. Environ. Contam. Toxicol.*, 36, 152, 1999.

19. U.S. Environmental Protection Agency, SW-846 Method 8330, Nitroaromatics and Nitramines by High Performance Liquid Chromatography (HPLC), Office of Solid Waste and Emergency Response, Washington, DC, 1998.

20. Brannon JM and Pennington JC, Environmental Fate and Transport Process Descriptors for Explosives, ERDC/EL TR-02-10, Engineer Research and Development Center, prepared for US Army Corps of Engineers, Washington, DC, 2002.

21. U.S. Environmental Protection Agency, Methods for Measuring the Toxicity and Bioaccumulation of Sediment-Associated Contaminants with Freshwater Invertebrates, 600/R-99/064, Office of Research and Development, Washington, DC, 2000.

22. U.S. Environmental Protection Agency, Methods for Measuring the Toxicity of Sediment-Associated Contaminants with Estuarine and Marine Amphipods, 600/R-94/025, Office of Research and Development, Washington, DC, 1994.

23. Haderlein SB and Schwarzenbach RP, Environmental process influencing the rate of abiotic reduction of nitroaromatic compounds in the subsurface, in *Biodegradation of Nitroaromatic Compounds and Explosives*, Spain JC, Hughes JB, and Knackmuss H-J,. Eds., CRC Press, Boca Raton, FL, 2000, 199.

24. Nipper M et al., Fate and effects of picric acid and 2,6-DNT in marine environments: Toxicity of degradation products, *Mar. Pollut. Bull.*, 50, 1205, 2005.

25. Hermens JLM et al., Application of negligible depletion solid-phase extraction (nd-SPE) for estimating bioavailability and bioaccumulation of individual chemicals and mixtures, in *Persistent, Bioaccumulative, and Toxic Chemicals II, Assessment and New Chemicals*, Lipnick RL, Jansson B, Mackay D, Petreas M, Eds., American Chemical Society, Washington, DC, 64, 2001.

26. Wells JB and Lanno RP, Passive sampling devices (PSDs) as biological surrogates for estimating the bioavailability of organic chemicals in soil, in *Environmental Toxicology and Risk Assessment: Science, Policy, and Standardization—Implications for Environmental Decisions*, Vol 10. STP 1403, Greenberg BM, Hull RN, Roberts MH Jr, Gensemer RW, Eds., American Society for Testing and Materials, Philadelphia, PA, 2001, 253.

27. Leslie HA et al., Biomimetic solid-phase microextraction to predict body residues and toxicity of chemicals that act by narcosis, *Environ. Toxicol. Chem.*, 21, 229, 2002.

28. Conder JM and La Point TW, Solid phase microextraction for predicting the bioavailability of TNT and its primary transformation products in sediment and water, *Environ. Toxicol. Chem.*, 24, 1059, 2005.

29. Pawliszyn J, *Solid Phase Microextraction: Theory and Practice*, Wiley-VCH, New York, 1997.

30. Conder JM et al., Nondestructive, minimal-disturbance, direct-burial solid-phase microextraction fiber technique for measuring TNT in sediment, *Environ. Sci. Technol.*, 37, 1625, 2003.

31. Conder JM et al., Solid phase microextraction fibers for estimating the toxicity and bioavailability of sediment-associated organic compounds, *Aquat. Ecosystem Health Manage.*, 7, 387, 2004.

6 Bioconcentration, Bioaccumulation, and Biotransformation of Explosives and Related Compounds in Aquatic Organisms

Guilherme R. Lotufo, Michael J. Lydy,
Gregory L. Rorrer, Octavio Cruz-Uribe,
and Donald P. Cheney

CONTENTS

6.1 Introduction 136
6.2 Bioconcentration, Bioaccumulation, and Biotransformation of
Explosives in Aquatic Animals 136
    6.2.1 Bioconcentration 136
    6.2.2 Body Distribution 142
    6.2.3 Bioaccumulation from Diet 142
    6.2.4 Toxicokinetics and Biotransformation 143
6.3 Uptake and Biotransformation of Explosives in Phototrophic
Organisms 145
    6.3.1 Bioconcentration of Explosives in Microalgae 145
    6.3.2 Uptake and Biotransformation of Explosives by Vascular
        Aquatic Plants 146
    6.3.3 Uptake and Biotransformation of Explosives by Nonvascular
        Marine Macrophytes 148
    6.3.4 Uptake and Biotransformation of Explosives by Nonvascular
        Freshwater Macrophytic Algae 150
6.4 Conclusions and Recommendations 151
References 152

6.1 INTRODUCTION

Despite their high environmental relevance, the presence of explosives and related compounds in aquatic systems is not well studied (see Chapter 4). Studies examining the bioaccumulation and biotransformation potential were conducted only for a few compounds using a small number of aquatic species.

Explosive compounds are weakly hydrophobic and therefore their bioaccumulative potential is expected to be low [1–3]. Bioconcentration (net accumulation of chemical from water) of explosives was low in a variety of aquatic animals [2–9]. Moreover, the dietary uptake of explosives in fish was minimal relative to aqueous uptake [7,8,10]. Although food web transfer of explosive compounds is expected to be insignificant, the demonstrated toxicity of those compounds to a wide variety of aquatic organisms (see Chapter 4) warrants pursuing an adequate understanding of their fate in relevant aquatic species.

Efficient biotransformation of TNT has been reported for fish [3,7,9] and an aquatic invertebrate [6], most of which accumulated biotransformed products including tissue-bound molecules, at higher concentrations relative to the parent compound at steady state [6,7,9]. The ability of aquatic animals to biotransform other explosives and related compounds is unknown. Recent studies examining the uptake from water and elimination kinetics of nitroaromatic and cyclic nitroamine explosives in aquatic invertebrates and fish [2,3,5–9] revealed that elimination of those compounds is very efficient, leading to steady-state levels within hours.

Aquatic vascular plants and macroalgae can take up TNT dissolved in water very efficiently as indicated by removal rates determined for several species [11,12]; some of which are promoted for use in phyto-treatment of explosives-contaminated water [11]. Less efficient removal of dissolved RDX was reported for wetland and aquatic plants [13,14]. Efficient biotransformation and elimination mechanisms in aquatic vascular plants and macroalgae resulted in a lack of bioconcentration of TNT and its solvent-extractable transformation products [11,12]. This chapter summarizes and discusses the bioconcentration, bioaccumulation, biotransformation, and toxicokinetic processes of explosives in aquatic organisms.

6.2 BIOCONCENTRATION, BIOACCUMULATION, AND BIOTRANSFORMATION OF EXPLOSIVES IN AQUATIC ANIMALS

6.2.1 BIOCONCENTRATION

A bioconcentration factor (BCF) is the ratio of the concentration of a chemical in an aquatic organism to that in the surrounding water, and is the most frequently used indicator of a compound's propensity to bioconcentrate in aquatic organisms [15]. Bioconcentration factors are determined primarily using steady-state concentrations, or by using kinetically derived uptake and elimination rates [3]. A summary of bioconcentration factors reported for explosives and related compounds in invertebrate and fish species is provided in Table 6.1.

TABLE 6.1

Empirically Determined Bioconcentration Factor (BCF) Values for Fish and Aquatic Invertebrates

Compound	Species	Exposure Duration (Days)	BCF (ml g^{-1})	Reference
2-NT	*Carassius auratus*	20	0.1	5
3-NT	*Carassius auratus*	20	0.06	5
4-NT	*Carassius auratus*	20	0.01	5
2,6-DNT	*Carassius auratus*	20	1.7	5
2,4-DNT	*Cyprinus carpio*	3	9.2	4
2,4-DNT	*Carassius auratus*	20	4.5	5
TNT	*Chironomus tentans*	1	2.6	7
TNT	*Lumbriculus variegatus*	1	4	7
TNT	*Tubifex tubifex*	2.2	2.5	6
TNT	*Ictalurus punctatus*	0.3	0.8	9
TNT	*Cyprinodon variegatus*	0.25	9.7[a]	3
TNT	*Pimephales promelas*	0.25	2.3, 5.0	2
2-ADNT	*Tubifex tubifex*	2.2	10.2	6
2-ADNT	*Cyprinodon variegatus*	0.25	13.1[a]	3
4-ADNT	*Tubifex tubifex*	2.2	12.4	6
2,4-DANT	*Tubifex tubifex*	2.2	2.8	6
2.4-DANT	*Cyprinodon variegatus*	0.25	0.5[a]	3
NB	*Carassius auratus*	20	0.01	5
1,2-DNB	*Carassius auratus*	20	0.07	5
1,3-DNB	*Carassius auratus*	20	8.5	5
1,4-DNB	*Carassius auratus*	20	12.4	5
RDX	*Lumbriculus variegatus*	0.7	2.1	8
RDX	*Ictalurus punctatus*	0.7	2	8
RDX	*Cyprinodon variegatus*	0.25	1.7[a]	3
HMX	*Cyprinodon variegatus*	0.25	0.5[a]	3

[a] Kinetically derived BCF (k_u k_e^{-1}).

The BCFs for TNT ranged widely from 0.8 to 9.7 ml g^{-1} for a variety of invertebrates and fish species [2,3,6,7,9]. Comparison of the bioconcentration of TNT and its major biotransformation products in the oligochaete *Tubifex tubifex* yielded lower BCFs for TNT and 2,4-DANT than for 2-ADNT and 4-ADNT [6]. A similar comparative study with sheepshead minnows (*Cyprinodon variegatus*) revealed similar BCFs for TNT and 2-ADNT, and a lower BCF value for 2,4-DANT [3]. The BCFs for nitrotoluenes (NTs), determined using goldfish (*Carassius auratus*) [5], were at least eight times lower than the values determined for TNT and DNTs. Likewise, the BCF for nitrobenzene (NB) was exceedingly low for goldfish [5], but the nitrobenzene BCF (2 ml g^{-1}) for golden orfe (*Leuciscus idus*) exposed to [^{14}C]-compound [16] was 200 times higher. The BCF for 1,2-dinitrobenzene (1,2-DNB) for goldfish

TABLE 6.2

Bioconcentration Factor (BCF) Values Predicted from log K_{ow} Values Using a Regression Model and Range of Experimentally Determined Values

Compound	log K_{ow}	Predicted BCF[a] (ml g^{-1})	Measured BCF (ml g^{-1})
2-NT	2.3[b]	38.7	0.1
3-NT	2.8[b]	104.2	0.06
4-NT	2.3[b]	38.7	0.01
2,4-DNT	1.98[c]	20.6	4.5 and 9.2
2,6-DNT	2.10[c]	26.1	1.7
TNT	1.60[c]	9.7	0.8–14.2
2-ADNT	1.95[d]	19.3	10.2 and 13.1
4-ADNT	2.1[d]	26.1	12.4
2,4-DANT	0.79[d]	2.0	0.5 and 2.8
NB	1.8[b]	17.2	0.01
1,2-DNB	1.55[b]	8.8	0.07
1,3-DNB	1.52[b]	8.3	8.5
1,4-DNB	1.45[b]	7.2	12.4
RDX	0.87[e]	2.3	2 and 2.1
HMX	0.18[f]	0.6	0.5

[a] log BCF = 0.86 log K_{ow} − 0.39 [15].
[b] cited in [5]
[c] [58]
[d] [59]
[e] [60]
[f] [61]

was also very low when compared to the over 100 times higher BCFs for 1,3-DNB and 1,4-DNB determined in the same study [5]. The RDX BCFs were similar for channel catfish [8] and sheepshead minnows [3] and in the oligochaete *Lumbriculus variegatus* [8], ranging from 1.7 to 2.1 ml g^{-1}. The only published study of HMX bioconcentration in aquatic organisms reported a BCF of 0.5 ml g^{-1} for juvenile sheepshead minnows [3].

Various predictive models have been proposed to estimate BCFs for organic compounds based on their hydrophobicity as represented by their n-octanol/water partition coefficient (K_{ow}) [15]. Explosive compounds are weakly hydrophobic (low K_{ow} values) and therefore BCFs predicted using these models should be low [3,6,17]. The equation derived by Meylan et al. [15] for nonionic compounds in the log K_{ow} range of 1 to 7 (log BCF = 0.86 log K_{ow} − 0.39) was used to calculate predicted BCFs for explosives and related compounds (Table 6.2) for empirically derived BCF values reported in Table 6.1. Comparison of predicted and empirically derived BCFs revealed that most explosives and related compounds bioconcentrate in aquatic

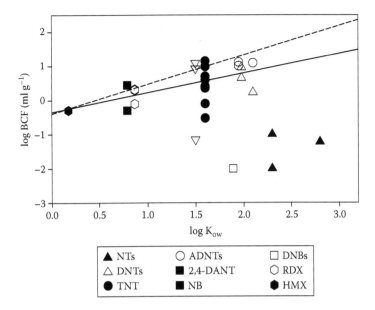

FIGURE 6.1   Relationship between the log bioconcentration factors (BCFs) for explosive and related compounds reported in Table 6.1 and log K_{ow}. The solid line represents the best-fit linear equation (log BCF = 0.53 log K_{ow} − 0.23, r^2 = 0.37) for all values except those with log BCF <−0.9. The dashed line represents the linear equation from Meylan et al. [15].

invertebrates and fish at concentrations lower than those predicted using their hydrophobicity (Table 6.2). Most measured BCFs were similar to model-predicted BCFs within a factor of 3. However, measured bioconcentration was dramatically lower than predicted for NTs, and NB and 1,2-DNB using goldfish [5], as the ratio between predicted and measured values ranged from 125 to 1733. Overall, the bioconcentration of explosives and related compounds from different chemical classes (nitrotoluenes, nitrobenzenes, cyclic nitramines) appear to be adequately predicted using a generic relationship derived for nonionic compounds (Table 6.2) suggesting that the bioconcentration of those compounds is a thermodynamically driven partitioning process between the water and the lipid phase of the exposed organisms. Further studies are necessary to confirm the unexpectedly low bioconcentration of NB, NTs, and 1,2-DNB in goldfish, and to determine the bioconcentration of those and other poorly studied compounds in other species of aquatic animals.

The linear relationship between log K_{ow} and the empirically derived log BCFs reported in Table 6.1 was nonsignificant (P = 0.372). When the exceedingly low BCFs reported for NB, NTs, and 1,2-DNB for goldfish (0.01–0.1 ml/mg) [5] were excluded, a significant (P = 0.003) linear relationship (log BCF = 0.53 log K_{ow} − 0.23, r^2 = 0.37) was obtained (Figure 6.1). For compounds in the log K_{ow} range of 0.5 to 2.5, the latter relationship predicts BCFs values lower than those predicted using the relationship of Meylan et al. [15]. More studies of the bioconcentration of explosive and related compounds in other aquatic species are needed to establish more accurate predictive models for the bioconcentration of those compounds.

TABLE 6.3

Bioconcentration Factor (BCF) Values for Aquatic Animals Determined
Using the Sum Concentration of the Parent Compound and Major
Transformation Products, or Radioactivity in Extract, or Total
Radioactivity Surrogate for Body Residue

Compound	Species	Exposure Duration (Days)	BCF (ml g⁻¹)	Analytical Measurement	Reference
TNT	*Chironomus tentans*	1	76	Total radioactivity	7
TNT	*Chironomus tentans*	1	8	Extractable radioactivity	7
TNT	*Daphnia magna*	4	209	Total radioactivity	17
TNT	*Lumbriculus variegatus*	4	202	Total radioactivity	17
TNT	*Lumbriculus variegatus*	1	216	Total radioactivity	7
TNT	*Lumbriculus variegatus*	1	38	Extractable radioactivity	7
TNT	*Tubifex tubifex*	2.2	12.3	Sum TNT, ADNTs, DANTs	6
TNT	*Ictalurus punctatus*	0.3	16	Total radioactivity	9
TNT	*Ictalurus punctatus*	0.3	10.5	Extractable radioactivity	9
TNT	*Pimephales promelas*	0.25	4.7, 8.4	Sum TNT and ADNTs	2
TNT	*Pimephales promelas*	4	5.5, 9.2	Sum TNT and ADNTs	2
2,4-DNT	*Daphnia magna*	4	13	Total radioactivity	18
2,4-DNT	*Lumbriculus variegatus*	4	58	Total radioactivity	18
RDX	*Daphnia magna*	4	3	Total radioactivity	17
RDX	*Lumbriculus variegatus*	4	1.6	Total radioactivity	17

In the studies listed in Table 6.1, BCFs were determined using the parent compound concentrations determined by high performance liquid chromatography
(HPLC) analysis of solvent extracts of whole-body homogenates. In early studies
of the bioconcentration of explosives in aquatic invertebrates and fish, BCFs were
derived using radioactivity measurements of whole-body homogenates as a surrogate for the sum molar concentration of parent and all transformation products.
Bioconcentration factors derived using total radioactivity measurements for body
residues were reported for TNT [17], 2,4-DNT [18], and RDX [17] (Table 6.3).
Recent studies of the bioconcentration of TNT parent compound in aquatic invertebrates and fish compared the BCFs calculated for the parent compound with BCFs
calculated using the sum concentration of TNT parent compound and its transformation products in the tissues (Table 6.3). In those studies, BCFs for the mixture
of parent and transformation products were calculated from body residues determined using different analytical techniques, which generated concentrations for

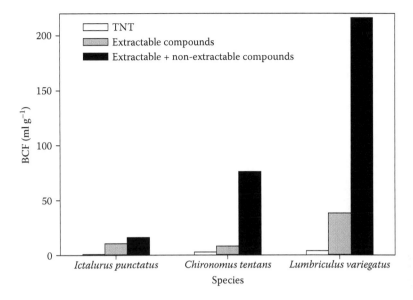

FIGURE 6.2 Bioconcentration factors (BCF) for aquatic animals exposed to TNT determined for body residues for parent compound, sum concentration of extractable compounds, and for the sum concentration of extractable and nonextractable compounds.

different compound mixtures. The different analytical techniques used included the following: (1) HPLC analysis of solvent extracts providing body residue for the sum molar concentration of TNT, ADNTs, and DANTs [2,6]; (2) radioactivity in solvent extracts providing body residue for the sum molar concentration of TNT and all its extractable transformation products [7,9]; and (3) total radioactivity of whole-body homogenates providing body residue for the sum molar concentration of TNT and all its extractable and nonextractable transformation products [7,9] (Table 6.3).

Comparison of BCF for parent compound (Table 6.1), for total extractable compounds, and for total extractable and nonextractable compounds for the same exposed organisms (Figure 6.2) reveals that the nonextractable (i.e., solvent-resistant and presumably tissue-bound) compounds are accumulated at substantially higher concentrations than extractable transformation products at steady state in fish and aquatic invertebrates. The tissue concentration of parent compounds represented only a fraction, sometimes minimal, of the total extractable concentration in all the species investigated.

Bioconcentration values determined using total radioactivity in the tissue were also reported for 2,4-DNT [18] and RDX [17] (Table 6.3). Comparison of BCFs determined for total radioactivity (Table 6.3) with those determined for parent compound only (Table 6.1) for RDX suggests dominant accumulation of parent compound relative to transformation products in aquatic animals. Similar comparison for 2,4-DNT suggests substantial accumulation of transformation products of that compound in aquatic animals.

6.2.2 BODY DISTRIBUTION

The body distribution of TNT and its transformation products as represented by radioactivity was examined in fathead minnows exposed to [^{14}C]TNT [2]. The highest concentrations of [^{14}C]TNT occurred in the internal visceral tissues, which greatly exceed the whole-body concentration [2]. The visceral tissues contained 57% of the radioactivity accumulated in the whole fish, while this body compartment represented only 12% of the mass of the fish. Conversely, the muscular, skeletal, and skin tissues contained only 34% of the total radioactivity while representing 82.4% of the fish mass. The [^{14}C]TNT concentration in viscera was 34 times higher than in somatic muscles in juvenile bluegill sunfish (*Lepomis macrochirus*) exposed to [^{14}C]TNT [17]. Similar results were obtained when juvenile channel catfish were exposed to [^{14}C]TNT, as [^{14}C]TNT concentrations were substantially higher in the visceral fraction (digestive tract, liver, reproductive system) than in other body parts [9]. However the concentration of TNT parent compound was greatest in the gills than in the viscera in fish used in the latter study. The bioaccumulation pattern for [^{14}C]TNT in fish was likely related to the lipid content of the tissue groups but was not solely influenced by that parameter [2]. Studies examining the bioconcentration of [^{14}C]RDX in juvenile bluegill sunfish reported higher accumulation in the viscera than in somatic muscles by factors of <3 [17,19], revealing that the RDX is more uniformly distributed in fish tissues compared to TNT.

6.2.3 BIOACCUMULATION FROM DIET

Aquatic animals accumulate xenobiotics through direct absorption from contaminated water mostly through respiratory surfaces and through the ingestion of contaminated food [20]. Dietary exposure to nonionic organic contaminants such as chlorinated pesticides and polychlorinated biphenyls resulted in significant bioaccumulation in fish [21,22]. Significant bioaccumulation of TNT transformation products in the salamander *Ambystoma tigrinum* through exposure to contaminated prey [23] encouraged recent investigations of the relative importance of dietary accumulation of explosives in fish [7,8,10].

The dietary uptake of TNT and RDX was examined using two fish species and different routes of contaminant delivery. Juvenile channel catfish that were force-fed with [^{14}C]TNT-spiked food pellets [7] daily for up to 10 d accumulated radioactivity, mostly as extractable and nonextractable unknown compounds. Steady-state bioaccumulation factors (BAFs, the concentration in fish relative to the concentration in the diet) for TNT (0.000024 g g^{-1}), sum of extractable compounds (0.00048 g g^{-1}), and total compounds (0.00091 g g^{-1}) were exceedingly low. In a separate study, juvenile fathead minnows (*Pimephales promelas*) were fed *Lumbriculus variegatus* that were preexposed to [^{14}C]TNT or [^{14}C]RDX [10]. The fathead minnows were fed the worms twice daily for up to 14 d and the worms were frozen prior to use. *Lumbriculus variegatus* preexposed to [^{14}C]-dichlorodiphenyltrichloroethane ([^{14}C] DDT) was used as a positive control for dietary uptake in the latter study because that compound is substantially more hydrophobic, and hence will bioaccumulate to a greater extent than explosives. Extensive biotransformation of the parent compound in worms occurred for TNT, but not for RDX or DDT. The body residues in the fish remained relatively unchanged over time for [^{14}C]TNT and [^{14}C]RDX, but did not approach steady state in the tissue for [^{14}C]DDT during the 14-d exposure period. The mean BAF values, determined using total radioactivity in worms and fish, were 0.018, 0.010, and 0.422 g g^{-1} for [^{14}C]TNT, [^{14}C]RDX, and [^{14}C]DDT, respectively, confirming the expected low bioaccumulative potential for TNT and RDX relative to DDT through the dietary route.

To directly assess the relative importance of aqueous and dietary uptake of RDX, juvenile channel catfish were fed *L. variegatus* freshly exposed to RDX, directly exposed to RDX-spiked water, or exposed to both for 7 d [8]. The same concentration was used for the worm and fish aqueous exposures. The mean BAF of 0.04 g g^{-1} obtained from that study for parent RDX was higher than the mean BAF (0.01 g g^{-1}) determined using radioactivity as a surrogate for RDX and its transformation products using fathead minnows [10]. Results from the comparative exposure scenario study [8] as well as from other dietary uptake investigations [8,10] indicated that aqueous exposure is likely the dominant route of exposure for TNT, RDX, and similar compounds, with dietary uptake providing only minimal contribution when both pathways are present.

6.2.4   TOXICOKINETICS AND BIOTRANSFORMATION

The uptake kinetics of TNT, 2-ADNT, 4-ADNT, and 2,4-DANT were compared in *T. tubifex* [6] with all four compounds reaching steady-state concentrations within 1 h of exposure. Using total radioactivity as a surrogate for whole-body concentration, the uptake rate coefficient (k_u) for TNT into the midge *Chironomus tentans*, the oligochaete *L. variegatus* [7] and juvenile channel catfish (*Ictalurus punctatus*) [7,9] were 26.7, 39.2, and 10.1 ml g^{-1} h^{-1}, respectively. Using analytically measured body residues, the k_u for TNT, 2-ADNT and 2,4-DANT into juvenile sheepshead minnows were 7.34, 12.55, and 1.27 ml g^{-1} h^{-1}, respectively [3]. The rates of RDX uptake in *L. variegatus* [8], *C. variegatus* [3], and juvenile channel catfish [8] were more variable (5.17, 0.15, and 1.28 ml g^{-1} h^{-1}, respectively), but were consistently less efficient than the uptake for TNT. The uptake clearance rate for 2,4-DNT in carp (*Cyprinus carpio*) (0.41 ml g^{-1} h^{-1}) [4] was slower than the rates derived for TNT using juvenile channel catfish or sheepshead minnows, and may be due to species-specific differences and/or a decrease in uptake efficiency with increasing organism size [24], because the carp (5–18 g) were substantially larger than the juvenile minnows and catfish (0.1–0.4 g) used in the latter studies. The k_u for HMX (0.06 ml g^{-1} h^{-1}) was only reported for sheepshead minnows [3] and was slower than the k_u determined for RDX (0.15 ml g^{-1} h^{-1}) in that study.

The nitro groups of TNT characteristically undergo biochemical reduction in living systems (see Chapter 2). Biotransformation of TNT to ADNTs during aqueous exposures occurred very quickly in *T. tubifex* (0.212 and 0.187 h^{-1} for 2-ADNT and 4-ADNT, respectively) [6], but was markedly slower (0.06 h^{-1} for ADNTs) in juvenile sheepshead minnows [3]. Substantial biotransformation of TNT to ADNTs in fathead minnows exposed to TNT for periods as short as 10 min was observed [2]. Efficient biotransformation of TNT to ADNTs was also reported for soil invertebrates (see Chapter 10). Formation of DANTs resulting from exposure to TNT or ADNTs was nonquantifiable or minimal [3,6,25]. In addition to the metabolic formation of ADNTs, exposure of *T. tubifex* [6], *L. variegatus*, *C. tentans* [7], and juvenile channel catfish [9] to aqueous solutions of [^{14}C]TNT resulted in the formation of unidentified extractable polar compounds as well as nonextractable (i.e., solvent-resistant) compounds. The fraction of total whole-body radioactivity corresponding to nonidentified extractable compounds accounted for the largest fraction of the total whole-body radioactivity in *T. tubifex* [6] and juvenile channel catfish [9], while nonextractable compounds made up the dominant fraction in *C. tentans* and *L. variegatus* [7] following aqueous exposures to [^{14}C]TNT. Bioaccumulation of large fractions of nonidentified compounds was also found in the marine amphipod *Eohaustorius estuarius* following exposure to sediment spiked with [^{14}C]TNT [26]. For RDX, comparisons of the steady-state BCF reported for total radioactivity (3.0 ml g^{-1}) [17] and parent compound (2.4 ml g^{-1}) for *L. variegatus* suggests that minimal biotransformation of RDX occurs in that species. Based on information from previous studies, Conder et al. [6] speculated that portions of the tissue-associated extractable nonidentifiable compounds in *T. tubifex* may exist as distinct TNT transformation products such as hydroxylaminodinitrotoluenes and azoxytoluenes, as reported for *E. albidus* [27].

Nonextractable radioactivity in invertebrate and fish tissues likely represents covalent conjugates (i.e., macromolecular bound residues) [6,7]. However, neither the biological half-life nor the chemical nature of nonextractable transformation products in invertebrates and fish has been investigated to date. Formation of covalent bonds with proteins in mammalian systems has been reported [28–30]. Investigations on the biotransformation of other explosives or related products in aquatic animals were not found in the available literature.

Elimination of TNT and its major biotransformation products was relatively fast for all of the aquatic organisms tested. For example, when *T. tubifex* or sheepshead minnows were transferred to noncontaminated water following exposure to TNT, 2-ADNT, 4-ADNT (*T. tubifex* only), or 2,4-DANT, measurable amounts of those compounds were completely eliminated from the tissues in less than 3 h [3,6]. The tissue sampling schedule used in the *T. tubifex* study precluded the calculation of the exceedingly fast elimination of TNT or 2,4-DANT in that species [6]. The elimination rate (k_e) for TNT in juvenile fathead minnows (2.2 h^{-1}) was exceedingly fast, corresponding to a half-life of 0.3 h. The k_e of TNT in juvenile sheepshead minnows was slower, 0.80 h^{-1}, corresponding to a half-life of 0.81 h [3]. The k_e values for *T. tubifex* for 2-ADNT and 4-ADNT were 1.07 h^{-1} and 0.91 h^{-1}, respectively. The k_e for 2-ADNT and 2,4-DANT were 0.96 and 2.74 h^{-1}, respectively [3]. The rates of elimination of DNTs in goldfish [5] were very similar (0.47 and 0.60 h^{-1} for 2,4-DNT and 2,6-DNT, respectively) to the rates determined for TNT and 2-ADNT in *T. tubifex* and sheepshead minnows, as expected given the similar hydrophobicity of those nitroaromatic compounds. However, the rate of elimination of 2,4-DNT in carp was substantially slower (0.06 h^{-1}), likely due to the larger size of the *C. carpio* compared to the organisms used in the other investigations.

Elimination of RDX was relatively fast in juvenile channel catfish and in the oligochaete *L. variegatus* (0.64 and 2.51 h^{-1}, respectively) [8], but was substantially slower in juvenile sheepshead minnows (0.09 h^{-1}) [3]. Bentley et al. [19] reported temporal changes in the concentration of RDX (as total radioactivity) in the viscera and edible tissues in three species of fish over time upon transfer from spiked water to clean water. Using the mean tissue burdens, the estimated elimination rates were 0.010 h^{-1} for both edible parts and viscera in juvenile bluegill sunfish; 0.006 h^{-1} and 0.008 h^{-1} for edible parts and viscera, respectively, in juvenile channel catfish; and 0.0002 h^{-1} and 0.0028 h^{-1} for edible parts and viscera, respectively, in fathead minnows [19]. The elimination of RDX from juvenile channel catfish and fathead minnows [19] was significantly slower compared to the elimination of RDX from juvenile channel catfish and sheepshead minnows reported in recent studies [3,8]. The reasons for the discrepancy are unknown. The k_e for HMX (0.123 h^{-1}) was only reported for sheepshead minnows [3]. Investigations on the elimination of other explosives or related products were not found in the available literature.

The elimination rate of ADNTs metabolically formed in *T. tubifex* [6] and sheepshead minnows [3] during exposure to TNT was substantially slower than the rate of elimination of 2-ADNT directly absorbed from water. For *T. tubifex*, absorbed ADNTs were quickly eliminated within 3 h, whereas the concentration of metabolically formed ADNTs remained unchanged for 54 h following transfer of TNT-exposed animals to uncontaminated water [6]. It has been recently demonstrated [31] that metabolically generated ADNTs are less likely available within tissues because of the lower extractability into polyacrylate-coated solid phase microextraction fibers relative to absorbed ADNTs. The reduction in extractability and rate of elimination of metabolically formed ADNTS may be due to the physiological location in which ADNTs are incorporated into tissue during metabolism of TNT to ADNTs. Absorbed ADNTs may concentrate in the dermis, whereas metabolically generated ADNTs may be bound to macromolecules within lipid bilayers, and therefore are more recalcitrant to elimination [31].

Sublethal biochemical effects can also occur due to exposure of organisms to explosive compounds. For example, significant inhibition of ethoxyresorufin-*O*-deethylase (EROD) activity was observed in juvenile rainbow trout (*Onchorhynchus mykiss*) dosed interperitoneally with TNT at different concentrations [32]. In a similar study, *O. mykiss* treated with TNT had increased glutathione *S*-transferase and glutathione reductase activities, and a decreased percentage of oxidized glutathione compared to control fish [33], indicating that TNT oxidizes macromolecules and activates antioxidant defense systems. The compounds TNT, 2-ADNT, and 4-ADNT were found in the hydrolyzed bile of the TNT-treated trout, indicating that this fish species can detoxify and excrete TNT. The presence of TNT and its transformation products in the bile may be suitable as a direct marker of TNT exposure in fish in contaminated sites [33].

6.3 UPTAKE AND BIOTRANSFORMATION OF EXPLOSIVES IN PHOTOTROPHIC ORGANISMS

6.3.1 BIOCONCENTRATION OF EXPLOSIVES IN MICROALGAE

Bioconcentration factors were determined for the freshwater microalga, *Pseudokirchneriella subcapitata* (formerly *Selenastrum capricornutum*) using radioactivity as a surrogate for body residues of parent compounds and all their transformation products [17]. The BCF of radioactivity-based TNT (453 ml g^{-1}) using *P. subcapitata* was approximately twice as high as similarly derived values for aquatic invertebrates and fish, whereas the BCF of radioactivity-based RDX (123 ml g^{-1}) was substantially higher (factor of 41 or higher) than those for aquatic invertebrates and fish. Studies of the bioconcentration of explosives in phytoplanktonic algae using analytical quantification of parent and transformation products are needed for achieving an adequate understanding of the movement of explosives in pelagic food chains. The relatively high BCF values measured in *P. subcapitata* suggest that TNT and RDX tissue-bound transformation products may be present at relatively high concentrations in microalgae.

6.3.2   UPTAKE AND BIOTRANSFORMATION OF EXPLOSIVES BY VASCULAR AQUATIC PLANTS

Aquatic vascular plants have been extensively studied for the phytoremediation of TNT from contaminated soil and groundwater using engineered wetland systems [34,35]. Removal of TNT from water has been extensively investigated using species belonging to the submerged aquatic plant genus *Myriophyllum* [11,36–42]. Several other species of aquatic and wetland plants also efficiently remove TNT from water [11,41,42]. Those studies have reported the "disappearance" of TNT from aqueous solutions using well-mixed suspensions of plant biomass and contaminant-spiked water. The concentration of dissolved TNT in the water decreased over time as a first-order rate process described by the decaying exponential relationship

$$C_{TNT} = C_{TNT,o} \cdot e^{-k \cdot X \cdot t}$$

where C_{TNT} is the concentration of TNT in the contaminated water at a given time (mg L^{-1}), $C_{TNT,o}$ is the initial TNT concentration in the contaminated water (mg L^{-1}), k is the specific first-order rate constant for TNT removal (L g^{-1} fresh weight [FW] h^{-1}), t is time (h), and X is the biomass density (g FW L^{-1}).

The specific first-order rate constants for TNT removal from water by *Myriophyllum* species ranged from 3.6×10^{-4} to 4.3×10^{-3} L g^{-1} FW h^{-1} [11,38,42]. In those studies, TNT was not recovered stoichiometrically from plant tissue because of extensive biotransformation.

A recent literature review [13] listed 17 wetland plants capable of RDX removal from water, or removal accompanied by RDX incorporation into plant biomass. The emergent wetland plant *Phalaris arundinacea* (reed canary grass) and the submerged wetland plant *Elodea canadensis* were particularly suitable for RDX bioremediation using constructed wetlands in terms of RDX removal efficiency and relative persistence [39]. Studies with the model aquatic plant *M. aquaticum* demonstrated that RDX removal from water was a first-order rate process, similar to TNT removal [14,43]. However, the first-order rate constants for RDX removal were 3.0×10^{-5} L g^{-1} FW h^{-1} versus 2.2×10^{-3} L g^{-1} FW h^{-1} for TNT removal at the same conditions [14], suggesting that RDX biotransformation was nearly two orders of magnitude slower than TNT biotransformation in *Myriophyllum*.

Terrestrial and aquatic vascular plants promote the biological reduction of nitro groups on TNT to amine groups, yielding 2-ADNT and 4-ADNT, similar to invertebrates and vertebrates [13]. In vascular plants, the putative reduction enzyme is a nitroreductase [44]. During reductive TNT biotransformation, ADNT products are accumulated within the plant tissue or excreted to the surrounding culture medium. For *Myriophyllum* species, the excreted ADNT products accounted for less than 20% of the initial TNT, and only trace levels of free ADNT remained in the biomass [37,38,42]. One study [40] also showed that *M. aquaticum* was capable of oxidative transformation of TNT via methyl oxidation or aromatic hydroxylation, with oxidation products accounting for nearly 36% of the TNT initially added.

The intracellular endpoints of TNT biotransformation in aquatic vascular plants per se have not been thoroughly characterized. [^{14}C]TNT studies showed that *Myriophyllum* did not mineralize TNT to CO_2 and that a significant fraction, approximately 30%, of the initial radioactivity was nonsolvent extractable and therefore was strongly associated with plant tissue [36,37]. Studies with terrestrial vascular plants have provided valuable information on TNT metabolism that likely applies to all vascular plants. For ADNT products that remained within the cell, *N*-glycosylation of ADNTs to ADNT-glyco-conjugates occurred [45,46]. Burken et al. [41] claimed that these ADNT-glyco-conjugates were sequestered within the plant vacuole or incorporated into "bound residues" within the plant cell wall. The TNT entering plant cells was incorporated into water-soluble, high molecular weight biopolymers, presumably formed by continued polymerization of the glycosyl residue on the ADNT-glyco-conjugate with other plant sugars [47]. The TNT transformation products were also incorporated into the lignin fraction of the cell wall [48].

The products of the initial biotransformation of RDX in aquatic vascular plants have not yet been identified. However, [^{14}C]RDX uptake studies with root cultures of the model terrestrial vascular plant *Catharanthus roseus* confirmed that about 25% of the initial RDX was incorporated into the intracellular bound residues. One recent study [49] provided evidence that RDX taken up by leaves of the wetland plant *P. arundinacea* was abiotically photodegraded to 4-nitro-2,4-diazabutanal. Just and Schnoor [49] further claimed that plant chloroplasts shuttled electrons to facilitate this reaction, a process they termed "phytophotolysis." One RDX biotransformation study [47] provided preliminary evidence for the formation of the nitroso product hexahydro-1-nitroso-3,5-dinitro-1,3,5 triazine (MNX) in the land plant *Cyperus esculentus* (yellow nutsedge), and so aquatic vascular plants may possess this biotransformation step as well.

At present, the ability of aquatic vascular plants to remove HMX from water has not been conclusively established. Preliminary experiments performed by Bhadra et al. [43] suggested that the vascular aquatic plant *M. aquaticum* did not take up or biotransform HMX, whereas root cultures of the terrestrial vascular plant *C. roseus* removed HMX from water inefficiently, as the removal rate was only slightly higher than those for the biomass-free and heat-killed biomass controls.

6.3.3   UPTAKE AND BIOTRANSFORMATION OF EXPLOSIVES
BY NONVASCULAR MARINE MACROPHYTES

The potential of nonvascular macrophytic marine plants (macroalgae or seaweeds) to take up and transform TNT has recently been demonstrated [12,50]. Marine macroalgae possess several enzymes known to detoxify xenobiotic compounds [51]. Macroalgae are multicellular phototrophic organisms that live in the benthic near-shore marine environments. They possess distinct anatomical features (blade, stipe, holdfast, and reproductive structures) and heteromorphic life cycles [52]. Photosynthetic tissue cultures for marine macroalgae established through plant cell and tissue culture techniques have many desirable features for studying the ability of these organisms to take up and biotransform compounds dissolved in seawater under controlled conditions. Unlike macroalgae in the natural environment, photosynthetic tissue cultures derived from marine macroalgae grown as a liquid suspension, propagate asexually, and maintain a consistent morphology during vegetative growth [53]. Furthermore, macroalgal tissue cultures are axenic. In contrast, field-collected macroalgae contain epiphytes and microscopic organisms, which also could potentially take up and biotransform TNT.

Tissue cultures from marine macroalgae are capable of taking up and metabolizing TNT dissolved in seawater [12,50]. Specific macroalgal cultures that remove TNT from seawater include a plantlet culture of the temperate red macroalga *Porphyra yezoensis* [54], a semidifferentiated shoot tissue culture (microplantlet culture) of the tropical red macroalgal *Portieria hornemannii* [55], and a filamentous tissue culture of the temperate green macroalgal *Acrosiphonia coalita* [56]. Batch TNT uptake studies were conducted within well-mixed bubbler flasks at an algal biomass suspension density of ~1 g FW L^{-1} in seawater containing TNT at initial concentrations ranging from 1 to 50 mg L^{-1}. For an initial TNT concentration of <10 mg L^{-1}, complete TNT removal from seawater was achieved within 72 h.

Removal of dissolved TNT from seawater by the aforementioned tissue cultures was a first-order process, similar to the TNT removal processes described for vascular plants. Specific first-order rate constants for TNT uptake are presented in Table 6.4. At initial TNT concentrations at or below 10 mg L^{-1}, first-order rate constants for TNT removal ranged from 0.016 to 0.062 L g^{-1} FW h^{-1} (Table 6.4), with *A. coalita* having the slowest rate and *P. hornemannii* having the fastest rate. Specific rate constants for TNT uptake by macroalgal tissue cultures were typically 5–10 times higher than those for TNT uptake by vascular plants. The difference in rates of TNT uptake between vascular plants and nonvascular macroalgae may reflect the differences in morphology and putative TNT uptake mechanism for these two types of phototrophic organisms. For the nonvascular macroalgal tissue cultures, TNT is most likely taken up directly by the submerged tissue surfaces exposed to seawater, where the dissolved TNT diffused directly into the algal thallus tissue. In contrast, for vascular aquatic plants, dissolved TNT most likely enters in the roots and then passes through the vascular elements to the stem and leaf [36]. The residual concentrations of TNT within the biomass are presented in Table 6.5. Generally, approximately 0.003 mg g^{-1} FW of TNT was found within acetonitrile extracts of the fresh macroalgal biomass, which was only 0.5%, on a molar basis, of the TNT initially

TABLE 6.4

Specific Rate Constants for TNT Removal from Seawater by Tissue Cultures of Marine Macroalgae

Culture	Temperature (°C)	Illumination (μE m^{-2}s^{-1}) Photoperiod (h light:h dark)	Initial TNT Concentration (mg L^{-1})	Specific Rate Constant for TNT Removal k (L g^{-1} FW h^{-1})
Acrosiphonia coalita filaments	13	35 (16:8)	1.13 ± 0.14	0.018 ± 0.0011
			10.5 ± 0.99	0.016 ± 0.0005
Porphyra yezoensis plantlets	15	35 (14:10)	2.54 ± 0.13	0.062 ± 0.0028
			10.6 ± 0.41	0.047 ± 0.0036
Portieria hornemannii microplantlets	22	150 (14:10)	1.03 ± 0.08	0.037 ± 0.0021
			9.69 ± 0.44	0.049 ± 0.0083

Note: Values are mean ±1 standard deviation.
Source: Data taken from Refs. 12 and 50.

TABLE 6.5

Comparison of Metabolite Levels from TNT Biotransformation by Three Tissue Cultures of Marine Macroalgae

TNT or Metabolite	Phase	*Acrosiphonia coalita* Filaments	*Porphyra yezoensis* Plantlets	*Portieria hornemannii* Microplantlets
Initial TNT conc. (mg L^{-1})	Liquid	1.13 ± 0.14	2.54 ± 0.13	1.03 ± 0.075
Biomass conc. (g FW L^{-1})	Suspension	1.10 ± 0.041	1.25 ± 0.12	1.15 ± 0.028
TNT removed (mg TNT g^{-1} FW)	Biomass	1.03 ± 0.13	2.04 ± 0.23	0.90 ± 0.069
Residual TNT conc. (mg g^{-1} FW)	Biomass	~0	0.00020 ± 0.00003	0.0031± 0.0016
Residual 2-ADNT conc. (mg g^{-1} FW)	Biomass	~0	0.00074 ± 0.00002	0.0022 ± 0.0006
Residual 4-ADNT conc. (mg g^{-1} FW)	Biomass	~0	0.00098 ± 0.00009	0.00096 ± 0.0007
Aqueous 2-ADNT conc. (mg g^{-1} FW)	Liquid	0.118 ± 0.0029	0.088 ± 0.0048	0.075 ± 0.0081
Aqueous 4-ADNT conc. (mg g^{-1} FW)	Liquid	0.034 ± 0.0046	0.061 ± 0.011	0.077 ± 0.013
Total metabolite recovery (mole percent initial TNT)	Liquid	15.5	6.77	17
	Biomass	0	0.11	0.74
	Suspension	15.5	6.88	17.7

Note: Values are mean ± 1 standard deviation.
Source: Data taken from Refs. 12 and 50.

added to the batch uptake experiment. Therefore, TNT was not simply adsorbed onto the tissue surface or absorbed into the biomass, but instead was converted to transformation products.

Illuminated control experiments containing seawater medium were conducted in parallel with the batch TNT uptake and metabolism experiments. The TNT concentration decreased over time due to transformation. The abiotic TNT oxidation product 1,3,5-trinitrobenzene was found in the seawater liquid medium, consistent with previous studies for the photolytic degradation of TNT in water [57]. The apparent first-order rate constants for abiotic degradation of TNT in seawater under ~150 µE m^{-2} s^{-1} light intensity were 0.0023–0.0028 h^{-1} at initial TNT concentrations ranging from 1.0 to 50 mg L^{-1}, which were 15–25 times smaller than the apparent first-order rate constants for TNT removal in the uptake and metabolism experiments using 1 g of biomass.

All three species of marine macroalgae investigated biologically reduced TNT to 2-ADNT, 4-ADNT, or both. Typical amounts of 2-ADNT and 4-ADNT found in the biomass and in the liquid medium at the conclusion of the batch uptake experiment are presented in Table 6.5. Both 2-ADNT and 4- ADNT were detected in the liquid culture medium of the entire batch TNT uptake and metabolism experiments, suggesting that these products were produced within the seaweed tissue and then excreted. At initial TNT concentrations of ~1 mg L^{-1}, the final concentrations of both 2-ADNT and 4-ADNT in the liquid medium were below 0.1 mg L^{-1}. The time course for the excretion of ADNTs to the liquid paralleled the time course for TNT removal from the seawater.

The total metabolite recovery shown in Table 6.5 was based on the molar amount of ADNT products measured divided by the molar amount of TNT initially added to the batch uptake experiment. Although 100% removal of TNT from the liquid culture medium was observed, the total molar-equivalents amounts of free TNT transformation products 2-ADNT and 4-ADNT in both the biomass and the liquid medium never represented more than 20% of the TNT initially added to the liquid medium, consistent with studies of TNT biotransformation by vascular plants. Only trace levels of ADNTs were detected within the biomass, all near or below 0.003 mg g^{-1} FW (<0.4 mole percent recovery, based on initial TNT), suggesting that these transformation products were further transformed to nonidentified products, such as the ADNT-glyco-conjugates described Section 6.3.1. Further studies with [14C]- or [15N]-labeled TNT are needed to accurately elucidate the endpoints of TNT biotransformation in marine macroalgae. The uptake and biotransformation of RDX, HMX, and other energetic compounds in marine macroalgae have not been reported and their study is therefore encouraged.

6.3.4   UPTAKE AND BIOTRANSFORMATION OF EXPLOSIVES BY NONVASCULAR FRESHWATER MACROPHYTIC ALGAE

The freshwater macrophytic alga *Nitella* spp. (stonewort) is capable of removing TNT from water [35]. The specific first-order rate constant for TNT uptake by this organism is 2.94 × 10^{-3} L g^{-1} FW h^{-1}, which is comparable to vascular freshwater aquatic plants of genus *Myriophyllum*, but is still at least five times lower than nonvascular macrophytic marine algae. After 7 h of TNT exposure, the transient products of TNT transformation by *Nitella* spp. identified in the aqueous phase were a mixture ADNTs and DANTs. However, all of these compounds disappeared after 48 h of TNT exposure.

6.4 CONCLUSIONS AND RECOMMENDATIONS

Bioconcentration and bioaccumulation studies have been conducted with a variety of environmentally significant explosives and relevant compounds, but information concerning other compounds like trinitrophenylmethylnitramine (tetryl) and trinitrophenol (picric acid) were not found in the literature. Available experimental data demonstrate that explosives and related compounds have low potential to bioconcentrate in aquatic organisms as expected given their weak hydrophobicity. Moreover, high biotransformation rates in aquatic organisms likely explain the observed bioconcentration of most nitroaromatic compounds at levels substantially lower than those predicted by their hydrophobicity. Spatially limited exposure to low chemical concentrations is the likely exposure scenario for biological receptors potentially at risk at underwater explosives-contaminated sites. For example, underwater unexploded ordnance (UXO) casings eventually breach and slowly release constituent compounds through mechanical stress, corrosion, and low-level remedial detonations. Several explosive and related compounds are efficiently eliminated from organisms upon removal from exposure. Therefore, mobile organisms inhabiting UXO-contaminated areas are unlikely to bioaccumulate common explosive compounds and their degradation products at high enough tissue concentrations and for long enough periods to result in deleterious biological effects. Direct uptake from water is expected to be the primary route of exposure of explosives in higher tropic level animals, as dietary uptake of TNT and RDX was substantially less than aqueous uptake.

Uptake of TNT in fish and invertebrates resulted in substantial bioaccumulation of nonidentified extractable and nonextractable compounds. Contrasting to the parent compound, metabolically formed transformation products of TNT appear to be eliminated at much slower rates. Neither the biological half-life nor the chemical nature of nonextractable transformation products of TNT in organisms has been investigated to date. Investigations on the biotransformation of explosives other than TNT in aquatic animals were not found in the available literature and are therefore warranted. Further studies of the fate of explosives in aquatic animals are necessary to reveal the identity of their transformation products present in the tissues of exposed organisms, to further characterize species-specific differences in the bioconcentration of transformation products, and to elucidate the mechanism of toxic action.

Phototrophic organisms take up and biotransform TNT. It is well established that aquatic vascular plants remove TNT from terrestrial waters. Recent studies show that nonvascular marine macroalgae (seaweeds) remove TNT from seawater. All phototrophic organisms initially reduce the nitro groups on TNT to form 2-ADNT and 4-ADNT. Some of the ADNTs are secreted back to the surrounding liquid milieu. The biochemical fate of the intracellular ADNTs is unknown, but it seems that conjugation of

ADNTs with sugars, followed by their sequestration into the cell vacuole or incorporation into bound residues with the cell wall are likely endpoints. The explosives RDX and HMX are also taken up by aquatic vascular plants, but their initial biotransformation products and endpoints are not known at this time. Fundamental studies on RDX and HMX metabolism by aquatic vascular plants and nonvascular macroalgae are needed.

REFERENCES

1. Talmage SS et al., Nitroaromatic munition compounds: Environmental effects and screening values, *Rev. Environ. Contam. Toxicol.*, 161, 1, 1999.
2. Yoo JL et al., Toxicity and bioaccumulation of TNT in fathead minnow (*Pimephales promelas*), *Environ. Toxicol. Chem.*, 25, 3253, 2006.
3. Lotufo GR and Lydy MJ, Comparative toxicokinetics of explosive compounds in sheepshead minnows, *Arch. Environ. Contam. Toxicol.*, 49, 206, 2005.
4. Lang PZ et al., Bioconcentration, elimination and metabolism of 2,4-dinitrotoluene in carps (*Cyprinus carpio* L.), *Chemosphere*, 35, 1799, 1997.
5. Wang Y et al., Uptake of weakly hydrophobic nitroaromatics from water by semipermeable membrane devices (SPMDs) and by goldfish (*Carassius auratus*), *Chemosphere*, 38, 51, 1999.
6. Conder JM, La Point TW, and Bowen AT, Preliminary kinetics and metabolism of 2,4,6-trinitrotoluene and its reduced metabolites in an aquatic oligochaete, *Aquat. Toxicol.*, 69, 199, 2004.
7. Belden JB et al., Accumulation of trinitrotoluene (TNT) in aquatic organisms: Part 2—Bioconcentration in aquatic invertebrates and potential for trophic transfer to channel catfish (*Ictalurus punctatus*), *Chemosphere*, 58, 1161, 2005.
8. Belden JB, Lotufo GR, and Lydy MJ, Accumulation of RDX in channel catfish (*Ictalurus punctatus*) and aquatic oligochaetes (*Lumbriculus variegatus*), *Environ. Toxicol. Chem.*, 24, 1962, 2005.
9. Ownby DR et al., Accumulation of trinitrotoluene (TNT) in aquatic organisms: Part 1—Bioconcentration and distribution in channel catfish (*Ictalurus punctatus*), *Chemosphere*, 58, 1153, 2005.
10. Houston JG and Lotufo GR, Dietary exposure of fathead minnows to the explosives TNT and RDX and to the pesticide DDT using contaminated invertebrates, *Int. J. Environ. Res. Public Health*, 2, 286, 2005.
11. Medina VF et al., Phyto-removal of trinitrotoluene from water with batch kinetic studies, *Water Res.*, 34, 2713, 2000.
12. Cruz-Uribe O and Rorrer GL, Uptake and biotransformation of 2,4,6-trinitrotoluene (TNT) by microplantlet suspension culture of the marine red macroalga *Portieria hornemannii*, *Biotechnol. Bioeng.*, 93, 401, 2006.
13. Hannink NK, Rosser SJ, and Bruce NC, Phytoremediation of explosives, *Crit. Rev. Plant. Sci.*, 21, 511, 2002.
14. Medina VF et al., Treatment of munitions in soils using phytoslurries, *Int. J. Phytoremediat.*, 4, 143, 2002.
15. Meylan WM et al., Improved method for estimating bioconcentration/bioaccumulation factor from octanol/water partition coefficient, *Environ. Toxicol. Chem.*, 18, 664, 1999.
16. Freitag D, Lay JP, and Korte F, Environmental hazard profile—Test results as related to structures and translation into the environment, in *QSAR in Environmental Toxicology: Proceedings of the Workshop on Quantitative Structure-Activity Relationships (QSAR) in Environmental Toxicology*, Kaiser KLE, Ed., D. Reidel Publishing Company, Dordrecht, Holland, 1984, 111.

17. Liu DH et al., Toxicity of TNT wastewaters to aquatic organisms, vol. I. Acute toxicity of LAP wastewater and 2,4,6-trinitrotoluene, Report AD-A142144, SRI Int., Menlo Park, CA, 1983.

18. Liu DH, Bailey HC, and Pearson JG. Toxicity of a complex munitions wastewater to aquatic organism, in *Aquatic Toxicology and Hazard Assessment: Sixth Symposium*, ASTM STP 802, Bishop WE, Cardwell RD, and Heidolph BB, Eds., American Society for Testing and Materials, Philadelphia, PA, 1983, 135.

19. Bentley RE et al., Laboratory evaluation of the toxicity of cyclotrimethylene trinitramine (RDX) to aquatic organisms, US Army Medical Research and Development Command, Fort Detrick, MD, 1977.

20. Arnot JA and Gobas FAPC, A food web bioaccumulation model for organic chemicals in aquatic ecosystems, *Environ. Toxicol. Chem.*, 23, 2343, 2004.

21. Hellou J, Mackay D, and Banoub JH, Dietary and aqueous exposure of finfish to organochlorine compounds—A case study, *Arch. Environ. Contam. Toxicol.*, 34, 280, 1998.

22. Gutjahr-Gobell RE et al., Feeding the mummichog (*Fundulus heteroclitus*) a diet spiked with non-ortho- and mono-ortho-substituted polychlorinated biphenyls: Accumulation and effects, *Environ. Toxicol. Chem.*, 18, 699, 1999.

23. Johnson MS et al., Bioaccumulation of 2,4,6-trinitrotoluene and polychlorinated biphenyls through two routes of exposure in a terrestrial amphibian: Is the dermal route significant?, *Environ. Toxicol. Chem.*, 18, 873, 1999.

24. Tarr BD, Barron MG, and Hayton WL, Effect of body size on the uptake and bioconcentration of di-2-ethylhexyl phthalate in rainbow trout, *Environ. Toxicol. Chem.*, 9, 989, 1990.

25. Lachance B et al., Toxicity and bioaccumulation of reduced TNT metabolites in the earthworm *Eisenia andrei* exposed to amended forest soil, *Chemosphere*, 55, 1339, 2004.

26. Rosen G and Lotufo GR, Toxicity and fate of two munitions constituents in spiked sediment exposures with the marine amphipod *Eohaustorius estuarius*, *Environ. Toxicol. Chem.*, 24, 2887, 2005.

27. Dodard SG, Powlowski J, and Sunahara GI, Biotransformation of 2,4,6-trinitrotoluene (TNT) by enchytraeids (*Enchytraeus albidus*) in vivo and in vitro, *Environ. Poll.*, 131, 263, 2004.

28. Liu YY et al., In vivo covalent binding of [^{14}C]trinitrotoluene to proteins in the rat, *Chem. Biol. Interact.*, 82, 1, 1992.

29. Leung KH et al., Mechanism of bioactivation and covalent binding of 2,4,6-trinitrotoluene, *Chem. Biol. Interact.*, 97, 37, 1995.

30. Bakhtiar R et al., Evidence for a novel heme adduct generated by the in vitro reaction of 2,4,6-trinitrotoluene with human hemoglobin using electrospray ionization mass spectrometry, *J. Inorg. Biochem.*, 68, 273, 1997.

31. Bowen AT, Conder JM, and La Point TW, Solid phase microextraction of aminodinitrotoluenes in tissue, *Chemosphere*, 63, 58, 2006.

32. Ek H et al., Acute effects of 2,4,6-trinitrotoluene (TNT) on haematology parameters and hepatic EROD-activity in rainbow trout (*Oncorhynchus mykiss*), *Aquat. Ecosystem Health Manage.*, 6, 415, 2003.

33. Ek H et al., Tentative biomarkers for 2,4,6-trinitrotoluene (TNT) in fish (*Oncorhynchus mykiss*), *Aquat. Toxicol.*, 72, 221, 2005.

34. Jacobson ME et al., Transformation kinetics of trinitrotoluene conversion in aquatic plants, in *Phytoremediation: Transformation and Control of Contaminants*, McCutcheon SC and Schnoor JL, Eds., John Wiley & Sons, Hoboken, NJ, 2003, 409.

35. McCutcheon SC, Medina VF, and Larson SL, Proof of phytoremediation for explosives in water, in *Phytoremediation: Transformation and Control of Contaminants*, McCutcheon SC and Schnoor JL, Eds., John Wiley & Sons, Hoboken, NJ, 2003, 429.

36. Vanderford M, Shanks JV, and Hughes JB, Phytotransformation of trinitrotoluene (TNT) and distribution of metabolic products in *Myriophyllum aquaticum*, *Biotechnol. Lett.*, 19, 277, 1997.

37. Hughes JB et al., Transformation of TNT by aquatic plants and plant tissue cultures, *Environ. Sci. Technol.*, 31, 266, 1997.

38. Pavlostathis SG et al., Transformation of 2,4,6-trinitrotoluene by the aquatic plant *Myriophyllum spicatum*, *Environ. Toxicol. Chem.*, 17, 2266, 1998.

39. Best EPH et al., Environmental behavior of explosives in groundwater from the Milan army ammunition plant in aquatic and wetland plant treatments. Removal, mass balances and fate in groundwater of TNT and RDX, *Chemosphere*, 38, 3383, 1999.

40. Bhadra R et al., Characterization of oxidation products of TNT metabolism in aquatic phytoremediation systems of *Myriophyllum aquaticum*, *Environ. Sci. Technol.*, 33, 3354, 1999.

41. Burken JG, Shanks JV, and Thompson PL, Phytoremediation and plant metabolism of explosives and nitroaromatic compounds, in *Biodegradation of Nitroaromatic Compounds and Explosives*, Spain JC, Hughes JB, and Knackmuss H-J, Eds., Boca Raton, FL, Lewis Publishers, 2000, 239.

42. Wang CY et al., Role of hydroxylamine intermediates in the phytotransformation of 2,4,6-trinitrotoluene by *Myriophyllum aquaticum*, *Environ. Sci. Technol.*, 37, 3595, 2003.

43. Bhadra R et al., Studies on plant-mediated fate of the explosives RDX and HMX, *Chemosphere*, 44, 1259, 2001.

44. Hannink N et al., Phytodetoxification of TNT by transgenic plants expressing a bacterial nitroreductase, *Nat. Biotechnol.*, 19, 1168, 2001.

45. Bhadra R et al., Confirmation of conjugation processes during TNT metabolism by axenic plant roots, *Environ. Sci. Technol.*, 33, 446, 1999.

46. Vila M et al., Metabolism of [C-14]-2,4,6-trinitrotoluene in tobacco cell suspension cultures, *Environ. Sci. Technol.*, 39, 663, 2005.

47. Larson SL et al., Classification of explosives transformation products in plant tissue, *Environ. Toxicol. Chem.*, 18, 1270, 1999.

48. Sens C, Scheidemann P, and Werner D, The distribution of C-14-TNT in different biochemical compartments of the monocotyledonous *Triticum aestivum*, *Environ. Poll.*, 104, 113, 1999.

49. Just CL and Schnoor JL, Phytophotolysis of hexahydro-1,3,5-trinitro-1,3,5-triazine (RDX) in leaves of reed canary grass, *Environ. Sci. Technol.*, 38, 290, 2004.

50. Cruz-Uribe O, Cheney DP, and Rorrer GL, Comparison of TNT removal from seawater by tissue cultures of three marine macroalgae, *Chemosphere*, 67, 1469, 2007.

51. Pflugmacher S, Wienke C, and Sandermann H, Activity of phase I and phase II detoxication enzymes in Antarctic and Arctic macroalgae, *Mar. Environ. Res.*, 48, 23, 1999.

52. Kain JM and Norton TA, Marine ecology, in *Biology of the Red Algae*, Cole KM and Sheath RG, Eds., Cambridge University Press, Cambridge, UK, 1990, 377.

53. Rorrer GL, Cell and tissue cultures of marine seaweeds, in *Encyclopedia of Cell Technology*, Spier RE, Ed., New York, John Wiley & Sons, 2000, 1105.

54. Waaland JR, Stiller JW, and Cheney DP, Macroalgal candidates for genomics, *J. Phycol.*, 40, 26, 2004.

55. Barahona LF and Rorrer GL, Isolation of halogenated monoterpenes from bioreactor-cultured microplantlets of the macrophytic red algae *Ochtodes secundiramea* and *Portieria hornemannii*, *J. Nat. Prod.*, 66, 743, 2003.

56. Rorrer GL, Polne-Fuller M, and Zhi C, Development and bioreactor cultivation of a novel semi-differentiated tissue suspension derived from the marine plant *Acrosiphonia coalita*, *Biotechnol. Bioeng.* 49, 559, 1996.

57. Schmelling DC and Gray KA, Photocatalytic transformation and mineralization of 2,4,6-trinitrotoluene (TNT) in TiO_2 slurries, *Water Res.*, 29, 2651, 1995.

58. Hansch C, Leo A, and Hoekman D, *Exploring QSAR, Hydrophobic, Electronic, and Steric Constants,* ACS Professional Reference Book, American Chemical Society, Washington, DC, 1995.

59. Elovitz MS and Weber EJ, Sediment mediated reduction of 2,4,6-trinitrotoluene and fate of the resulting aromatic (poly)amines, *Environ. Sci. Technol.*, 33, 2617, 1999.

60. Banerjee S, Yalkowsky SH, and Valvani SC, Water solubility and octanol/water partition coefficients of organics. Limitations of the solubility-partition coefficient correlation, *Environ. Sci. Technol.*, 14, 1227, 1980.

61. Yoon JM et al., Uptake and leaching of octahydro-1,3,5,7-tetranitro-1,3,5,7-tetrazocine by hybrid poplar trees, *Environ. Sci. Technol.*, 36, 4649, 2002.

7 Toxicity of Energetic Compounds to Wildlife Species

Mark S. Johnson and Christopher J. Salice

CONTENTS

7.1 Introduction 157
7.2 Effects of Energetic Compounds in Mammals.. 158
7.3 Effects of Energetic Compounds in Birds 163
7.4 Effects of Other Energetic Compounds in Birds.. 166
7.5 Effects of Energetic Compounds in Reptiles... 168
7.6 Effects of Energetic Compounds in Amphibians 169
7.7 Summary 171
Acknowledgments.. 172
References... 172

7.1 INTRODUCTION

Energetics is a class of compounds that includes explosives, pyrotechnics, and propellants. The predominant manufacturing and use of these substances is specific to the military, particularly in regard to training. Consequently, environmental contamination from the use of energetic compounds within the U.S. Department of Defense is widespread. Causes include practices associated with manufacturing, loading, and packing operations; testing/firing ranges; and demilitarization of outdated supplies. Explosives, such as 2,4,6-trinitrotoluene (TNT) and 1,3,5-trinitrohexahydro-1,3,5-triazine (RDX), are considerably more prevalent in environmental media given the nature of their use and source for contamination (e.g., through unexploded ordnance and production).

Depending on the activity and mission, areas containing these substances in soil can be extensive, as some artillery ranges are several square miles in area. However, the residues found in these areas are very heterogeneous in distribution [1]. Human access to these areas is often highly restricted, making them relatively attractive to many wildlife species. This relative "attractiveness" is heightened through human encroachment in areas surrounding military installations. Nevertheless, as environmental stewards of these lands, the military has supported research in understanding the potential for adverse effects from exposure to these substances.

Much of the interest in the effects of exposure to energetic compounds was initially fostered by human health concerns resulting from reports of adverse workplace effects and the potential for exposure by military personnel [2]. This interest was directed toward understanding these effects through controlled laboratory toxicity investigations using rodents and other domestic laboratory animal species (see Agency for Toxic Substance Disease Registry [ATSDR] for a review [3]). Because the present review involves the effects on wildlife, toxicity results from the use of domestic laboratory animals intended for human health applications will not be reviewed in this chapter. The reader is guided to Chapter 8 for a description of the genotoxic effects of explosives. Wildlife is defined here as animal species of the phylum Chordata within the classes Mammalia, Aves, Amphibia, or Reptilia, or those species generally considered as terrestrial vertebrates.

Only circumstantial evidence is available that supports the supposition that wildlife species are exposed to explosive compounds. Studies conducted at U.S. Army ammunition plants and other areas of known soil contamination have failed to detect body burdens of suspected explosive compounds in mice, deer, and some bird species [4–7] (see Chapter 10 in this book for a more complete review). Given the relatively rapid metabolic potential of many explosives in vivo, the heterogeneous distribution of these substances in the environment, and the potential for bioaccumulation of some nitramines in plants, body burden analysis may not adequately describe exposure potential. Therefore, the data reviewed in this chapter will focus on controlled laboratory toxicity studies conducted to evaluate the effects in wildlife species, many of which were designed for specific risk assessment applications.

Primary nitroaromatics are: TNT; the environmental breakdown products, including 1,3,5-trinitrobenzene (TNB), 1,3-dinitrobenzene (DNB), 2,4-and 2,6-dinitrotoluene (DNTs); and the primary reduction products 2-amino 4,6-dinitrotoluene and 4-amino 2,6-dinitrotoluene (ADNTs). Nitramines include RDX and octahydro-1,3,5,7-tetranitro-1,3,5,7-tetrazocine (HMX). Additional energetic compounds discussed in this chapter include nitroglycerin, white phosphorus, and ammonium perchlorate. Other energetic compounds are not discussed due to a lack of information regarding toxicity to wildlife species.

7.2   EFFECTS OF ENERGETIC COMPOUNDS IN MAMMALS

In mammals, short- and long-term oral exposures to nitroaromatics have been reported to cause hematological effects (e.g., transient methemoglobinemia), complications resulting from red blood cell lysis, and, at high levels, neurological effects [8–11]. Other associated targets include the main blood conditioning organs such as the liver, kidneys and spleen, and the testes causing tissue degeneration and sperm abnormalities [8,11,12]. Mechanisms for these effects are not clearly understood, though hypotheses include: the generation of radical oxygen species during reduction of the nitro subgroups; inhibition of α-aminolevulinic acid dehydratase activity; formation of Heinz bodies; and mitochondrial damage [13–15]. The aromatic subgroup may be involved in manifestation of neuromuscular effects [16].

Although not a direct indicator of physiological damage, chromaturia or red-orange-colored urine is often coincident with exposures associated with adverse effects for many nitroaromatics, especially TNT [17–20]. The specific metabolites responsible for this coloration are not well understood, though it has been postulated that the coloration was due to the formation of 2,4,6-trinitrobenzyl alcohol (TNB alcohol) through oxidation of the methyl group [11]. However, this coloration is also consistent with the formation of "pink water" found in surface water associated with TNT manufacturing activities [21] and may be due to photolytic abiotic processes.

Generally, wild mammals show more variability in response to oral TNT exposures than laboratory-bred animals. Laboratory studies conducted with wild cotton rats (*Sigmodon hispidus*) resulted in oral TNT LD_{50} (median lethal dose) estimates of 607 and 767 mg TNT kg^{-1} body weight (bw) for males and females, respectively (mean for both sexes = 683 mg kg^{-1}; 95% CI = 477–978 mg kg^{-1}) [22]. Animals exhibited an increased respiratory rate within 90 min after dosing. Orange-colored urine (chromaturia) and urinary bladder distention were observed in all animals at necropsy. These results suggest that the oral lethal dose for cotton rats is lower than for Sprague-Dawley rats (LD_{50} = 1320 and 795 for males and females, respectively [8]). Seven-day repeated dose treatments of 1/8, 1/4, and 1/2 the LD_{50} for male (75.9, 151.8, and 303.5 mg TNT kg^{-1} bw d^{-1}) and female cotton rats (96, 192, and 384 mg TNT kg^{-1} bw d^{-1}) were conducted using a corn oil vehicle [22]. This study reported evidence of hemolytic anemia (i.e., decreased red blood cell counts, hematocrit, and hemoglobin) in addition to leucocytosis and other compensatory responses to red blood cell loss including increased numbers of reticulocytes and variation in color and size of red blood cells (polychromasia and anisocytosis, respectively). Rats exposed to the lowest dose (75.9 and 96 mg kg^{-1} d^{-1} for males and females, respectively) were reported to show histological evidence of extravascular hemolytic anemia through observations of hemosiderin-laden macrophages, reticulocytes, and indications of extramedullary hematopoiesis (EMH) [22]. Consistent with the accumulation of lytic red blood cells, splenic weights were increased in females dosed with 192 and the 384 mg kg^{-1} d^{-1}, and liver weights were increased in males dosed with 151.8 and 303.5 mg kg^{-1} d^{-1}. Testicular lesions were subtle and occurred at higher dose levels (152 and 304 mg kg^{-1} d^{-1}). These lesions were characterized as dilated tubules with detached spermatozoa and spermatids free in the lumen [22].

Results of three feeding studies using white-footed mice suggest that Nearctic mice may be more resistant to oral exposures of TNT than Palearctic (Old World) species (e.g., *Mus* sp.). Two 14-d feeding studies using TNT were conducted at levels approaching and exceeding the LD_{50} in laboratory mice (*Mus* sp.) yet produced no mortalities in white-footed mice (*Peromyscus leucopus*) [19,20]. White-footed mice (10 per group per sex) were exposed to one of five treatments of 0, 170, 330, 660, and 2640 ppm TNT in feed for 14 d [19]. These concentrations were determined to approximate daily exposures of 0, 41, 82, 165, and 660 mg kg^{-1} bw d^{-1} for both sexes. Limited health criteria were evaluated (organ to body weight ratios of liver, kidney, spleen, gonads; histopathological suite) and the induction of bone marrow micronuclei was investigated. Only the liver showed increased weight relative to controls in the high dose group. No other changes in organ or body mass were reported. There were no dose-related findings with regard to the presence of micronuclei [19].

This study was repeated in white-footed mice using higher levels of TNT, which were 0, 420, 830, 1650, and 3300 ppm TNT in feed for 14 d [20]. Daily exposures were calculated to be equivalent to 66, 145, 275, and 602 mg TNT kg^{-1} bw d^{-1} for males and 70, 142, 283, and 550 mg kg^{-1} d^{-1} for females for the 420, 830, 1650, and 3300 ppm TNT, respectively. Levels of radical oxygen intermediate production (H_2O_2) were impaired in isolated splenic preparations at levels exceeding 275 mg $kg^{-1} d^{-1}$ [20]. Males and females also showed signs of hemolytic anemia including splenic congestion and EMH being most severe and consistent in animals from the highest exposure group. Chromaturia was also most consistently evident in the high dose group. No blood-specific parameters were measured.

McCain and Ferguson [19] exposed 100 male and female white-footed mice to concentrations of 0, 660, 1320, and 2640 ppm TNT in feed for 90 d. The calculated oral dose was 0, 165, 330, and 660 mg $kg^{-1} d^{-1}$, respectively. The highest concentration used in this study (2640 ppm; 660 mg $kg^{-1} d^{-1}$) was equivalent to the LD_{50} of 660 mg kg^{-1} reported by Dilley et al. [8] in *Mus musculus*, yet there was no mortality during the study. Initial animal body mass reduction consistent with reduced feed intake was reported, yet all groups gained body mass over time. McCain and Ferguson [19] found only exposures to 330 and 660 mg $kg^{-1} d^{-1}$ were associated with adverse physiological changes (e.g., organ weight differences), and reported an no observable adverse effect level (NOAEL) of 660 ppm (165 mg $kg^{-1} d^{-1}$).

A 14-d feeding study was used to investigate the effects of TNB exposures on least shrews (*Cryptotis parva*) [23]. Ten animals per sex per treatment were exposed to cat chow containing 0, 5, 10, 20, and 40 ppm of 1,3,5-TNB [23]. The calculated oral dose estimates were 0, 11, 22, 38, and 98 mg $kg^{-1} d^{-1}$ for males and 0, 11, 22, 45, and 99 mg $kg^{-1} d^{-1}$ for females. Food and water consumption, body weight, and organ weights were measured, and hematological and histopathological changes were evaluated. There were no significant differences in food and water consumption between control and TNB-exposed shrews. Data indicating erythrolysis were mixed generally due to variability between and within treatments. Males, but not females, had significantly lower red blood cell count and mean corpuscular hemoglobin indices at the high dose treatment only (40 ppm, 98 mg $kg^{-1} d^{-1}$) compared with controls. In addition, there was a positive trend in splenic weight in females, but not in males relative to treatment. Given the lack of clear dose–response relationships and the variation between treatments, the biological significance of these results is unclear and suggests that shrews may be tolerant to TNB under these conditions.

The subchronic effects of TNB were evaluated in white-footed mice in a 90-d feeding study [24]. Mice (10 animals per sex per treatment) were exposed to a 0, 150, 375, or 750 ppm diet or 0, 24, 67, and 114 mg $kg^{-1} d^{-1}$ in males and 0, 20, 65, and 108 mg $kg^{-1} d^{-1}$ in females (calculated). Data collection included weekly body mass and food consumption, standard blood indices, organ weights, and histopathology. There was a general lack of clear dose–response relationships in many parameters. Spleen to body weight values were higher in high-dose males but not females, and kidney to body weight values were higher only for the 375 ppm female group. White blood cell counts were higher for high-dose males, and reticulocyte percentages were higher for the 150 and 750 ppm group only compared with controls [24]. Hematology results for females were equivocal. The only histopathological effects attributed to treatment were erythroid cell hyperplasia of the spleen and seminiferous tubule degeneration in the testes of males. The incidence of these findings was low but was not evident in the control group. Given these data, the authors attribute biologically significant adverse effects to only the high-dose (750 ppm) treatment. The authors reported a NOAEL of 20 mg kg^{-1} d^{-1} for females and 24 mg kg^{-1} d^{-1} for males. These values differ from the NOAEL of 67.4 mg kg^{-1} d^{-1} that was chosen by Talmage et al. [21] from data on the TNB-induced testicular effects observed in high-dose animals in the same study.

Two 14-d feeding studies were conducted separately with 2,4- or 2,6-DNT in white-footed mice to determine the practical use of this species as an environmental sentinel. Ferguson [17] exposed 100 mice (10 animals per sex per group) to 0%, 6.25%, 12.5%, 25%, and 50% of the oral LD$_{50}$ for 2,4-DNT in *Peromyscus leucopus*, which calculated to target feed concentrations of 0, 493, 988, 1975, and 3950 ppm DNT in feed, based on a 4 g per day ingestion rate. This resulted in calculated exposures of 0, 74, 158, 286, and 512 mg DNT kg bw^{-1} d^{-1} for males and 0, 75, 149, 303, and 499 mg DNT kg^{-1} bw d^{-1} for females in the 0, 493, 988, 1975, and 3950 ppm DNT in feed, respectively. Food consumption, body mass, organ to body mass ratios, and general daily observations were recorded. Although there were three deaths during the study, none was considered to be a result of 2,4-DNT exposure. Males, but not females, had a decrease in feed consumption in the 1975 and 3950 ppm groups; however, both sexes had normal body masses, which ranged from 21 to 33 g. Significant increases in the liver to body weight ratios were observed in the two high exposure groups of males (1975 and 3950 ppm) and three high exposure groups of females (ranging from 988 to 3950 ppm for the females). The highest exposure group of males had significantly higher spleen to body weight ratios. There was a decrease in testes weight in the high exposure group only. None of these observations was supported by positive histological observations. Chromaturia was observed more often in males than females and restricted to the high exposure groups [17]. It was concluded that white-footed mice were more resistant to the effects of 2,4-DNT exposure than *M. musculus* and attributed this to a higher metabolic rate and faster food transit times in white-footed mice.

Ferguson [18] conducted a feeding study in white-footed mice using 2,6-DNT under the same experimental conditions. Nominal dietary concentrations were 0, 388, 776, 1552, and 3105 ppm that calculated to 0, 44, 103, 238, and 342 mg kg^{-1} bw d^{-1} for males and 0, 44, 103, 231, and 343 mg kg^{-1} bw d^{-1} for females, respectively. There were no deaths attributed to 2,6-DNT exposure. All animals maintained a normal range in body mass (19 to 23 g), yet there were variations in feed consumption in the initial stages of exposure [18]. Liver to body weight ratios were increased in males and females in the two high exposure groups (1552 and 3105 ppm) and females only in the 776 ppm group. Adrenal weights were lower for males, but not females in the two high exposure groups. No adverse histological lesions were found to be associated with treatment. Chromaturia and alopecia was observed in the high exposure groups (3105 ppm) only on day 14 of the study. The author reports a NOAEL of 103 mg kg^{-1} d^{-1} for males and 44 mg kg^{-1} d^{-1} for females, yet these endpoints are of uncertain biological and ecological significance.

No studies were found that evaluated the toxicity of 1,3-DNB or any of the primary reduction products of TNT (2-ADNT and 4-ADNT) in mammalian wildlife species; however, data for laboratory rodents for 1,3-DNB suggest similar etiology with other similar nitroaromatic compounds [25,26].

Effects of nitramines in wild mammal species are poorly understood; however, there is a robust database of common laboratory mammal tests from which some generalizations can be made. The primary targets in mammals from oral RDX exposure primarily include the central nervous, hepatic, and the hematological systems. RDX is generally considered the most toxic of the explosives to mammals, with LD_{50} values ranging from 59 to 200 mg kg^{-1}. Symptoms coincident with death include convulsions, tremors, and hyperactivity [27–29].

Ferguson and McCain [30] evaluated the effects of RDX exposure in feed for 14 d in white-footed mice using the methods previously described. Dietary levels were based on 0%, 6.25%, 12.5%, 25%, and 50% of the reported LD_{50} values for *Mus musculus* (LD_{50} = 118 mg kg^{-1}) [31] assuming an average food intake rate of 4 g d^{-1}. This yielded dietary concentrations of 0, 50, 100, 200, and 400 ppm RDX in feed. Actual intake was estimated at 0, 8, 16, 31, and 59 mg kg^{-1} d^{-1} for males and 0, 8, 15, 32, and 68 mg kg^{-1} d^{-1} for females. There were no deaths and all animals maintained normal body mass (range from 17.7 to 22.2 g for both sexes). Spleen weights were elevated in the two high exposure groups (200 and 400 ppm) and liver weights were elevated for the 50 and 200 ppm groups in females only. Ovary weights were increased in all RDX treatments except the 100 ppm group. Histopathological examination of these and other tissues found no adverse effects attributable to treatment. These results suggest that white-footed mice are more tolerant of oral RDX exposures than laboratory mice; however, the biological significance of the increased organ weights and lack of adverse histological findings associated with these organs is of uncertain biological significance.

Toxicity from oral exposures to HMX is generally low in mammals. Rodent LD_{50} values range from 2 to 7.6 g kg^{-1}. The low toxicity of HMX is likely due to its low gastrointestinal absorption in monogastric species. Metabolic studies conducted with [^{14}C]HMX in rats compared levels of radioactivity in the urine following intravenous versus oral administration [32]. These data suggest that less than 5% of the oral dose of the compound crosses the gastrointestinal absorption barrier. This result is consistent with the low oral lethality reported by others [33]. Additionally, very little tissue deposition of HMX was found in the liver, kidney, and brain following oral administration [32].

However, there are questions regarding the gastrointestinal absorption of HMX in ruminants and hind-gut fermenting species. The approximate lethal dose in rabbits of the New Zealand strain (*Oryctologus cuniculus*) was evaluated where one animal of each sex was exposed to a single oral dose of 50, 100, 250, 428.5, 1000, or 2000 mg HMX kg^{-1} bw [33]. All females and males exposed to ≥250 mg kg^{-1} died or were moribund following the administration of compound. Major clinical signs included hyperkinesia, hypokinesia, clonic convulsions, cyanosis, and mydriasis. Postmortem observations suggested systemic exposure. Although few animals were used, mortality occurred in rabbits at levels far below that which have caused death in rodent species yet are consistent with intravenous rodent LD_{50} data [33]. No studies conducted with HMX using mammalian wildlife species were found.

Effects of oral white phosphorus exposures in laboratory mammals suggest that the liver, bone, and uterus are the targets of toxicity. Hepatic fat deposition has been described in rats from single exposures [34,35], possibly as a result of phosphorus-induced protein synthesis inhibition. A limited subacute study [36] reported changes in bone resorption at 0.065 mg rat^{-1} d^{-1}. Two one-generation reproductive studies conducted in rats suggest that oral phosphorus exposures can affect parturition dam mortality [37,38]. Precise etiology was unknown, but trends in pup survival [37] and severe liver necroses were suggested as influences. NOAELs and LOAELs were reported as 0.015 and 0.075 mg kg^{-1} d^{-1}, respectively. No data for mammalian wildlife species were found.

7.3 EFFECTS OF ENERGETIC COMPOUNDS IN BIRDS

Birds respond differently than mammals to oral nitroaromatic and nitramine exposure. Although hematological parameters are often affected by high exposures, the resiliency of the avian hematopoietic system and the refractory nature of the nucleated red blood cells suggest that the blood is not a sensitive target. Explosive compounds appear to predominantly affect the central nervous system (CNS) and cause neuromuscular effects (e.g., convulsions, lethargy, and tremors). The moiety as well as the potential mode of action responsible for these effects is largely unknown.

The approximate lethal dose (ALD) of oral TNT exposure using northern bobwhite (*Colinus virginianus*) was determined in a set of experiments using one individual of each sex at eight different oral exposures ranging from 263 to 4508 mg kg^{-1} [39]. Birds were fasted overnight prior to exposure, and distilled water was used as a vehicle. Birds receiving more than 2000 mg kg^{-1} bw were lethargic and had reddish-brown feces. All birds receiving >2000 mg kg^{-1} died or were moribund within eight days. The ALD was determined to be 2003 mg kg^{-1} based on these results.

Following ALD determination, a 14-d range finding and 90-d subchronic feeding study using TNT were conducted with northern bobwhite [39,40]. A serial mixture procedure and the birds' ability to discriminate TNT-tainted particles in the 14-d study led to time-varying feed concentrations, therefore the results were considered unreliable. A 90-d study was then conducted using 70 northern bobwhite (seven birds per sex per treatment). Treatment TNT feed concentrations for the 90-d study were batch prepared at 0, 100, 750, 1500, and 3000 ppm and checked for consistency. This resulted in estimated daily oral exposures of 0, 7, 48, 98, and 178 mg kg^{-1} d^{-1} [40]. Body mass, feed intake, hematological parameters (red blood cell count, white blood cell count, five-part differentials), several functional immune assays, mass of main blood conditioning organs, and histopathology were assessed. Within the first 30 d of exposure, 4 out of 14 birds in the 3000 ppm treatment group exhibited symptoms of convulsions, ataxia, and muscular weakness prior to death. No other birds exhibited these symptoms. Variation within and between groups contributed to a finding of no difference in blood indices and immune parameters; however, there were some statistically significant trends associated with increasing exposure. Exposure to TNT was positively correlated with a general decrease in red blood cell count, decrease in packed cell volume, decline in total plasma protein, decline in prolymphocyte and leucocyte counts, and an increase in indeterminate apoptotic/necrotic leucocytic cells [40]. Most of these indices were within the normal ranges for this species. A nonsignificant trend was observed in an increased presence of hemosiderin in the Kupffer cells of the liver. No other adverse histological effect was associated with treatment [40].

Given the variation associated with the former study, another was designed to evaluate the effects of TNT in the rock pigeon (*Columba livia*) using a controlled dosing protocol. Sixty male and female common pigeons of the White Carneaux strain ($N = 120$) were assigned to one of five treatments where birds were exposed to 0, 20, 70, 120, or 200 mg TNT kg^{-1} bw d^{-1} in corn oil via oral gavage for 60 d [41]. Birds in the 200 mg kg^{-1} d^{-1} group displayed neuromuscular symptoms consistent with the earlier study including tremors, ataxia, lethargy, and death after 2 to 3 weeks of exposure. Liver and kidney to body weight ratios were depressed in the 70 mg kg^{-1} d^{-1} treatments and above. Ovary mass was reduced relative to body weight in females in the two high dose groups only. Hemoglobin and packed cell volumes were reduced in 70, 120, and 200 mg kg^{-1} d^{-1} treatments, and red blood cell counts were lower in the 120 and 200 mg kg^{-1} d^{-1} groups only. However, these blood indices were not considered biologically relevant because they were within the normal ranges for this species. Histopathological evaluations including fluoro-jade staining of brain sections failed to reveal any treatment-related effects.

Using a similar protocol, the effects of oral 2,4-DNT exposure were evaluated using northern bobwhite. Using the up/down method, the LD_{50} was determined to be 55 mg kg^{-1} [42]. Following a 14-d range finding study, twelve northern bobwhite of each sex ($N = 120$) received 0, 1, 5, 15, or 25 mg DNT kg^{-1} d^{-1} via gavage in corn oil for 60 d. All the females and 10 of the 12 males in the high dose group, and 4 females and 3 males of the 15 mg kg^{-1} d^{-1} group died or were moribund mostly within the first week of exposure. Symptoms included lethargy, scant feces and diarrhea, and a drastic reduction in feed intake. Gross postmortem examination found feed in the gizzard and often the crop, suggesting a cessation of peristalsis of potential neurological origin.

Changes in packed cell volume (PCV), hemoglobin (Hb), and red blood cell count (RBC) were all affected by treatment for females but not for males. Triglycerides were higher in the male 5 and 15 mg kg^{-1} d^{-1} groups compared with the control and 1 mg kg^{-1} d^{-1} groups. There was a dose-related increasing tendency in plasma uric acid concentrations, which was consistent with the kidney to body weight data. Incidence of multifocal lymphocytic infiltration and urate accumulation in the kidney was also suggestive of adverse kidney effects [42]. These data suggest that the mechanism of DNT toxicity is quite different from that of TNT and more potent in the manifestation of effects.

Few data on the effects of other nitroaromatic compounds in birds are available. The LD_{50}s for 1,3-DNB of >100 and 42 mg kg^{-1} for European starlings (*Sturnus vulgaris*) and red-winged blackbirds (*Agelaius phoeniceus*), respectively, have been reported [43]. Toxicity in birds from exposures to nitramines is different from what can be predicted from the mammalian data. Gogal et al. [44] conducted an ALD, a 14-d range finding, and a 90-d feeding study using northern bobwhite with RDX. Using a distilled water vehicle, eight groups consisting of one male and one female received doses of 125, 187, 280, 420, 630, 945, 1417, and 2125 mg RDX kg^{-1} bw. Birds receiving doses greater than 630 mg kg^{-1} died within 72 h displaying CNS effects such as convulsions and respiratory distress. Birds receiving more than 280 mg kg^{-1} often became moribund and were euthanized. Based on these results, the ALD was determined to be 187 and 280 mg kg^{-1} for males and females, respectively.

Feed concentrations of RDX for the 14-d and the 90-d study were formulated at 0, 83, 125, 187, 280, and 420 ppm. Treatments included six birds of each sex for the 14-d and ten birds of each sex for the 90-d feeding regimes. Initial feed intake and body mass were depressed in birds in a dose-dependent manner coinciding with a decrease in egg production at levels greater than 125 mg kg^{-1} d^{-1} for the 14-d study (Figure 7.1) [44]. Liver weights were lower for all birds on RDX-tainted feed. However, in the 90-d feeding study, feeding resumed in the third week of exposure and no significant differences were observed in body mass, feed intake, or egg production. This suggests that body mass and egg production changes of the 14-d regime were likely due to the palatability of the treatment (i.e., concentration of

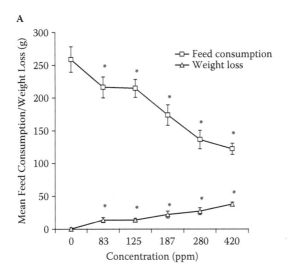

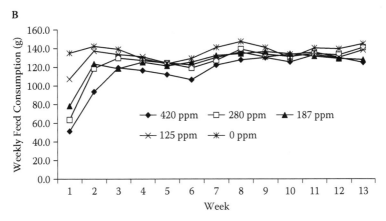

FIGURE 7.1   Changes in body mass and feed consumption of northern bobwhites (*Colinus virginianus*) to dietary exposures of RDX for (a) 14 days and (b) 90 days. Egg production for (c) 14 days as other data are presented as means ±SEM. Asterisks (*) denote different from controls at *p* > 0.05. (Data taken from Gogal et al. [44], with permission.)

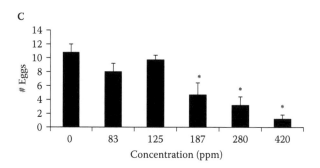

FIGURE 7.1   (continued).

RDX in feed). During the 90-d feeding study, four females (three at 420 and one at 187 ppm) and one male (exposed to 420 ppm) died within 56 d of exposure. It is likely the exposures to 420 ppm are of consequence, and that the moribund individual at 187 ppm was an anomaly, particularly since there were no statistical differences in organ weights, body mass, egg production, hematology, histopathology, and various functional and descriptive immune assays that were also conducted as part of the 90-d study [44]. The feed consumption rates and mean body masses resulted in calculated RDX intakes of 0, 6.0, 8.7, 10.6, 12.4, and 18.4 mg RDX kg^{-1} bw d^{-1} for both sexes for the 0, 83, 125, 187, 280, and 420 ppm feed treatments, respectively.

Absorption of HMX from oral exposures in birds appears to be low, consistent with the information reported from studies using monogastric mammalian species. To determine an ALD, northern bobwhite were dosed with HMX in distilled water to concentrations of up to 10 g kg^{-1} without consistent results [45]. Sixteen birds (1 male and 1 female per group) were gavaged with concentrations of HMX ranging from 125 to 2125 mg kg^{-1}. A single female died on day 6 at 187 mg kg^{-1}; all others survived without incident. A limited repeated ALD was conducted at concentrations ranging from 3188 to 10,760 mg kg^{-1}. A single female that received 7173 mg kg^{-1} died on day 12. Following these trials, limited tests were conducted that incorporated fasting and corn oil as a vehicle, none of which resulted in an outcome different than seen in previous studies. As a result, an ALD could not be determined [45].

A limited 28-d feeding study using northern bobwhite ensued conducted at concentrations of 0, 100, 1000, and 10000 ppm HMX in feed [45]. These results suggest a strong dose-related feeding aversion whereby feed consumption, body mass, and egg production were markedly decreased in the two high exposure groups. No other adverse effects were observed.

7.4   EFFECTS OF OTHER ENERGETIC COMPOUNDS IN BIRDS

Three- to four-day old northern bobwhite were used to evaluate the influence of ammonium perchlorate (AP) on thyroid function, as well as thyroidal hormone level systemic effects on the hypothalamic–pituitary–thyroid axis [46]. The perchlorate ion competitively inhibits iodide uptake, thus with time AP has the potential to reduce thyroid hormone levels. Systemic thyroid hormone levels (T_3 and T_4) were measured by radioimmunoassay. Birds were exposed to AP through the drinking water at high concentrations (250–4000 mg L^{-1}) for two weeks and low concentrations (0.05–250 mg L^{-1}) for eight weeks. Thyroid weight, body mass, hindlimb growth, and systemic and colloidal thyroid hormone levels were evaluated. Plasma T_4 levels were decreased relative to controls in birds exposed to $\geq$500 mg L^{-1} after two weeks of exposure, though this was variable [46]. Plasma T_4 levels were less than those of controls at AP concentrations $\geq$1000 mg L^{-1} after eight weeks, yet there were trends that suggested it was important at lower exposures (~32% lower for other treatments, but not statistically significant). Thyroid weight was increased at exposures greater than 1000 mg L^{-1} at one week and greater than 500 mg L^{-1} at two weeks, but only greater at 1000 mg L^{-1} after eight weeks. All AP exposures resulted in lower thyroidal T_4 concentrations for shorter durations. Differences in growth (body weight, tibia and femur length) were not seen except at 4000 mg L^{-1} [46]. Thyroidal hormone content was the most sensitive parameter measured in relation to exposure to AP. Plasma T_4 levels were not considered useful given the variability in the assay between animals and exposure groups. Overall, these data suggest that AP will only adversely affect northern bobwhite after sustained exposures to relatively high concentrations.

A limited 8-d dietary study using nitroglycerin was conducted using northern bobwhite (N = 150) in exposures ranging from 15.9 to 5620 ppm, including a control. No adverse effects were reported with exposures up to 5620 ppm, yet some decreases in body mass were noted. Therefore, the NOAEL was reported as 3160 ppm and the LOAEL at 5620 ppm [47].

Mass mortalities of ducks and swans have been described resulting from ingestion of particulate white phosphorus at Eagle River Flats in Alaska [48]. Particle sizes were consistent with those chosen by waterfowl species and were a result of incomplete combustion of tracer rounds used in training exercises. Secondary poisoning of predators that fed upon diseased waterfowl has also been reported [49].

An LD_{50} of 6.46 mg kg^{-1} for male and 6.96 mg kg^{-1} for female mallards (*Anas platyrhynchos*) was derived using a single oral dose of phosphorus dissolved in corn oil [50]. However, phosphorus was more toxic when administered in pelletized form (LD_{50} = 4.05). Other pathological responses included weight loss, enlarged spleens, formation of hyaline droplets in the kidney, and necrosis/fat accumulation in the liver. Adverse observations from acute exposures to mallards, mute swans (*Cygnus olor*), and domestic chickens (*Gallus domesticus*) in the range of 3 to 5 mg kg^{-1} include erosion of gizzard lining, petechial hemorrhages in the pancreas, hepatomegaly and hemorrhage in the liver, and reductions in egg production [51–53].

Effects from longer-term continuous oral exposures include changes in hematological and cytological blood parameters [52], and severe developmental changes in embryos as a result of exposure to the egg-laying female [54]. Immature chickens were dosed with phosphorus and then fed to juvenile American kestrels (*Falco sparverius*), some containing an implanted phosphorus pellet [55]. Exposures continued in the kestrels until half the birds in either exposure group (those with or without the phosphorus pellet) died. Hematocrit and hemoglobin levels decreased, and plasma lactate dehydrogenase, L-alanine aminotransferase, and glucose levels were increased in birds receiving phosphorus-containing chicks compared with controls [55]. Mortality and a reduction in feed consumption were most pronounced in kestrels receiving chicks containing a phosphorus pellet.

7.5   EFFECTS OF ENERGETIC COMPOUNDS IN REPTILES

There are few toxicity data available on the effects of munition compounds on reptiles. In fact, the class Reptilia has largely been ignored by the field of wildlife toxicology, despite the fact that reptiles are fairly widespread and can represent a significant portion of the biomass in certain habitats [56,57]. In addition, there is evidence that lizard populations are declining and environmental contaminants are cited as a contributing factor [58]. Reptiles may have been excluded from consideration because it was believed that benchmarks protective of mammalian and avian receptors were likely also protective of reptiles. Moreover, given the life history variation in the class Reptilia (e.g., relatively long life spans, complex reproduction strategies) and special husbandry requirements, controlled laboratory toxicity studies were considered prohibitory [59]. Other considerations may include perceived societal values of reptile species relative to more prominent mammalian and avian species. Regardless, at present there is a paucity of reptilian toxicity data available, particularly for energetic compounds.

Efforts to identify and develop a reptile toxicity model resulted in several lines of Eastern (*Sceloporus undulatus*) and Western (*Sceloporus occidentalis*) fence lizards that can be reared and bred successfully in the laboratory [59]. The system was modeled after standard mammalian toxicity tests and follows a hierarchical approach. An ALD procedure is used as an acute range-finding assay. One individual of each sex is used for each of several exposures (usually eight) to determine the ALD. Although not a statistically robust value, the ALD provides an easy, convenient means of determining the lowest level at which mortality occurs while using a minimum number of animals. Although longer-term sublethal data are required for risk assessment purposes, some inferences can be made between vertebrate classes.

In an effort to test the system and to generate preliminary data on the toxicity of munition compounds to reptiles, acute oral toxicity studies on TNT, DNT, and RDX were conducted using Western fence lizards (*S. occidentalis*) [60]. All three compounds were administered via gavage in a corn oil vehicle. Table 7.1 shows the median lethal dose (LD_{50}) for lizards orally exposed to these common munition compounds. Based on these data, RDX is relatively more toxic to fence lizards than TNT and DNT. Males appear more sensitive to RDX than females, while females are more sensitive to DNT than males, although generalizations on the relative sensitivity of males and females should be approached with caution. Overall, these data suggest that RDX is most toxic, and DNT and TNT are less toxic.

Toxicity studies with other compounds involving *S. occidentalis* have indicated that the species is amenable to longer-term studies of at least 60 d. The model has proven to be a viable reptilian toxicity testing system and will be used to further evaluate the toxicity of military-related energetic compounds.

TABLE 7.1

Acute Oral Median Lethal Dose (LD$_{50}$) of Specific Explosive Compounds to Western Fence Lizards (*Sceloporus occidentalis*)

	RDX[a] (mg kg^{-1})	DNT[b] (mg kg^{-1})	TNT[c] (mg kg^{-1})
Males	72 (49, 106)	380 (149, 515)	1038 (332, 2360)
Females	88 (65, 119)	577 (406, 785)	1579 (593, 3356)

Note: 95% Confidence intervals in parentheses.

[a] McFarland et al. [68].

[b] Suski et al. [69].

[c] McFarland et al. [70].

7.6 EFFECTS OF ENERGETIC COMPOUNDS IN AMPHIBIANS

The unique physiology and life histories of amphibians require a more adaptive approach to toxicity testing. The applicability of any outcome depends upon the relevance of the exposure pathway or conditions to the specific species evaluated. For example, there is considerable life history variation among amphibian species; some species exist almost exclusively in aquatic environments throughout their entire life cycle (e.g., sirens, neotenic salamanders, *Xenopus* sp.) and others exist almost exclusively in terrestrial environments (e.g., *Plethodon cinereus*). Moreover, there are also unique physiological attributes of amphibians that require consideration. Sensitivities associated with a biphasic life history (metamorphosis) require an understanding of the amphibian endocrine/exocrine system and how xenobiotics may influence it. This also requires knowledge of the environmental influences of different media and how amphibian species are exposed. In some species, gas exchange is primarily cutaneous resulting in a relatively thin and unique integument. This is particularly important because many energetic compounds have properties that can result in significant dermal absorption [60].

Although it is often assumed that the oral pathway contributes most to contaminant exposure for wildlife species, it may not be true for amphibian species exposed to energetic compounds. A study that evaluated the relative importance of the oral and dermal pathways of a lipophilic substance (a mixture of polychlorinated biphenyls) and TNT found that the dermal pathway contributed most to the TNT body burdens in the tiger salamander (*Ambystoma tigrinum*) exposed in a terrestrial microcosm design [61]. Another study found that the water-absorbing ventral surface of the American toad (*Bufo americanus*) was an important pathway contributing to uptake of methoxychlor, a water-soluble organochlorine pesticide [62]. These studies suggest that the dermal pathway should be considered in amphibian toxicity studies, particularly if the results are to assist in making risk management decisions.

Few data are available on the effects of energetic compounds on amphibian species. The Frog Embryo Teratogenesis Assay–*Xenopus* (FETAX) was used to

TABLE 7.2

Developmental Toxicity Values of Several Selected Substances Using the FETAX Assay

Compound	96-h LC$_{50}$[a] and 95% CI[b] (µM)	96-h EC50[c] and 95% CI (µM)	MCIG[d] (µM) and MCIG/96-h LC$_{50}$ (%)
2,4,6-Trinitrotoluene (TNT)	16.7 (13.0–25.0)	9.78 (7.80–14.6)	5.50 (33)
2-Amino-4,6-dinitrotoluene (2-ADNT)	166 (141–210)	16.9 (13.8–21.1)	12.4 (7.5)
4-Amino-2,6-dinitrotoluene (4-ADNT)	115 (103–130)	85.8 (73.1–106)	53.5 (47)

[a] LC$_{50}$ = median lethal concentration.
[b] CI = confidence interval.
[c] EC$_{50}$ = median effective concentration for malformation.
[d] MCIG = minimum concentration to inhibit growth.
Source: Saka [64].

test TNT, 2-ADNT, and 4-ADNT in an aqueous solution [63]. Concentrations of nitroaromatics (TNT: 0, 13.8, 27.5, 55.1, and 110 µM; 2-ADNT or 4-ADNT: 0, 8.88, 17.8, 35.5, 71.1, and 142 µM) were diluted into the FETAX solution. Microplates with replicates were used that exposed individual embryos under these conditions using a 24 h water renewal interval for 96 h. Mortality, malformation incidence, and growth (measured by head-to-tail length) were quantified. TNT was found to be more toxic than either of the primary reduction products (Table 7.2). The most typical observed terata include axial and eye malformations, edema, and irregular gut coiling [64].

Effects of soil exposures to TNT were evaluated in adult, field-collected tiger salamanders (*A. tigrinum*) using a microcosm design [64,65]. TNT was mixed in soil, hydrated, and analyzed during the course of the 14-d exposure period. TNT was significantly reduced to 2-ADNT and 4-ADNT during the course of the exposure period (from a target TNT concentration of 1200 µg g^{-1} to an initial concentration of 280 µg g^{-1} to a final concentration of 59 µg g^{-1}). Concentrations of 2-ADNT and 4-ADNT approached but did not exceed that of parent compound in soil. However, concentrations of ADNTs did exceed that of TNT in food tissue (worms; TNT range = 0.25–0.79 µg g^{-1}; 2-ADNT = 2.1–2.6 µg g^{-1}; 4-ADNT = 2.1–2.5 µg g^{-1}). Only one exposure group with a concurrent control was used. Controls gained weight with time but TNT-exposed salamanders did not. Mass of main blood conditioning organs, hematological indices, and histopathology were inconclusive as were the functional immune assays. No overt toxicological symptoms or behaviors were associated with TNT exposure.

Field-collected red-backed salamanders (*Plethodon cinereus*) were exposed to 2,4-DNT in soil for 28 d [66]. A sandy loam soil was characterized and spiked with 2,4-DNT at concentrations of 0, 75, 200, 800, and 1500 mg kg^{-1} dry weight (dw) soil. Food consisted of adult mutant fruit flies (*Drosophila melanogaster*; flightless) that were not exposed to DNT. All animals exposed to analytical concentrations greater than 1000 mg DNT kg^{-1} and not soil died within the first week of exposure.

Prior to death, salamanders exhibited signs of lethargy, buccopharyngeal ventilation, and ataxia. Salamanders exposed to analytical concentrations greater than 345 mg kg⁻¹ lost body weight compared with controls. This loss of body mass was coincident with a reduction in feeding efficiency and an increase in buccopharyngeal ventilation observations. Overall, these observations suggest that high soil concentrations of 2,4-DNT may affect the chemosensory ability of salamanders under these conditions.

Effects from soil exposures to RDX were evaluated in red-backed salamanders using a similar protocol as discussed in the previous study [67]. An oak–beech forest soil was spiked with target concentrations of RDX at 0, 10, 100, 1000, and 5000 mg kg⁻¹ dry weight and was characterized according to organic carbon content, pH, cation exchange capacity, and the proportion of sand, silt, and clay. Half of the salamanders in the 5000 mg kg⁻¹ group, and two in the 1000 mg kg⁻¹ group exhibited symptoms of hyperactivity, rolling and gaping motions, and lethargy following 12 d of exposure. One animal from the high-dose group was euthanized after losing 22% of its body mass. Salamanders of the 5000 mg kg⁻¹ group lost weight and had a reduced feed consumption rate compared with controls. No other adverse effects were associated with exposure; however, these CNS-related symptoms were similar to those described for mammals exposed to RDX [27].

7.7 SUMMARY

Among classes of wildlife species, effects from exposures to energetic compounds are dependent largely upon physiological and phenotypical attributes. Mammals are largely more sensitive to exposures to nitroaromatic explosive compounds through lytic effects on mature anucleated red blood cells, whereas birds, reptiles, and amphibians that have nucleated blood cells are least affected. Targets in birds tend to be neurologic in origin and may be due to the accumulation of metabolic intermediates in subchronic exposure regimes. Data for amphibians suggest that dermal exposures are most important and may foster a greater appreciation of the differences among wildlife species that sense the environmental changes.

The significant variation in effects observed between substances; particularly in birds (e.g., TNT and DNT) suggest that the use of quantitative structure–activity relationships (QSARs) may have questionable relevance. As we gain a greater understanding of avian distribution, metabolism, and excretion in regard to explosive compounds, extrapolation and estimation of effects among similar compounds and species may be possible. Moreover, as we understand more about the toxicokinetics of these compounds, we can begin to develop strategies to investigate the mechanisms of toxicity in regard to these observed effects. The understanding of mechanisms will provide the optimal information to begin to predict effects between explosive compounds and avian species.

With the exception of white phosphorus, there are few data that suggest environmental exposures of energetics to wildlife are problematic. However, through the use of controlled toxicity data and the evolution of more accurate risk assessment techniques, exposures can be more accurately characterized and the probability of risk to these valued resources can be used to make wise land-use decisions.

ACKNOWLEDGMENTS

We thank Mark Michie, Mick Major, James Boles, and three other anonymous reviewers for their helpful comments and review of this manuscript. The views expressed in this chapter are those of the authors and do not necessarily reflect the views and policies of the U.S. Army. Mention of trade names or commercial products does not constitute endorsement or recommendation for use.

REFERENCES

1. Jenkins TF et al., Characterization of Explosives Contamination at Military Firing Ranges, ERDC TR-01-5, Final/Technical Report, Engineer Research and Development Center, U.S. Army Corps of Engineers, Hanover, NH, 2001.
2. Hathaway JA, Trinitrotoluene: A review of reported dose-related effects providing documentation for a workplace standard, *J. Occup. Med.,* 19, 341, 1977.
3. Agency for Toxic Substances and Disease Registry (ATSDR), *Toxicological Profiles for HMX, RDX, TNT, and DNTs,* Public Health Service, U.S. Department of Health and Human Services, Atlanta, GA, 1977–2002.
4. Shugart LR, Dinitrotoluene in Deer Tissues, Oak Ridge National Laboratory Final Report, ORNL/M-1765, Oak Ridge National Laboratory, Oak Ridge, TN, 1991.
5. Shugart LR et al., TNT Metabolites in Animal Tissues, Final Report, Oak Ridge National Laboratory, ORNL/M-1336, Oak Ridge National Laboratory, Oak Ridge, TN, 1990.
6. Whaley JE and Leach GJ, Health Risk Assessment for Consumption of Deer Muscle and Liver from Joliet Army Ammunition Plant, Joliet, IL, Final Report, Project #75-51-YF23, U.S. Army Environmental Hygiene Agency, Aberdeen Proving Ground, MD, 1994.
7. Whaley JE, personal communication, 1999.
8. Dilley JV et al., Short-term oral toxicity of 2,4,6-trinitrotoluene in mice, rats, and dogs, *J. Toxicol. Environ. Health,* 9, 565, 1982.
9. Levine BS et al., Subchronic toxicity of trinitrotoluene in Fischer 344 rats, *Toxicology,* 32, 253, 1984.
10. Chandra AM, Qualls SCW Jr, and Reddy G, Testicular effects of 1,3,5-trinitrobenzene (TNB), 1. Dose response and reversibility studies, *J. Toxicol. Environ. Health,* 50, 365, 1997.
11. Yinon J, *Toxicity and Metabolism of Explosives,* CRC Press, Boca Raton, FL,1990.
12. Linder RE et al., Acute effects and long-term sequelae of 1,3-dinitrobenzene on male reproduction in the rat, I. Sperm quality, quantity and fertilizing ability, *J. Androl.,* 9, 317, 1988.
13. Zitting A et al., Acute toxic effects of trinitrotoluene on rat brain, liver, and kidney: Role of radical production, *Arch. Toxicol.,* 51, 53, 1982.
14. Tenhunen R et al., Trinitrotoluene-induced effects on rat heme metabolism, *Exp. Molec. Pathol.,* 40, 362, 1984.
15. Savolainen H, Tenhunen R, and Harkonen H, Reticulocyte haem synthesis in occupational exposure to trinitrotoluene, *Br. J. Ind. Med.,* 42, 354, 1985.
16. Bruckner JV and Warren DA, Toxic effects of solvents and vapors, in *Casarett and Doull's Toxicology,* Klaassen CD, Ed., McGraw-Hill, New York, 2001, 869.
17. Ferguson JW, Fourteen Day Feeding Study of 2,4-Dinitrotoluene in the White-Footed Mouse, *Peromyscus leucopus,* Toxicological Study No. 4152-31-96-06-14, U.S. Army Center for Health Promotion and Preventive Medicine, Aberdeen Proving Ground, MD, 1996.
18. Ferguson JW, Fourteen Day Feeding Study of 2,6-Dinitrotoluene in the White-Footed Mouse, *Peromyscus leucopus,* Toxicological Study No. 4152-31-96-06-15, U.S. Army Center for Health Promotion and Preventive Medicine, Aberdeen Proving Ground, MD, 1998.

19. McCain WC and Ferguson JW, Fourteen Day Range Finding and Ninety Day Feeding Study of 2,4,6-Trinitrotoluene in the White Footed Mouse (*Peromyscus leucopus*), Toxicological Study No. 2340-38-95-6-1, U.S. Army Center for Health Promotion and Preventive Medicine, Aberdeen Proving Ground, MD, 1998.

20. Johnson MS, Ferguson JW, and Holladay SD, Immune effects of oral 2,4,6-trinitrotoluene (TNT) exposure to the white-footed mouse, *Peromyscus leucopus*, *Int. J. Toxicol.*, 19, 5, 2000.

21. Talmage SS et al., Nitroaromatic munition compounds: Environmental effects and screening values, *Rev. Environ. Contam. Toxicol.*, 161, 1, 1999.

22. Reddy G et al., Toxicity of 2,4,6-trinitrotoluene (TNT) in hispid cotton rats (*Sigmodon hispidus*): Hematological, biochemical, and pathological effects, *Int. J. Toxicol.*, 19, 169, 2000.

23. Reddy G et al., Fourteen day toxicity evaluation of 1,3,5-trinitrobenzene (TNB) in the shrew (*Cryptotis parva*), Abstract, Twenty-First Meeting of the American College of Toxicology, San Diego, CA, 2000.

24. Reddy TV et al., Ninety-day toxicity evaluation of 1,3,5-trinitrobenzene (TNB) in *Peromyscus leucopus*, Abstract, Society of Environmental Toxicology and Chemistry Second World Congress, Vancouver, BC, 1995.

25. Reddy TV et al., Subchronic Toxicity Studies on 1,3,5-Trinitrobenzene, 1,3-Dinitrobenzene and Tetryl in Rats, Fourteen-Day Toxicity Evaluation of 1,3-Dinitrobenzene in Fischer 344 Rats, AD A290641, prepared by the U.S. EPA Environmental Monitoring Systems Laboratory, Cincinnati, OH, for the U.S. Army Medical Research and Development Command, Fort Detrick, Frederick, MD, 1994.

26. Reddy TV et al., Subchronic Toxicity Studies on 1,3,5-Trinitrobenzene, 1,3-Dinitrobenzene and Tetryl in Rats: Ninety Day Evaluation of 1,3-Dinitrobenzene in Fischer 344 Rats, AD A297458, prepared by the U.S. EPA Environmental Monitoring Systems Laboratory, (Cincinnati, OH) for the U.S. Army Medical Research and Development Command, Fort Detrick, Frederick, MD, 1995.

27. Levine BS et al., Thirteen week toxicity study of hexahydro-1,3,5-trinitro-1,3,5-triazine (RDX) in Fischer 344 rats, *Toxicol. Lett.*, 8, 241, 1981.

28. Levine BS et al., Determination of the Chronic Mammalian Toxicological Effects of RDX: Twenty-Four Month Chronic Toxicity/Carcinogenicity Study of Hexahydro-1,3,5-Trinitro-1,3,5-Triazine (RDX) in the Fischer 344 Rat, AD A160774, prepared by ITT Research Institute (Chicago, IL) for the U.S. Army Medical Research and Development Command, Fort Detrick, MD, 1983.

29. Crouse CBL, unpublished data, 2005.

30. Ferguson JW and McCain WC, Fourteen Day Feeding Study of Hexahydro-1,3,5-Trinitro-1,3,5-Triazine (RDX) in the White-Footed Mouse, *Peromyscus leucopus*, Toxicological Study No. 6955-31-97-05-02, U.S. Army Center for Health Promotion and Preventive Medicine, Aberdeen Proving Ground, MD, 1999.

31. Cholakis JM et al., Mammalian Toxicological Evaluation of RDX, AD A092531, prepared by the Midwest Research Institute (Kansas City, MO) for the U.S. Army Medical Research and Development Command, Frederick, MD, 1980.

32. Wilson AB, Determination of Acute and Subchronic Mammalian Toxicity of HMX, AD A173743, U.S. Army Medical Research and Development Command, Fort Detrick, MD, 1985.

33. Cuthbert JA, D'Arcy-Burt KJ, and Carr SMA, HMX: Acute Toxicity Tests in Laboratory Animals, AD A171598, U.S. Army Medical Research and Development Command, Fort Detrick, Frederick, MD, 1985.

34. Seakins A and Robinson DS, Changes associated with the production of fatty livers by white phosphorus and by ethanol in the rat, *Biochem. J.*, 92, 308, 1964.

35. Pani P et al., On the mechanism of fatty liver in white phosphorus-poisoned rats, *Exp. Mol. Pathol.*, 16, 201, 1972.

36. Whalen JP et al., Pathogenesis of abnormal remodeling of bones: Effects of yellow phosphorus in the growing rat, *Anat. Rec.*, 177, 15, 1973.

37. Monsanto, One-Generation Reproductive Study in Rats with Elemental Phosphorus with Attachments and Cover Letter Dated 010990, Monsanto Co., EPAOTS0518525-1, 1990.

38. Monsanto, Follow-Up Information: The One-Generation Reproduction Study in Rats with Elemental Phosphorus with Attachments and Cover Letter Dated 080591, Monsanto Co., EPAOTS0518525-4, 1991.

39. Gogal RM Jr, Influence of Feed Exposure with TNT Derivatives on Quail (Northern Bobwhite, *Colinus virginianus*) Immunity as Measured by National Toxicology Program Immune Endpoints, Final Report, Avian Toxicology Study No. 990503G, 981124G, prepared for U.S. Army Center for Health Promotion and Preventive Medicine, Aberdeen Proving Ground, MD, 2000.

40. Gogal RM Jr et al., Influence of dietary 2,4,6-trinitrotoluene exposure in the northern bobwhite (*Colinus virginianus*), *Environ. Toxicol. Chem.*, 21, 81, 2002.

41. Johnson MS et al., Responses of oral 2,4,6-trinitrotoluene (TNT) exposure to the common pigeon (*Columba livia*): A phylogenic and methodological comparison, *Int. J. Toxicol.*, 24, 221, 2005.

42. Johnson MS, Michie MW, and Gogal RM Jr, Influence of oral 2,4-dinitrotoluene exposure to the northern bobwhite (*Colinus virginianus*), *Int. J. Toxicol.*, 24, 265, 2005.

43. Schafer EW, The acute oral toxicity of 369 pesticidal, pharmaceutical, and other chemicals to wild birds, *Toxicol. Appl. Pharmacol.*, 21, 315, 1972.

44. Gogal RM Jr et al., Dietary oral exposure to 1,3,5-trinitro-1,3,5-triazine in the northern bobwhite (*Colinus virginianus*), *Environ. Toxicol. Chem.*, 22, 381, 2003.

45. Johnson MS, Gogal RM Jr, and Larsen CT, Food avoidance behavior to dietary octahydro-1,3,5,7-tetranitro-1,3,5,7-tetrazocine (HMX) exposure in the northern bobwhite (*Colinus virginianus*), *J. Toxicol. Environ. Health, Part A*, 68, 1349, 2005.

46. McNabb FMA, Larsen CT, and Pooler PS, Ammonium perchlorate effects on thyroid function and growth in bobwhite quail chicks, *Environ. Toxicol. Chem.*, 23, 997, 2004.

47. Hercules International, unpublished data.

48. Racine CH et al., White phosphorus poisoning of waterfowl in an Alaskan salt marsh, *J. Wildlife Dis.*, 28, 669, 1992.

49. Roebuck BD et al., Predation of ducks poisoned by white phosphorus: Exposure and risk to predators, *Environ. Toxicol. Chem.*, 13, 1613, 1994.

50. Sparling DW et al., Toxicity of white phosphorus to waterfowl: Acute exposure in mallards, *J. Wildlife Dis.*, 33, 187, 1997.

51. Nam S-I, MacMillan DL, and Roebuck BD, The translocation of white phosphorus from hen (*Gallus domesticus*) to egg, *Environ. Toxicol. Chem.*, 15, 1564, 1996.

52. Sparling DW, Vann S, and Grove RA, Blood changes in mallards exposed to white phosphorus, *Environ. Toxicol. Chem.*, 17, 2521, 1998.

53. Sparling DW, Day D, and Klein P, Acute toxicity and sublethal effects of white phosphorus in mute swans, *Cygnus olor, Arch. Environ. Contam. Toxicol.*, 36, 316, 1999.

54. Vann SL, Sparling DW, and Ottinger MA, Effects of white phosphorus on mallard reproduction, *Environ. Toxicol. Chem.*, 19, 2525, 2000.

55. Sparling DW and Federoff NE, Secondary poisoning of kestrels by white phosphorus, *Ecotoxicol.*, 6, 239, 1997.

56. Campbell KM and Campbell TS, A logical starting point for developing priorities for lizard and snake ecotoxicology: A review of available data, *Environ. Toxicol. Chem.*, 21, 894, 2002.

57. Hopkins W, Reptile toxicology: Challenges and opportunities on the last frontier in vertebrate ecotoxicology, *Environ. Toxicol. Chem.*, 19, 2391, 2000.

58. Gibbons JW et al., The global decline of reptiles, déjà vu amphibians, *Bioscience*, 50, 653, 2000.

59. Talent LG et al., Evaluation of western fence lizards (*Sceloporus occidentalis*) and eastern fence lizards (*Sceloporus undulatus*) as laboratory reptile models for toxicological investigations, *Environ. Toxicol. Chem.*, 21, 899, 2002.

60. Reifenrath WG et al., Percutaneous absorption of explosives and related compounds: An empirical model of bioavailability of organic nitro compounds from soil, *Toxicol. Appl. Pharmacol.*, 182, 160, 2002.

61. Johnson MS et al., Bioaccumulation of 2,4,6-trinitrotoluene and polychlorinated biphenyls through two routes of exposure in a terrestrial amphibian: Is the dermal route significant? *Environ. Toxicol. Chem.*, 18, 873, 1999.

62. Hall RJ and Swineford D, Uptake of methoxychlor from food and water by the American toad (*Bufo americanus*), *Bull. Environ. Contam. Toxicol.*, 23, 335, 1979.

63. Saka M, Developmental toxicity of *p,p'*-dichlorodiphenyltrichloroethane, 2,4,6-trinitrotoluene, their metabolites, and benzo[*a*]pyrene in *Xenopus laevis* embryos, *Environ. Toxicol. Chem.*, 23, 1065, 2004.

64. Johnson MS et al., Effects of 2,4,6-trinitrotoluene in a holistic environmental exposure regime to a terrestrial salamander: *Ambystoma tigrinum, Toxicological Pathol.*, 28, 334, 2000.

65. Johnson MS et al., Fate and biochemical effects of 2,4,6-trinitrotoluene exposure to tiger salamanders (*Ambystoma tigrinum*), *Ecotoxicol. Environ. Saf.*, 46, 186, 2000.

66. Johnson MS et al., Toxicological responses of red-backed salamanders (*Plethodon cinereus*) to subchronic soil exposures of 2,4-dinitrotoluene. *Environ. Pollut.* 147, 604, 2007.

67. Johnson MS et al., Toxicologic and histologic response of the terrestrial salamander *Plethodon cinereus* to soil exposures of 1,3,5-trinitrohexahydro-1,3,5-triazine, *Arch. Environ. Contam. Toxicol.*, 47, 496, 2004.

68. McFarland CA et al., Toxicity of RDX (hexahydro-1,3,5-trinitro-1,3,5-triazine) in western fence lizards (*Sceloporus occidentalis*). *Environ. Toxicol. Chem.* 27, 1102, 2008.

69. Suski JG et al., Dose-related effects following oral exposure of 2,4-dinitrotoluene on the western fence lizard, *Sceloporus occidentalis*, *Environ. Toxicol. Chem.*, 27, 352, 2008.

70. McFarland CA et al., Toxicity of oral 2,4,6-trinitrotoluene exposure in the western fence lizard (*Sceloporus occidentalis*), *Environ. Toxicol. Chem.*, 27, 1102, 2008.

8 Genotoxicity of Explosives

Laura Inouye, Bernard Lachance, and Ping Gong

CONTENTS

8.1 Introduction 177
8.2 In Vitro Methodologies 178
    8.2.1 Microorganism-Based Mutagenicity Assays 178
        8.2.1.1 2,4,6-Trinitrotoluene (TNT) and Its Products 180
        8.2.1.2 Dinitrotoluenes (DNT) and Their Products 182
        8.2.1.3 Cyclic Nitramines and Their Products 183
        8.2.1.4 Other Compounds 183
        8.2.1.5 Mixtures and Environmental Samples 184
    8.2.2 Mammalian Cell Line–Based Genotoxicity Assays 185
        8.2.2.1 2,4,6-Trinitrotoluene (TNT) and Its Products 185
        8.2.2.2 Dinitrotoluenes and Their Products 187
        8.2.2.3 Cyclic Nitramines and Their Products 188
        8.2.2.4 Other Compounds 188
        8.2.2.5 Mammalian Cell Line Testing of Environmental
                 Samples 189
8.3 In Vivo Methodologies 189
    8.3.1 Mammalian In Vivo Genotoxicity Investigations 190
        8.3.1.1 TNT and Its Products 190
        8.3.1.2 Dinitrotoluenes and Their Products 191
        8.3.1.3 Cyclic Nitramines and Their Products 192
        8.3.1.4 Other Compounds 193
    8.3.2 Other In Vivo Genotoxicity Investigations 193
8.4 Structure–Activity Relationships (SARS) 194
8.5 Application to Ecotoxicology 195
8.6 New Approaches for Assessing Genotoxicity of Explosives 198
8.7 Summary 199
8.8 Data Gaps and Future Directions 201
References 202

8.1 INTRODUCTION

Genotoxicity is a specialized case of biological effects in which the toxicological endpoint is an alteration of the information content, structure, or segregation of DNA in an organism. Genotoxicants have been defined as compounds that covalently bind and/or cause gene mutations or chromosomal changes in in vitro or in vivo systems, as opposed to nongenotoxic carcinogens that act via altering enzyme levels or inducing cell proliferation. Typical in vitro assays for genotoxicity include microbial reversion assays (e.g., *Salmonella* assay), mammalian cell-based assays (e.g., chromosomal aberration), as well as measures of interactions at the DNA level (adduct formation, strand breakage). In vivo assays share many of the endpoints monitored in vitro, such as adducts, strand breakage, chromosomal aberrations, micronuclei formation, and mutations (e.g., alterations in the hypoxanthine guanine phosphoribosyl transferase [HGPRT or HPRT] gene sequence), and endpoints that cannot be monitored in vitro such as the formation of neoplastic lesions and increases in cancer incidence. Data for genotoxicity of explosives are limited as compared to those available for common environmental contaminants (e.g., benzo[a]pyrene). This chapter will present both in vitro and in vivo mutagenicity methodologies including microorganism assays, mammalian cell-based assays, plant assays, and available carcinogenicity data, as well as a short discussion on structure–activity relationships for mutagenicity of nitroaromatics.

Although concerns regarding genotoxicity of compounds are generally related to survival of an individual (cancer) rather than effects on ecosystems, linkages between the two have been made. However, the paucity of data for genotoxicity of explosives makes it difficult to support the linkage of their genotoxicity and effects at the population level. Evidence available for other classes of compounds will be used to demonstrate the potential linkage.

Although many other explosives and explosives-related compounds exist, only the data for 2,4,6-trinitrotoluene (TNT) and its breakdown products, the cyclic nitramines (RDX and HMX), tetryl, trinitrobenzene, and nitroglycerin will be reviewed in this chapter. Two dinitrotoluene isomers are also included, as they are still used as explosives in addition to being side-products of TNT synthesis.

8.2   IN VITRO METHODOLOGIES

In vitro methodologies include microbial and mammalian cell-based assays for mutagenesis. Although they do not address the complex toxicodynamic processes that affect genotoxicity in multicellular organisms, these methodologies have the advantage of being rapid screening assays that minimize the cost and difficulties associated with live animal testing. The use of an S9 liver homogenate as an exogenous source of enzymes helps to address the metabolic activation pathways missing in microbial systems. However, in many cases, addition of the S9 fraction actually reduces the observed mutagenic response, possibly due to either metabolism to nonmutagenic compounds or to mutagenic compounds reacting with the proteins and other macromolecules in the S9 fraction.

8.2.1   MICROORGANISM-BASED MUTAGENICITY ASSAYS

Several microorganism-based mutagenicity assays have been used to assess genotoxic potential of explosives as summarized in Table 8.1. The *Salmonella typhimurium* reverse mutation assay is the most common test. Various tester strains have been developed to detect base-pair substitution and frameshift mutations.

TABLE 8.1

In Vitro Microorganism-Based Systems Used to Assess Mutagenic Potential for Explosives and Explosives-Related Compounds

Microorganism	Basis of Assay	Endpoint	Detects	Reference
Salmonella typhimurium	Bacterial, reverse mutation	Reversion to histidine independence	Frameshift and base-pair substitutions	1–21
Vibrio fischeri	Bacterial, reverse mutation	Reversion from dark variant to photoluminescent wild-type	Frameshift, base-pair substitution, SOS induction, and DNA intercalation	12, 22, 23
Escherichia coli	Bacterial, DNA repair	Comparison of growth inhibition zones for wild-type and repair-deficient strains	Lethal DNA damage to bacteria	25
		SOS induction	Induction of DNA repair	27
Neurospora crassa	Fungal, reverse mutation	Reversion to adenine or tryptophan independence	Frameshift and base-pair substitutions	24
Saccharomyces cerevisiae	Yeast, intragenic recombination	Phenotypic alteration (colony color) or reversion to adenine or tryptophan independence	Frameshift and base-pair substitutions	24–26

TA98 and TA100 are the most common strains to detect base-pair substitution and frameshift mutations, respectively. Several versions of the *Salmonella* assay (standard plate method, spot tests, modified plate methods, fluctuation test) have also been used to test explosives [1–21]. Two of the modified versions include increased incubation times, potentially resulting in increased sensitivity for mutagenicity detection or increased susceptibility to cytotoxicity, which may impede the ability to detect mutagenicity. In one modification [1], the bacteria are preincubated for 90 min with an S9 homogenate before addition of top agar and plating; the standard assay uses no preincubation. The revertant colonies are enumerated after two days, just as with the standard plate assay. The other modified assay is a fluctuation test [2] in which the bacteria are exposed in liquid culture and their growth (positive or negative) monitored after five days. Due to the large number of strains available for the *Salmonella* assay, the discussions in this section are mostly limited to the TA98 and TA100 strains. Other assays based on the reverse mutation concept included in this section are the bacterial-based *Vibrio fischerii* assay (Mutatox) [12,22,23] and fungus-based *Neurospora crassa* assay [24]. The Mutatox assay, although based on reverse mutation (base-pair and frameshift), can also detect induction of enzymes of the SOS repair system and damage induced by DNA intercalation. The *Neurospora* assay is similar to the *Salmonella* assay, with several strains developed for detection of base-pair substitution or frameshift mutations. The yeast *Saccharomyces cerevisiae* was used to monitor a different endpoint, that of intragenic recombination events [24–26], whereas the bacteria *Escherichia coli* was used to monitor DNA repair occurring after DNA damage, with one assay through comparison of toxicity in DNA repair enzyme deficient-strains versus wild-type strains [25], and another through the monitoring of the SOS induction response [27].

8.2.1.1   2,4,6-Trinitrotoluene (TNT) and Its Products

TNT tests positive in both TA98 and TA100 *Salmonella* strains [1–10]. The reported lowest observed effect levels ranged from 2 μM [8] to 500 μM [3]. Negative results reported for TA100 were probably due to the use of lower dose ranges, as authors either tested lower concentrations (2 to 44 μM) [11] or did not report the concentrations tested [12]. There was no consistent trend in the differential ability to induce base-pair or frameshift mutations. With only one exception [1], the addition of S9 resulted in reduced mutagenic activity for both strains; the increase observed in TA98 upon addition of S9 may have been due to the use of the modified plate assay. The utilization of special *Salmonella* strains [7–9,13] has determined that the presence of bacterial nitroreductases and O-acetyltransferases increases mutagenicity of TNT and many of its known metabolites, emphasizing the importance of metabolic pathways other than the oxidative metabolism provided by the addition of S9. The *Vibrio* assay [12,22,23] confirmed the direct mutagenicity and lack of metabolic activation observed in the *Salmonella* assay. In contrast, the *E. coli*-based assay [27] indicated that TNT only induced DNA damage in the presence of S9. The differences observed between these assays may be due to the fact that these assays detect different genotoxic events (Table 8.1). Another point to be considered is that the S9 fraction contains not only phase I oxidative/reductive enzymes, which produce amino, nitroso, and hydroxylamine metabolites, but also phase II conjugating enzymes such as UDP-glucuronosyltransferases, sulfotransferases, methyltransferases, acetyltransferases, and glutathione S-transferases. In general, conjugation reactions yield inactive metabolites, as exemplified by the in vivo acetylation of DNT and its subsequent excretion, but this is not always the case [28]. Further discussion of this complex subject is, however, beyond the scope of this chapter.

In addition to the parent compound, many TNT reduction products that may not form under the microorganism assay conditions but have been detected either as environmental degradation products or in vivo metabolic products have also been tested. The two major aminodinitrotoluene (ADNT) isomers test positive in a variety of assays. The 2-amino isomer was mutagenic in both *Salmonella* strains at test concentrations ranging from 1 to 13 mM [1,2,5–7,12], the two negative reports tested lower concentrations (up to 500 μM for TA98 [1] or 25 μM for TA98 and TA100 [11]). The addition of S9 did not increase mutagenicity in either strain, with the exception of one report using the modified plate assay [1]. Results for the 4-amino isomer were similar for concentrations ranging from 0.5 to 13 mM [1,2,5–7,12], with nonmutagenic outliers either testing at lower concentrations (up to 40 μM for TA98 [2], and 25 μM for TA98 and TA100 [11]) or not reporting the maximal concentrations used [12]. Again, there was no increase in mutagenicity in the presence of S9 except in a modified assay [1] for the TA98 strain. The *Vibrio* assay was consistent with the *Salmonella* assay, with both isomers testing positive [12,23,27], although the addition of S9 resulted in increased mutagenicity for the 4-amino compound [12,23].

The diaminonitrotoluene (DANT) isomers are mutagenic in a variety of assay types, with the 2,6-DANT isomer being more potent than 2,4-DANT in the *Salmonella* assay. The 2,6- isomer was consistently positive for both TA98 and TA100 strains and was not metabolically activated by S9 homogenate [1,2,6]. The 2,4-DANT results were not consistent, possibly due to cytotoxicity of the compound. Karamova et al. [8] reported a maximal induction of mutants at 30 µM and an abrupt loss of mutagenicity at the next highest dose of 300 µM, which was probably due to toxicity. Thus, the lack of response reported by Tan et al. [6] may have been due to the use of concentrations toxic to the bacteria (150 to 749 µM). Similarly, the fluctuation assay [2] included concentrations (up to 64 µM) found to be nontoxic in the standard plate assay [8], but the extended exposure period may have increased sensitivity toward toxic effects. Alternatively, when the concentrations used were not high enough to detect mutagenicity, mutagenicity for the DANT isomers appears to be detectable only at cytotoxic or near cytotoxic concentrations. Won et al. [11] used a maximal concentration (25 µM) 10-fold to 40-fold lower than the investigations that reported positive mutagenicity in TA98 [1,8] or TA100 [1,2,8], whereas Honeycutt et al. [12] did not report the concentrations tested. Thus, the lack of reported mutagenicity in the majority of the investigations on 2,4-DANT may be due to the testing of concentrations above toxic levels or below detection levels. Addition of S9 either did not increase or decrease mutagenicity [1,2,8,11,12] in the TA100 strain, whereas in the TA98 strain, S9 increased in mutagenicity in one report, which used the modified fluctuation assay [2]. Both the *Vibrio* [12] and *E. coli* [27] assays support the direct mutagenicity of 2,4-DANT, and the lack of increased mutagenicity upon S9 addition.

The mutagenicity of hydroxylamino DNTs was determined in two studies. The earlier study [11] reported two isomers as nonmutagenic, but the concentrations tested (up to 2.4 µM) were 50-fold lower than the lowest dose (117 µM) of the later study [4], which reported both the 2- and 4-hydroxylamino-dinitrotoluenes as well as 2,4-dihydroxylamino-6-nitrotoluene as direct-acting mutagens in both TA98 and TA100 strains. All hydroxylamino compounds were about 10-fold more potent as base-pair mutagens than as frameshift mutagens, and increasing hydroxylation appeared to increase mutagenicity, despite the lack of evidence from *Salmonella* assays with ADNT and DANT isomers in the presence of S9 oxidative enzymes.

The mutagenicity of the condensation products of TNT reduction products, tetranitroazoxytoluenes, is isomer dependent. The 2,2',6,6'-tetranitro-4,4'-azoxytoluene isomer (4,4'-AZT) was not mutagenic in the TA98 strain at concentrations up to 300 µM, both with and without S9 [1,11,12]. It was mutagenic in the TA100 strain in two reports using test concentrations between 10 µM and 300 µM [1,7]; concentrations lower than this were not positive [11]. The 4,4',6,6'-tetranitro-2,2'azoxytoluene isomer (2,2'-AZT) tested positive for TA98 [1] and TA100 [1,7,12] strains at a concentration of 50 µM or greater, and exhibited similar potencies for the two bacterial test strains. Neither 2,2'-AZT or 4,4'-AZT was activated by addition of S9 [1,7,12,14]. The 2,4',6,6'-tetranitro-2',4-azoxytoluene (2',4-AZT) also tested positive in both strains, with increased mutagenicity in the presence of S9 (fivefold and eightfold increases in TA98 and TA100 strains, respectively) [1,7]. When activated by S9 homogenate, 2′,4-AZT was the most potent of all compounds tested, with or without S9 homogenate [1]. Both 4,4′-AZT and 2,2′-AZT were reported as suspected mutagens in the *Vibrio* assay, due to nonreproducible positive hits without the presence of a dose-related response [12]; the 2′,4-AZT isomer was not tested in this assay. For the azoxy compounds, the presence of nitro groups in the 4 positions increased the potency for inducing frameshift mutations. The isomer with no nitro groups in the 4 position (4,4′-AZT) was the least potent isomer and acted only as a base-pair mutagen. 2′,4-AZT has one nitro group in the 4 position and was active as both a frameshift and a base-pair mutagen (more potent as base-pair mutagen), and 2,2′-AZT that has nitro groups in both 4 positions was the most potent mutagen with similar potencies for base-pair and frameshift mutations [1].

8.2.1.2   Dinitrotoluenes (DNT) and Their Products

Both 2,4- and 2,6-DNT have been shown to be nonmutagenic in the TA98 *Salmonella* strain [1,7,15] indicating that they do not cause frameshift mutations. Although addition of S9 did not result in detectable mutagenicity for 2,6-DNT [1,7], 2,4-DNT was slightly mutagenic in the TA98 strain upon addition of S9 in a modified plate assay [1]. The results for the TA100 strain are less consistent. Positive results in the absence of S9 activating enzymes have been reported for 2,4-DNT [7] and 2,6-DNT [1,7], and those reporting negative results were consistently tested at lower concentrations [1,17]. In the presence of S9, 2,6-DNT was mutagenic [1,7] in the TA100 strain, whereas 2,4-DNT tested negative in the standard plate assay [7] but positive in the modified [1]. The importance of non-S9 homogenate in activating nitrotoluene compounds is emphasized through the use of *Salmonella* strains that overproduce nitroreductase and O-acetyl-transferase enzymes. Mutagenicity was as much as 40-fold greater for 2,4-DNT and 2,6-DNT in the strain that overproduced both enzymes as compared to that of the wild-type strain [15].

The DNT reduction products 2,4- and 2,6-diaminotoluene (DATs) are non-mutagenic in both base-pair and frameshift detection strains [1,17]. However, in the presence of S9 metabolic enzymes, both compounds induced frameshift mutations (2,6-DAT more so than 2,4-DAT) [1], and 2,6-DAT also induced base-pair mutations, although to a lesser extent than frameshifts (0.3 and 2.5 revertants μg^{-1}, respectively). The *Vibrio*-based assay indicated that 2,4-DAT was not mutagenic without metabolic activation, but 2,4-DAT with S9 and 2,6-DAT with or without S9 were all reported only as suspected mutagens due to positive responses lacking clear dose-related responses or inconsistent positive responses [22]. The hydroxylamino-nitrotoluene isomers were all direct-acting mutagens in the TA100 *Salmonella* strain [17], supporting reports of mutagenicity of DATs in the presence of S9 homogenate. Again, enzymes other than those found in S9 fractions are of importance to the mutagenicity of these compounds; although a wide range of metabolites were determined to be nonmutagenic, all compounds tested positive in the absence of S9 in at least one strain that overproduced nitroreductases and/or O-acetyltransferases [15]. Hydroxylamino compounds were more mutagenic than their parent compounds, but amino-nitro compounds were less mutagenic than their respective parent compounds. Interestingly, the frameshift-detecting strain YG1041 was more sensitive to azoxybenzenes than a corresponding strain that detects base-pair substitution (TG1042); azoxybenzenes were by far the most potent of the tested metabolites, ranging from 33 to 2417 revertants μg^{-1}. These results are similar to those observed for azoxy compounds derived from TNT, indicating that unlike most other explosives-associated compounds, azoxy compounds may be potent frameshift mutagens.

In summary, it appears that 2,4-DNT, 2,6-DNT and their breakdown products do not directly cause frameshift mutations, with the exception of azoxybenzenes. Evidence suggests that although they are metabolically activated by S9 homogenate, nitroreductases and O-acetyltransferases also play an important part in the activation of these compounds. Both DNTs are direct-acting base-pair mutagens when tested at high enough concentrations, and their activity is increased by metabolic activation; the latter is supported by data from both the modified S9 test and the direct base-pair mutagenicity activity of the hydroxylamino-nitrotoluene intermediates.

8.2.1.3   Cyclic Nitramines and Their Products

The cyclic nitramines hexahydro-1,3,5-trinitro-1,3,5-triazine (RDX) and octa-hydro-1,3,5,7-tetranitro-1,3,5,7-tetrazocine (HMX) consistently tested negative in the *Salmonella* assay for all strains; assay concentrations up to and exceeding solubility showed no evidence of mutagenicity with or without the presence of S9 [1,2,6,24]. Negative results for RDX in *Saccharomyces* strain D3 with and without metabolic activation have also been reported [26], although the concentrations tested were not reported. RDX was also tested in the *Vibrio*-based assay, with mixed results [22]. Without the presence of S9, two of three replicates resulted in positive mutagenic results, but no dose-related response was observed. In the presence of S9, the *Vibrio* assay results were positive, but a dose-related response was observed in only two of three replicates. Due to these inconsistencies, RDX was listed as a suspected rather than a confirmed mutagen. These results do not necessarily conflict with those of the *Salmonella* assay, as Mutatox detects a more diverse range of genotoxic mechanisms (Table 8.1). In contrast to the parent RDX, the trinitroso product (TNX) was slightly mutagenic in the base-pair detection strain (TA100) both with and without metabolic activation, whereas the mono- and dinitroso products gave negative results in all tests. The mutagenic activity was very low compared to other explosives, resulting in only 0.3 revertants μg^{-1}, compared to 3.4 revertants μg^{-1} for TNT or 219 revertants μg^{-1} for one of the TNT azoxy metabolites [1].

8.2.1.4   Other Compounds

Tetryl was shown to be a potent, direct-acting mutagen in the *Salmonella* assay, and was consistently more active in the TA100 strain (base-pair substitution) than in the TA98 strain (frameshift) [1,6,24,25]. This is supported by the *Neurospora* reverse mutation assay, in which the N23 mutant (detects base-pair mutations) was positive for tetryl but the 12-9-17 mutant (detects frameshift mutations) was not [24]. Tetryl also tested positive for mutagenicity in the *Saccharomyces* assay [24,25]. Tetryl appears to act directly, as S9 did not significantly increase mutagenicity regardless of the assay system. In the *E. coli* DNA repair assay [25], tetryl was found to be cytotoxic, but not genotoxic, to *E. coli*.

Trinitrobenzene (TNB) is also a direct-acting mutagen, causing both frameshift and base-pair mutations. There was no consistent trend between the differential sensitivity toward TA98 or TA100. Two studies [2,25] reported similar activity for the two strains. George et al. [1] used a modified plate method and reported the TA98 strain to be more sensitive than the TA100 strain, whereas Assmann et al. [14] reported significantly greater sensitivity of the TA100 strain. Addition of S9 reduced mutagenic potential in all cases. In other assay systems, TNB was cytotoxic rather than mutagenic, and did not induce either DNA repair in *E. coli* or mitotic recombination in *S. cerevisiae* [25].

Nitroglycerin (NG) is weakly mutagenic in *Salmonella* tests [16]. Although NG gave positive results in the base-pair substitution sensitive strain (TA1535), no significant reversions were observed for six other *Salmonella* strains. Addition of S9 reduced its mutagenicity. Analysis of the mutant spectra in the TA1535 strain indicated that NG has a similar mechanism to spermine and nitric oxide (C to T transitions), suggesting that nitric oxide may be the active metabolite of NG [16]. Coexposure with extracellular nitric oxide scavengers did not decrease mutagenicity, indicating that intracellular production of nitric oxide was responsible for this effect [16].

8.2.1.5   Mixtures and Environmental Samples

Several investigations conducted at Oak Ridge National Laboratory in Tennessee [6,18,19] have utilized the *Salmonella* assay to follow the mutagenicity of TNT-contaminated soils during degradation. Based on HPLC analytical data and relative potencies of the known products, it is predicted that: (1) the TA100 strain should have a higher mutagenic response to compost extracts than the TA98 strain, and (2) addition of S9 should decrease mutagenic activity in both strains. Neither of these hypotheses held true. In addition, the mutagenic activity predicted on the basis of mutagenic potency and concentrations in composts ranged from 13% to 33% of the measured values for TA100 and from 2% to 12% for TA98. Similar results have been reported for extracts of TNT and RDX-contaminated soils [20], which induced more mutations in the TA98 strain compared to the TA100 strain, caused greater than expected mutagenicity based on measured concentrations of compounds, and resulted in increased mutagenicity upon addition of S9. In these investigations, the authors hypothesized that unknown mutagenic compounds present in the samples were the basis of the unexplained results.

The formation of AZTs may be responsible for the unpredicted results. Although not detected in the extracts, AZTs were present in compost leachates and may have formed in vitro in the concentrated extracts. Several AZTs are direct mutagens [1,12] with potencies up to 164-fold higher than TNT. The presence of AZTs may also explain the increase in mutagenicity upon addition of S9 metabolic enzymes observed in initial compost extracts [19]. Although none of the metabolites detected in extracts are known to increase mutagenicity upon S9 addition, 2',4-AZT is metabolically activated to a high degree [1]. Alternatively, the higher than expected mutagenicity may also be due to mixture interactions between explosives-related compounds, or the presence of nonexplosives-related compounds that modify availability and reactivity of the explosives-related compounds. Kaplan and Kaplan [3] found synergistic (more than additive) effects of surfactants and TNT on mutagenicity rates in *Salmonella* strains TA1538 and TA98.

Although it is possible that the unexpectedly high mutagenicity may be due to compounds unrelated to TNT, in vivo rat exposure data supports the hypothesis that at least some of the mutagenicity is related to unknown TNT products. When urine from rats exposed to TNT was fractionated by high performance liquid chromatography (HPLC) prior to running the *Salmonella* assay, the majority of the mutagenic activity was associated with several unknown peaks [21], including a peak eluting near 2,6-DANT and a late-eluting nonpolar peak. Both peaks were associated with substantially higher revertants per plate in the TA98 strain than in the TA100 strain. No mutagenicity was detected in the unexposed rat urine. Hence, it has been shown that unknown products possess higher activity in TA98 than in TA100 tester strains and that they contribute significantly to the mutagenicity of environmental samples.

8.2.2 Mammalian Cell Line-Based Genotoxicity Assays

Most studies conducted on nitroaromatics in mammalian cells have dealt with polycyclic nitroaromatic hydrocarbons. Although a wide range of mutagenic potencies for various nitro arenes have been reported, recent reviews [29,30] did not include explosive compounds. The various assays used to assess genotoxicity of explosives are described in Table 8.2.

8.2.2.1 2,4,6-Trinitrotoluene (TNT) and Its Products

In contrast to the well-recognized mutagenic potency of TNT in bacterial assays, there is some disagreement on its genotoxicity in mammalian cells. In the P388 mouse lymphoma cell assay, exposure to TNT resulted in mutagenicity at the TK locus, but only in absence of S9 and at concentrations causing a high cell-death rate [31]. Additionally, the exposure period (30 min) was brief, and there was no mention of whether a carrier solvent was used to reach the high concentrations (up to 1000 μg ml^{-1}) employed. TNT was also reported to be mutagenic at the HGPRT locus in Chinese hamster ovary (CHO) cells [32], although the dose–response curve was atypical, with peak activity at 40 mg L^{-1} and much lower activity at all other concentrations. This response may be a result of the known ability of aryl amines to suppress their own conversion into hydroxylamino derivatives, the proximate genotoxic products [33]. The low reproducibility of the data led the authors to suggest that the results were due to relatively infrequent mutations, which were close to the limit of detection of the assay [32]. This observation would be consistent with the reported negative results at the HGPRT locus in V79 Chinese hamster lung cells [2]. Both the V79 and CHO HPRT assays are reported to be insensitive to clastogens due to the localization of the HPRT gene on the unique X chromosome [34], which may explain the apparent weak mammalian mutagenicity of TNT. In isolated rat lens epithelial cells exposed to TNT in glycerol, single-strand breaks were detected [35]. The cells were highly sensitive to TNT, indicating that primary cells still possess the metabolic capacity to activate TNT and/or are devoid of protective mechanisms active in other cells. Contrary to these results, TNT tested negative in absence of metabolic activation with the TK6 human lymphoblastoma cell assay at concentrations ranging

TABLE 8.2

In Vitro Mammalian Cell-Based Systems Used to Assess Mutagenic Potential for Explosives and Explosives-Related Compounds

Cell Type	Basis of Assay	Endpoint	Reference
P388 mouse lymphoma	TK locus, forward mutation	5-IdU$_r$	31
L5178Y mouse lymphoma		F$_3$T$_r$	49
Mouse lymphoma			50
K$_1$ – BH$_4$ Chinese hamster ovary	HPRT locus, forward mutation	6-TG$_r$	32
CHO K$_1$			41, 42
CHO cells			48
Chinese hamster ovary	DNA damage (clastogenicity)	CA	44
		CA + B	50
V79 Chinese hamster lung	HGPRT locus, forward mutation	6-TG$_r$	2, 43, 39
TK6 human lymphoblast	TK locus, forward mutation	F$_3$T$_r$	36
Human lymphocytes	DNA damage (clastogenicity)	CA	46
WI-38 human fibroblast	DNA damage	UDS	47
Primary rat hepatocyte	DNA damage	UDS	40, 50
Rat hepatocyte in vivo–in vitro assay	DNA damage	UDS	62, 65
Rat isolated epithelial lens cells	DNA damage	SSB	35
Human liver carcinoma HepG2 transgenic lines	CAT-Tox	DNA damage	38

Note: 5-IdU$_r$, resistance to 5-iodo-2-deoxyuridine; 6-TG$_r$, resistance to 6-thioguanine; F$_3$T$_r$, resistance to trifluorothymidine; CA, chromosomal aberrations; B, breaks; UDS, unscheduled DNA synthesis; SSB, single-strand breaks; CAT-Tox clones detect alterations in DNA sequence or helical structure.

from 40 to 112 mg L^{-1} [36]. For TNT, cell survival was only reduced to 65% at the highest concentration tested, suggesting that TK6 cells lack the necessary activation enzymes, or that higher concentrations are needed to detect the low level mutagenicity of TNT. Differential sensitivity between human and other mammalian cells [37], as well as organ specific toxicity, is another possible explanation of the contradictory results reported. Another likely explanation for the low mutagenicity of TNT in standard assays is the lack (or the oxygen sensitivity, resulting in inactivation) of the enzymes necessary for metabolic activation in some of these cells [30]. Based on newly developed transgenic cell lines, transcriptional gene activation was recently studied following TNT exposure [38]. TNT demonstrated the potential to alter DNA sequence or DNA helical structure by inducing DNA repair systems. The most specific effect of TNT was induction of the c-fos promoter that is implicated in the mammalian cell response to the well-known mutagens 4-nitroquinoline-1-oxide and methyl methane sulfonate [38].

The genotoxicity of TNT's known metabolites has also been investigated. In the CHO HPRT assay, 4,4′-AZT, 2′,4,6,6′-tetranitro-2,4′-azoxytoluene (2,4′-AZT), and triaminotoluene (TAT) were found to be directly mutagenic and 4-ADNT and 2′,4-AZT were mutagenic following metabolic (S9) activation [32]. The positive response of these compounds was characterized by an absence of a concentration–response relationship, indicating a weak activity or a significant suppressing activity in this in vitro model, as described earlier for TNT. The following compounds were statistically inactive in CHO cells: 2-ADNT, 2,4-DANT, 2,6-DANT, and 2,2′-AZT. The classification of 2,6-DANT as nonmutagenic in spite of a clear, linear, concentration–response relationship with S9 activation may be attributed to the fact that only a single experiment was conducted and the resulting numbers of mutant cells were close to the background level. None of the reduced TNT metabolites studied in the V79 Chinese hamster lung cells assay were mutagenic, including 2-ADNT, 4-ADNT, 2,4-DANT, 2,6-DANT, and TAT [2]. However, one positive result out of three experiments was seen for 4-ADNT in the absence of S9 activation. Following metabolic activation, high concentrations (70 mg L^{-1}) of 2,6-DANT were mutagenic and highly toxic (87%–89% cell growth inhibition) [39]. Both 2- and 4-ADNT were also tested in the TK6 human lymphoblastoma cell assay, giving negative results at concentrations of 52 mg L^{-1} [36]. No toxicity was observed at this concentration.

8.2.2.2 Dinitrotoluenes and Their Products

Both 2,4- and 2,6-DNT were first characterized as nongenotoxic in an unscheduled DNA synthesis (UDS) assay in primary rat hepatocytes [40]. The reduced metabolite 2,4-DAT (but not 2,6-DAT) produced a positive response in the same assay. Later, 2,4-DNT was found to be mutagenic in absence of metabolic activation in the P388 mouse lymphoma cell assay [31]. In the same assay, technical grade dinitrotoluene (76% 2,4-DNT, 19% 2,6-DNT, 5% other isomers) and pure 2,6-DNT were inactive with or without rat liver S9 activation. In a separate study, pure 2,4-DNT gave negative results in CHO cells at the HGPRT locus when tested in conjunction with Aroclor-induced rat liver metabolic activation [41]. Under reduced oxygen tension, exposure to technical grade 2,4-DNT resulted in increased toxicity and produced a weak but significant increase in 6-TG$_r$ mutant frequency [41]. A linear concentration–response could not be established, reminiscent of the data previously discussed for TNT and its metabolites [32]. The use of primary hepatocytes from Aroclor-treated rats and of *Salmonella* bacteria as activating systems for pure 2,4-DNT also resulted in the production of mutagenic metabolites in CHO cells [41]. Bacterial transformation of 2,4-DNT under anaerobic conditions produced 2-amino-4-nitrotoluene and 4-amino-2-nitrotoluene as well as azoxy compounds. Subsequently, the same team reported that technical grade 2,4-DNT as well as pure 2,4-DNT, 3,4-DNT and 2,6-DNT (at concentrations up to 2 mM) were nonmutagenic using standard conditions in the CHO-HPRT assay [42]. Both 2,4-DNT and 2,6-DNT isomers gave negative results in the HGPRT-V79 assay when tested in absence of metabolic activation [43].

2,4-DNT is considered nonclastogenic in the National Institute for Environmental Health Sciences (NIEHS) database, following tests for chromosomal aberrations in Chinese hamster lung and/or CHO cells but can cause chromatid exchanges in the same cells [44]. Furthermore, DNT isomers (2,3-DNT, 2,4-DNT, 3,4-DNT, 2,6-DNT) and technical grade DNT were unable to induce morphological transformation of Syrian hamster embryo (SHE) cells in culture, suggesting that some important activation step was missing in SHE cells [45]. An example showing that some primary cells may have the ability to metabolize nitro aromatics, or may be more sensitive to various metabolites, is given by the fact that 2,4-DNT was clastogenic in phytohemagglutinin-stimulated human peripheral lymphocytes [46]. Transcriptional gene activation was studied following DNT exposure in transgenic cell lines [38], and 2,4-DNT, but not 2,6-DNT, demonstrated the potential to alter DNA sequence or DNA helical structure by inducing DNA repair systems. This finding partly contradicts the results of other researchers who consider 2,4-DNT as mainly a promoter, and 2,6-DNT as an initiator/promoter [45]. However, this classification takes into consideration complex metabolic bioactivation in animals, a phenomenon that cannot be duplicated in vitro.

8.2.2.3 Cyclic Nitramines and Their Products

RDX was shown to be inactive in the UDS in WI-38 human fibroblasts, with and without S9 metabolic activation [47]. Reported test concentrations were in the range of 250 to 4000 μg ml^{-1}, well over the aqueous solubility of the compound (approximately 70 μg ml^{-1}) [48]. Similarly, in the HGPRT-V79 assay with or without metabolic activation, both RDX and HMX were inefficient in inducing 6-thioguanine resistant cells. The test concentrations used were up to 40 μg ml^{-1} for RDX and 10 μg ml^{-1} for HMX, close to the solubility limit of these compounds in water [2]. Another confirmation of the lack of mutagenic activity of RDX was provided recently when the compound was found inactive in the mouse lymphoma forward mutation assay using the L5178Y cell line [49].

The mutagenic potential of RDX at the TK locus was evaluated at concentrations ranging from 3.93 to 500 μg ml^{-1} with and without metabolic activation. None of the treatments induced mutant frequency that exceeded the minimum criteria for a positive response. MNX, the mononitroso metabolite of RDX, has been tested using a battery of in vitro assays and found to be genotoxic in mammalian cells. MNX was found mutagenic at the TK locus in the mouse lymphoma assay, with and without Aroclor–S9 activation [50]. It was also found to be clastogenic following S9 activation in the CHO chromosomal aberration assay, inducing breaks, chromosome number changes and chromatid interchanges [50]. No chromosomal aberrations occurred without S9 activation. In addition, in the primary rat hepatocyte unscheduled DNA synthesis assay, MNX significantly increased DNA repair [50]. Effective concentrations were not stated in the cited report.

8.2.2.4 Other Compounds

In spite of its high bacterial mutagenicity, TNB failed to elicit a significant number of 6-thioguanine mutants at the HGPRT locus. This may be related to its higher toxicity for mammalian cells. TNB was inactive in the CHO assay with or without metabolic (S9) activation [32], at concentrations (330 μM) causing up to 40% cell death. In the CHO assay, survival was reduced following metabolic activation. TNB was also inactive in the V79 Chinese hamster lung cells assay, with or without S9 metabolic activation [2]. In the V79 cell system, a 4-h exposure at a concentration of 300 μM with metabolic activation caused no decrease in cell survival (clonal assay). TNB was also nonmutagenic in the absence of metabolic activation in the TK6 human lymphoblast cell assay [36].

Results from earlier studies conducted in mammalian cells on NG were all negative. NG did not induce chromosomal aberrations in lymphocytes obtained from treated animals, and NG was reported to be toxic but nonmutagenic in CHO cells [48]. Later testing revealed that NG has a weak mutagenic potency, causing increased mutants (20 net mutants per 10^5) in MN-11 murine tumor cells [51]. NG's weak mutagenicity is attributed to nitric oxide production derived from NG via metabolic reduction. It is not known if nitric oxide production can occur following exposure to other nitro compounds, such as nitroaromatics or nitramines. Nitric oxide-mediated mutagenic effects of NG may be implicated in its carcinogenicity in animals [52,53].

8.2.2.5 Mammalian Cell Line Testing of Environmental Samples

Mammalian cells are not generally used to assess the genotoxicity of environmental samples, probably because of the complexity and increased cost compared to microbial assays. In a recent review, only three studies using mammalian cells were located, of which one dealt with explosives [54]. The V79 Chinese hamster cell assay was used to evaluate the mutagenic potential of organic extracts of explosives-contaminated soil to provide data on mammalian genotoxic potential for risk assessment purposes [20]. Interestingly, extracts were mutagenic at nontoxic concentrations although the authors could not identify the responsible component(s), because at the time no data were available for explosives in this test system. Given the fact that all individual chemicals have since tested negative in V79 cells (without S9 activation), one may conclude that this observation was a mixture effect.

8.3   IN VIVO METHODOLOGIES

Whereas in vivo assays are more difficult to conduct, they include toxicodynamic factors that cannot be readily accounted for in vitro, such as uptake, distribution, metabolism (includes both activation and deactivation of genotoxicants), repair, and excretion. The pitfalls of relying strictly on in vitro methodologies for assessment of genotoxic potential are exemplified by DNT. Ingestion of technical grade DNT resulted in 100% incidence of hepatocellular carcinomas in male Fischer rats, even though in vitro assays indicated DNTs are not potent mutagens. Contrary to an in vitro UDS hepatocyte assay, hepatocytes isolated from DNT-exposed rats indicated strong UDS induction [55]. Exposures of axenic (germ-free) rats proved intestinal flora were responsible for metabolism to aminonitrotoluenes, which are products known to induce UDS in mammalian systems.

The activation of DNT has been shown to be a multistep process involving metabolism in the liver, excretion into the bile, deconjugation of metabolites and further metabolism by the intestinal flora, re-uptake (enterohepatic transport) of metabolites into liver, and finally activation and binding to cellular macromolecules in the liver [56]. More recent studies [57] involving rats pretreated with coal tar creosote, which potentiates the genotoxicity of 2,6-DNT, elucidated a complex interaction that balances metabolic activation, uptake, and detoxification. The study monitored intestinal flora enzyme activities, bacterial analysis, mutagenicity of urine samples, HPLC analysis, and hepatic DNA adducts over a five-week exposure period. The location of nitroreductase activity was an important factor in determining if increased activity would potentiate or reduce hepatic DNA adducts; activity in the small intestine resulted in reduction of DNT before it was absorbed, thus reducing genotoxicity; whereas activity in the lower intestinal tract bioactivated DNT metabolites. Optimal conditions for bioactivation of DNTs require low nitroreductase activity in the small intestine to allow uptake of DNT into the liver, elevated hepatic mixed function oxidases for increased metabolism and release of conjugated metabolites into the intestines by way of the bile, and elevated intestinal β-glucuronidase activity to increase release of hepatic DNT metabolites to allow reuptake of intermediate compounds. Although assessment of mutagenic potential of individual metabolites in in vitro assays has helped support this hypothesized complex metabolic activation pathway, it is clear that in vivo exposure studies were critical in the understanding of the process.

8.3.1   Mammalian In Vivo Genotoxicity Investigations

As compared to the amount of data available for in vitro genotoxicity assays, very few in vivo genotoxicity investigations have been conducted for explosives and their associated compounds. No data were available for lifetime in vivo exposures to tetryl or HMX. The most common in vivo genotoxicity methodology is the classical carcinogenicity assessment in which animals are exposed daily for six months to two years prior to sacrificing the animals for hematological and histopathological analyses. Another method that has been used to assess explosives is the dominant-lethal mutation assay, in which males are exposed prior to mating to unexposed females, and the number of surviving implanted embryos and dead implanted embryos are determined and compared to controls. This assay detects mutations in the sperm cell DNA that does not cause dysfunction in the sperm but is lethal to the egg or developing embryo; it is generally understood that dead implants (dominant lethals) are the result of chromosomal damage, although gene mutations cannot be excluded. Formation of micronuclei or unscheduled DNA synthesis in bone marrow and/or blood of rodents after in vivo exposure is yet another method that has been used for assessing DNA damage to rodents exposed to explosives in vivo; these assays are typically run for shorter exposure periods.

8.3.1.1   TNT and Its Products

TNT is regarded as a carcinogen in animal models with an oral slope factor for carcinogenic risk, listed in the Integrated Risk Information System (IRIS) as 0.03 per mg kg^{-1} d^{-1} [58]. Two-year carcinogenicity studies conducted with rats resulted in urinary bladder papilloma and carcinoma in female Fischer 344 rats [59]. In addition, hepatocellular (in male rats), renal and urinary bladder hyperplasia (in female rats) are seen at doses of 10 mg kg^{-1} d^{-1}. In contrast, TNT was not considered carcinogenic at doses up to 70 mg kg^{-1} d^{-1} over a two-year period in hybrid B6C3F1 mice [60]. Although initial data analysis indicated the combined incidence of lymphomas and leukemias was significantly elevated in female mice from the highest dose, later analysis [61] demonstrated no significant effects or trends, indicating that the neoplasms were not treatment related.

Endpoints for acute exposures to TNT included bone marrow micronucleus in rats and mice, and UDS in rat liver. TNT did not test positive for either mouse bone marrow micronuclei or rat liver UDS [62]. In a separate study, rats treated for 28 d with up to 190.4 mg TNT kg^{-1} d^{-1} in feed-generated negative genotoxicity results as assessed by in vivo cytogenetic analyses in bone marrow [47]. These negative results may be due to the selection of bone marrow as the target, since chronic TNT exposure in rats generates tumors in urinary bladder and liver. Negative results from the liver UDS assay may be due to the acute toxicity of TNT that may interfere with the detection of damage to DNA in vivo because doses in short-term testing are typically higher than those used in chronic studies. It is unlikely that the negative results for UDS in rat livers were due to the inability of TNT to form reactive metabolites in vivo, as TNT administered intraperitoneally (ip) to rats resulted in about 2% of a radiolabeled dose of TNT accumulating in the liver, and 30% covalently bound to proteins within a 4-h period [63]. Although DNA adducts were not quantified, the data clearly indicate that TNT and its reactive metabolites are found in the liver.

Although no dominant lethal studies have been reported for TNT, evidence exists that it can damage gametes in vivo. Following exposure of rats to a high dose of 300 mg TNT kg^{-1}, the amount of 8-oxo-7,8-dihydro-2′-deoxyguanosine (8-oxo-dG) increased in the caput epididymis, whereas the increase in the whole testis was not statistically significant [64]. Lack of alteration in plasma testosterone, combined with the ability of antioxidants (catalase and bathocuproine) to ameliorate the observed DNA damage, pointed to the involvement of hydrogen peroxide and copper (I). The increased formation of oxidized DNA in caput epididymis was attributed to the low DNA repair capacity of cells in this tissue. The authors also reported that among the compounds tested in vitro using isolated calf thymus DNA, 4-hydroxylamino-2,6-dinitrotoluene was found to be a potent inducer of 8-oxo-dG formation; TNT and 4-ADNT did not oxidize DNA in vitro.

8.3.1.2 Dinitrotoluenes and Their Products

The complexities involved in assessing the carcinogenic potential of DNT isomers have already been discussed. The U.S. Environmental Protection Agency (USEPA) classifies selected DNTs as animal carcinogens; the technical grade mixture has a listed oral slope factor for carcinogenic risk of 0.68 per mg kg^{-1} d^{-1} [58]. In addition to the previously discussed data [55–57], several studies have shown that genotoxic activity of DNT is isomer specific. Both 2,4- and 2,6-DNT gave positive results in the in vivo–in vitro rat hepatocyte UDS assay [65], with 2,6-DNT being 10 times more potent than 2,4-DNT. Dietary exposures of rats to 2,4-DNT at 95% purity or greater tended to result in benign tumors at lower doses [66] or elevated hepatocarcinomas only at high doses (700 ppm) [67], whereas technical grade DNT (76% 2,4-DNT, 19% 2,6-DNT) resulted in hepatocarcinomas at doses as low as 10 mg kg^{-1} d^{-1} [68]. Pure 2,4-DNT did not increase the incidence of hepatocarcinomas at doses up to 27 mg kg^{-1} d^{-1}, whereas technical grade DNT induced tumors (47% of the treated rats) at 35 mg kg^{-1} d^{-1}, and pure 2,6-DNT induced significant increases in tumors at doses as low as 7 mg kg^{-1} d^{-1} (85% of treated rats) [69]. In a series of tests [70–72], it was determined that 2,6-DNT acts as a complete carcinogen, whereas 2,4-DNT acts only as a promoter and thus cannot induce tumors without the presence of the 2,6-DNT or other initiators. In vivo studies with 2,4-DNT, 3,5-DNT, and their diamino counterparts in mice indicate that administration intraperitonally or by gavage did not result in genotoxicity as determined by dominant lethal assay; the reduced percentage of fertile matings between exposed mice were due to testicular damage via nongenotoxic mechanisms [73,74].

One interesting aspect of the various in vivo exposures is that susceptibility to the formation of tumors and the target organ varies by both sex and species. Rats are more sensitive than mice, and for rats, the tumors are formed primarily in the liver; whereas in mice, tumors are located primarily in the kidney. Additionally, females of both species tend to be significantly less susceptible to tumor formation. Both aspects are likely to be a result of the complex metabolic activation and deactivation processes controlling carcinogenicity of DNTs in vivo. In liver tissue, which possesses a high level of oxidative enzymes, N- acetylation can lead to a proximal carcinogen (N-acetylamine) that is activated to the ultimate carcinogen via oxidative metabolism. In kidneys, N-acetylation is primarily a detoxification mechanism, as the compounds become more water soluble and more readily excreted into the urine. In an epidemiological study of underground mining workers exposed to technical grade DNT [28], the importance of N-acetylation was highlighted as it was determined that all patients with urothelial tumors were identified as "slow acetylators."

8.3.1.3   Cyclic Nitramines and Their Products

The carcinogenic potential of RDX has been evaluated in Fischer 344 rats, Sprague-Dawley rats, and B6C3F1 mice [75–77]. RDX was not found to be carcinogenic when fed to either strain of rats. In mice, it was found to produce significant increases in the combined hepatocellular adenomas/carcinomas in B6C3F1 female mice. Based on this mouse study, the USEPA classifies RDX as a possible human carcinogen, and the IRIS database currently reports an oral slope factor for carcinogenic risk of 0.1 per mg kg^{-1} d^{-1} [58]. Because B6C3F1 mice are known for their high spontaneous tumor formation and the article by Lish et al. [77] reported an unusually low control tumor level, the RDX data set was recently reevaluated by a NIEHS Pathology Working Group. Comparison of exposed animals to historical controls indicated that the high dose animals did indeed have significantly elevated tumor formation. As a result of the reevaluation, RDX will still be listed as a possible human carcinogen, but the USEPA may reduce the oral slope factor for carcinogenic risk to one-third of the current value [78]. Results from an in vivo mouse micronucleus assay with RDX in the control CD-1 (ICR) BR mouse [79] showed that whereas RDX induced signs of clinical toxicity in all tested doses (31 to 250 mg kg^{-1}), it was neither cytotoxic nor capable of inducing chromosomal damage. No significant effects (number of corpora lutea, implants, or live or dead embryos) were observed for the dominant lethal assay conducted in F-344 rats [80] fed on a diet containing RDX at up to 50 mg kg^{-1} d^{-1}.

Regarding products of RDX, a single preliminary assessment of the mononitroso degradation product of RDX (MNX) is available, although details regarding doses and statistics are not available [50]. In this report, subacute doses to male mice mated to unexposed virgin females did not result in dominant lethal effects. Because RDX has been shown to result in tumors in some model organisms and products have been shown to be mutagenic in in vitro assay conditions, it is possible that the compounds either do not reach the sperm in genotoxic concentrations or the testes may not metabolize RDX to genotoxic metabolites within the organ.

8.3.1.4   Other Compounds

Although TNB is positive in several in vitro assays, it does not appear to be genotoxic in vivo [81]. Slaga et al. [82] found TNB tested negative in a skin initiation assay. Chronic two-year oral exposure of male and female F344 rats to doses ranging from 5 to 300 mg kg^{-1} d^{-1} also did not reveal any carcinogenic potential [83,84].

Nitroglycerin is not listed as carcinogenic by USEPA [58], although there are reports of hepatic tumors in rats but not mice (doses >300 mg kg^{-1} d^{-1}) [52,53]. Chromosomal aberrations were not found in dogs given up to 5 mg kg^{-1} d^{-1} for nine weeks or in rats given up to 234 mg kg^{-1} d^{-1} for eight weeks [85]. A dominant lethal test in rats given up to 363 mg kg^{-1} d^{-1} in the diet for 13 weeks also showed no effect on male fertility and no genotoxic activity [85].

8.3.2   Other In Vivo Genotoxicity Investigations

Higher plants have long been used for testing the genotoxicity of pure chemicals and have been recognized as excellent indicators for detecting, monitoring, and assessing the mutagenicity of air, water, or soil [86,87]. Plant genetic assays are highly sensitive and are considered appropriate tests in the prediction of mutagenicity/carcinogenicity. The rapidly dividing cells found in pollen and roots are appropriate and sensitive tissues for cytological endpoints, such as chromosome/chromatid aberrations, sister-chromatid exchanges, and micronuclei. Among the more than 20 species used during the past half century, four plant genetic bioassays have proved to be highly sensitive, simple, and cost-effective. They are *Tradescantia* micronucleus in tetrad-stage pollen mother cells (Trad-MCN) assay, *Tradescantia* stamen hair mutation (Trad-SHM) assay, *Allium cepa* root tip micronucleus assay, and *Vicia faba* root tip micronucleus assay.

Although these assays are commonly used for assessing environmental contaminants, only one study was found in the current literature for explosives-related compounds. Gong et al. [88] tested the genotoxicity of 2,4-DNT and 2,6-DNT using the Trad-MCN assay. Plant cuttings bearing young inflorescences were exposed to aqueous solutions of the two DNT isomers. Micronuclei were scored in the tetrad-stage pollen mother cells after 6-h exposures. Results indicated that both 2,4-DNT and 2,6-DNT were genotoxic with minimum effective doses being 30 and 135 mg L^{-1}, respectively. The same authors also tested a Sassafras sandy loam soil amended with 25 to 2000 mg 2,4-DNT kg^{-1} soil, and plant cuttings were exposed to soil slurries made from two volumes of water and one volume of soil (unpublished results). After 6-h exposure followed by 24-h recovery, significant increases in micronuclei frequency were consistently observed in amended soils as compared to control soil. Although both DNTs require metabolic activation to exhibit genotoxicity [15], no metabolites were detected in the test solution. It is unclear whether the mechanisms of genotoxic activity are different between the microbial model and the plant model, or if the metabolites were retained in the plant tissues, which were not analyzed for

DNT or DNT products. Further research is required to determine and compare the underlying mechanisms of genotoxicity in different test systems.

8.4 STRUCTURE–ACTIVITY RELATIONSHIPS (SARS)

This section will focus on SARs in relation to mutagenicity of explosives. An excellent review of quantitative structure–activity relationship (QSAR) models for mutagenicity and carcinogenicity is available for a wide range of chemical classes [89]. Although several models based on carcinogenicity have been developed for aromatic amines, they include heterocyclics and multiple ring systems, and are thus not specific to nitroaromatic explosive compounds. To predict genotoxicity of compounds, it is helpful to understand the underlying mechanisms controlling genotoxicity. Chapter 9 provides an in-depth discussion of the mechanisms of cytotoxicity. Briefly, the classical mechanism through which nitro-containing compounds exert genotoxic effects is through the generation of the electrophilic metabolites, the reactive hydroxylamine metabolites, which form during nitro reduction or amino group oxidation. Alternatively, cyclic nitramines explosives are known to form N-nitroso compounds. N-nitroso compounds are generally believed to form reactive metabolites via oxidative metabolism at carbon atoms α or β to the nitroso.

The importance of nitroreductases was exemplified in a study testing 36 nitroaromatic compounds related to the manufacture of TNT, using five strains of *Salmonella* [5], in which the contribution of various structural characteristics to mutagenicity was determined. All isomers of DNT and ADNT, as well as 2,4,6-TNT and 2,3,4-TNT required nitroreductases to exhibit mutagenicity. Compounds possessing labile nitro groups that can readily undergo nucleophilic displacement are capable of directly reacting with DNA without the requirement of metabolic activation by nitroreductases. This category includes 3,4,5-TNT, 2,3,6-TNT, 2,4,5-TNT isomers, and 2A-3,6-DNT, all of which are active both with and without nitroreductases in the bacterial system. Within the group of compounds requiring nitroreductases to be mutagenic, potency appeared to be related to the ability to create an electron-deficient aromatic ring. Nitro groups in para relationships to each other enhance this electron deficiency and have been shown to increase mutagenicity.

Debnath et al. [90,91] published several articles on QSAR models for aromatic nitro compounds based on direct mutagenicity to the TA98 and TA100 bacterial strains. Two major factors were determined to be important predictors of mutagenicity for both groups. The water-octanol partition coefficient (log P or log K_{ow}) and water solubility were predictive of the ability of a compound to be absorbed and transported to a receptor site (metabolic enzymes/DNA). For nitro compounds, the optimal log P was 5.44. The other major factor was the LUMO (lowest unoccupied molecular orbital), which predicts the ability to undergo oxidative metabolism (electronic effects); mutagenicity increases inversely with LUMO. Benigni et al. [92] proposed that separate QSAR models are required for the prediction of whether a compound is mutagenic and prediction of mutagenic potency of compounds. Electronic and steric factors tended to discriminate whether a compound is mutagenic (the compound can be metabolized to an active compound), whereas log P better modeled the mutagenic potency (the extent to which the compounds can be metabolized). This may in part explain some of the negative results observed for N-nitroso degradation products of RDX. Although N-nitroso compounds are generally regarded as potent mutagens [93], RDX and its products are water soluble (low log K_{ow} of 0.81–0.87 for RDX) [94], which may impede both metabolic activation and the ability of the reactive metabolites to reach the DNA target.

Because QSARs developed for genotoxicity require large data sets for validating predictions, many are based on the *Salmonella* assay, the most populated data set of all the assays available to screen and rank chemicals for mutagenicity and carcinogenicity. Although the *Salmonella* assay is a good predictor of rat carcinogenicity (80% of compounds testing positive are also rat carcinogens), it is a poor predictor for the lack of genotoxicity. A compound that tests negative in the *Salmonella* assay has about equal chance of being carcinogenic or noncarcinogenic (e.g., DNTs) [95]. Thus, QSAR models based solely on *Salmonella* assays tend to be of limited utility due to the high false-negative rate. Benigni et al. [96], in developing a computer model to predict carcinogenicity from short-term tests, recommended a battery of tests that included in vivo micronuclei, *Salmonella*, and *Saccharomyces* for overall sensitivity and selectivity. Patlewicz et al. [89] concluded that for development of more accurate QSARs, efforts must focus on the development of appropriate data sets.

8.5   APPLICATION TO ECOTOXICOLOGY

The first major conference focusing on the application of genotoxicity to ecotoxicology (the 1993 Napa Conference on Genetic and Molecular Toxicology [97]) covered a wide variety of topics, ranging from markers of exposure to linking the markers to effects at population levels. A special issue of the journal *Ecotoxicology* was published on environmental population genetics [98]. In this special issue, genetic ecotoxicology was defined as the study of effects of xenobiotic compounds on DNA structure and function in indigenous organisms and their relationships to higher-order effects [99]. This includes information on DNA damage and repair, as previously discussed in this chapter, as well as population genetics in contaminated environments. The general concepts covered in these special issues are provided in Figure 8.1 that outlines a chain of events linking the interaction of a chemical with DNA to effects measurable at the population level. Differential species sensitivity is an important consideration for ecological impacts; even among rodents, mice and rats have very different susceptibilities (see Section 8.3.1). Differential susceptibility may profoundly affect community structure through alterations in predator–prey relationships and other interspecies interactions.

Although available data for explosives and explosives-related compounds are not sufficient to link genotoxicity to effects at the ecosystem level, the potential for genotoxic damage to cause effects observable at the population level in marine ecosystems has been studied for more common environmental contaminants with genotoxic potential, notably compounds found in fuel or oils. However, it is important to understand that explosives and related compounds tend to be far less genotoxic and bioaccumulative than the compounds discussed later, which will in turn reduce the risk to the ecosystem. For example, the common contaminant benzo[a]pyrene (BaP) has a 10-fold higher oral slope factor for carcinogenic risk than that of technical grade

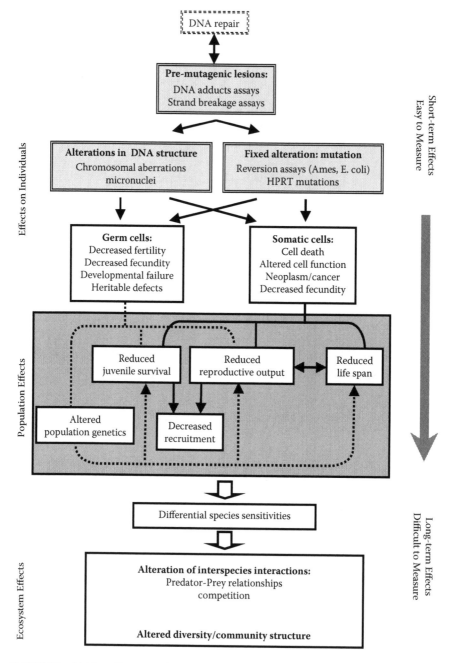

FIGURE 8.1 Linking effects on an individual level to an ecological scale. Effects measured in individuals are commonly utilized as biomarkers of exposure and can range from premutagenic lesions (adducts, etc.) that can be repaired, to mutations and clastogenic effects. Depending on the cell type (somatic or germ) mutated, different effects may be observed at the population level. Differential species sensitivity will contribute to effects observed at the ecosystem level; sensitive species are affected to a greater extent than resistant species, resulting in altered diversity and interspecies interactions.

DNT, the most carcinogenic of the compounds discussed in this chapter (7.3 vs. 0.68 per mg $kg^{-1}d^{-1}$, respectively [58]).

Sections 8.2 and 8.3 of this chapter demonstrate the capacity of explosives and explosives-related compounds to damage genetic material both in vitro and in vivo. However, positively linking assays for genetic damage to effects at the whole organism level has been controversial. Some of these issues have been discussed in Section 8.3. Additionally, many of the in vitro assays fail to consider the ability of organisms to repair DNA damage or to dispose of the damaged cell(s) via mechanisms such as apoptosis. Research with benthic fish populations has made progress in correlating markers of exposure to genotoxic compounds to actual effects at the whole organism level (e.g., tumor formation). DNA adducts have been correlated with the prevalence of liver tumors in flounder [100] and were shown to be a significant risk factor for lesions in feral English sole [101]. Significant correlations between visible lesions and increased allyl formate-induced hepatic micronuclei formation in field-exposed brown bullhead have also been recorded [102]. Thus, in vitro assays indicating that selected explosives form adducts or micronuclei are likely predictive of their ability to adversely affect individuals.

Although the ability of compounds to induce tumor formation is of grave concern for individuals, it is not a major issue when placed in context of ecological impacts. Unless the affected population is extremely small and/or possesses low reproductive rates, effects at the population level will not be observed unless reproduction and/or recruitment is significantly reduced, either directly (damaged gametes) or through the diminished reproductive capability of adult organisms. Most genotoxicity studies at the population level have been conducted with species that have adopted a life history strategy of early and prolific reproduction (r-strategy), as studies can be conducted with high enough numbers of organisms for statistical analysis within a reasonable amount of time. Populations of species that are long lived, reproduce later in life, and have fewer offspring (K-strategy) would be more susceptible to genotoxic effects, but are difficult to study.

Gametes have often been cited as having limited capability for DNA repair, thus they are more likely to retain DNA damage than somatic cells. TNT has been shown to damage sperm DNA in vivo [64], but there are no available dominant lethal studies to confirm whether the damaged sperm would result in reduced fertilization ability or reduced survival of the fertilized eggs. However, DNA damage has been linked to reduced survival of embryos for various test organisms. Exposure of the parental generation to doses of radiation or chemical concentrations below those that affect adult or gamete survival have been shown to cause embryonic abnormalities and decreased survival in early life stages in marine polychaetes [103], increased mutation frequency in nematodes [104], and anaphase aberrations in sea urchins [104].

If it can be shown that genotoxicity is indeed resulting in reduced reproductive output and/or reduced juvenile survival, the susceptibility of the population for a particular species will depend on its reproductive strategy. Failure of a low percentage of gametes to survive can severely affect species with K-strategy life history traits, but may only result in minor effects on species depending on the r-strategy. However, even populations with high reproductive rates can be adversely affected via genetic damage, as reviewed by Anderson et al. [104]. Analysis of marine fish allowed correlations to be made between chemical exposure to polycyclic aromatic hydrocarbons and DNA damage, and linked the resultant DNA damage to endpoints that can affect population density. Anaphase aberrations were correlated with embryo mortality and malformations in field populations as well as decreased adult survival and decreased juvenile survival and recruitment. Other studies correlated exposure to aromatic hydrocarbons, chlorinated hydrocarbons, and heavy metals to the formation of anaphase aberrations, and linked the anaphase aberrations and other mitotic abnormalities with embryo mortality and gross malformations. These investigations indicate that of all the genotoxicity endpoints tested, chromosomal aberrations best correlated to effects at a population level. Although the ability of explosives and related compounds to cause chromosomal aberrations has been proven in only a few instances, studies on the qualitative and quantitative relationships between various short-term tests for genotoxicity [105,106] indicate that the in vitro mutation assays may be predictive of the ability to cause in vivo chromosomal damage.

It has been hypothesized that genetic damage can potentially result in a mutation that confers adaptation to adverse environmental conditions, leading to increased ability to survive. Theodorakis et al. [107–111] have correlated strand breakage to both contamination in the environment and fecundity of the mosquitofish (*Gambusia*) exposed to radionuclides. Interestingly, fish at the radionuclide-contaminated sites also possessed higher population diversity as measured by random amplified polymorphic DNA (RAPD) and allozyme techniques. Several markers were correlated with increased resistance to radionuclide-induced damage, inferring that the fish have an evolutionarily selective advantage at the contaminated sites. However, most evidence suggests that adaptations occur through selection of naturally occurring variants, not chemically induced mutations. Although the best examples of this process are found with pesticides and antibiotics, it has also been shown to be true for a case involving a TNT-resistant algae variant [112]. Exposure of the microalga *Dictyosphaerium chlorelloides* to TNT led to the generation of a TNT-resistant variant. Fluctuation analysis indicated that the resistant strain was a result of rare spontaneous mutations and was not linked to exposure to TNT. However, this adaptation came at the cost of a decreased photosynthetic rate and diminished capacity to compete with nonresistant strains in the absence of TNT. In general, regardless of how they were generated, adaptations to adverse conditions typically come at a high cost to the organism in the form of reduced fitness (decreased reproductive capability, decreased immunocompetence/increased susceptibility to parasites, and so forth) [113,114], and are likely to result in a population that is less capable of adapting to other stressors.

8.6   NEW APPROACHES FOR ASSESSING GENOTOXICITY OF EXPLOSIVES

Recently, many new approaches have been pursued with the primary goal of enhancing genotoxicant detection and understanding the mechanisms of genotoxicity. Genetic engineering has resulted in the development of transgenic models for investigating in vivo genotoxicity, including mice and rats [115,116] harboring various reporter genes, and the plant, *Arabidopsis thaliana*, carrying a β-glucuronidase marker gene [117,118]. Molecular tools such as amplified fragment length polymorphism (AFLP) in combination with flow cytometry were recently used to assess DNA damage in white clover (*Trifolium repens*) induced by heavy metals-contaminated soils [119]. Although these newly developed assays have been claimed to be more sensitive than traditional methods, more efforts are required to validate these new assays before being adopted for routine use, and to make them as applicable and cost-effective as standard genotoxicity assays.

Driven by interests in mechanism-based risk assessment and the rapid developments in the field, a toxicogenomics approach has recently been applied to genotoxicity studies. This new subdiscipline of toxicology combines the emerging technologies of genomics, proteomics, and bioinformatics to identify and characterize mechanisms of action of known and suspected toxicants [120]. Starting in 1999, an international collaborative toxicogenomics program evaluated the utility of gene expression profile analysis for risk assessment of genotoxicants [121]. There are several advantages of the toxicogenomics approach: (1) toxicogenomics monitors global gene expression, unlike traditional promoter–reporter genotoxicity assays that cover only a limited number of biological pathways; (2) gene expression changes often precede changes at the cellular or tissue level; (3) DNA microarray allows monitoring of hundreds or thousands of genes simultaneously instead of a single gene at a time; (4) gene expression profiling provides mechanistic insight into the mode of action of a genotoxic compound; and (5) genotoxic stress-associated gene expression profiles or change patterns may be characteristic of specific classes of toxicants [122]. Using this approach, two cell lines commonly employed in standard genotoxiciy testing were exposed to model mutagenic compounds with diverse known mechanisms [123]. Both gene expression profiles and traditional genotoxicity endpoints were determined. Results indicated that patterns of altered gene expression could distinguish between indirect-acting and direct-acting genotoxicants [124], and could differentiate genotoxic and cytotoxic stresses [125]. Furthermore, alterations in gene expression as measured by microarrays were not as sensitive as traditional genotoxicity assays, diminishing an earlier concern that the technology might be overly sensitive [123].

The application of modern molecular biology methodologies in characterizing genotoxicity at the population level has also increased in recent years [126]. Techniques such as RFLP (restriction fragment length polymorphism), RAPD, SSR (simple sequence repeats, e.g., mini- and microsatellites), and AFLP have greatly increased the capacity to generate large numbers of both monomorphic and polymorphic genetic markers [127], which are used to characterize alterations of genetic diversity resulting from environmental toxicant exposure.

8.7 SUMMARY

Table 8.3 broadly summarizes results from the various genotoxicity assays. TNT and DNT (technical grade) were consistently genotoxic, whereas HMX tested negative in all assay systems. Pure 2,4-DNT tested positive in in vitro microorganism and in

TABLE 8.3

Summary for Genotoxicity Testing of Explosives Compounds

Compound	Microorganism	Mammalian In Vitro	Mammalian In Vivo	Plant In Vivo
TNT	+	+/–	+	NA
Pure 2,6-DNT				
Technical grade 2,4-DNT	+/–	+/–	+	+
Pure 2,4-DNT	+/–	–	–	+
Tetryl	+	NA	NA	NA
TNB	+	–	–	NA
RDX	–	–	+	NA
HMX	–	–	–	NA
NG	Weak	+/–	–	NA

Note: + denotes the compound was determined to be genotoxic; – denotes negative genotoxicity reported.; +/– denotes mixed results; NA denotes no data available; weak denotes the compound is a weak mutagen.

vivo plant assays but negative in both in vitro and in vivo mammalian assays. TNB and NG tested positive in in vitro assays but negative in in vivo mammalian assays, whereas the opposite was true for RDX. Overall, in vitro assays produced many false positives, overpredicting carcinogenicity. RDX was an exception, the only false negative, as it is classified as carcinogenic yet tests negative in in vitro assays.

In the *Salmonella* and *Neurospora* assays, the reported compounds invariably had a greater tendency toward base-pair substitution than frameshift mutation. Most compounds were not metabolically activated by the S9 homogenate, although there were some exceptions (AZTs and DNT products). Comparison of the relative mutagenic potency of the parent explosive compounds using various *Salmonella* strains without the presence of metabolic enzymes strains indicates that the cyclic nitramines are nonmutagenic, the DNTs have low to nonexistent mutagenicity, and TNT tends to be about tenfold more mutagenic than the DNTs. TNB and tetryl were consistently reported as more mutagenic than TNT, ranging from two-fold [6] to greater than tenfold more mutagenic [1,2]. Relative potencies of tetryl and TNB were inconsistent [1,25]. For the most part, the metabolites and breakdown products tended to be equally or less mutagenic than the parent compounds. One exception comes from the only available report for the breakdown products of RDX. Only the trinitroso product tested positive, with a very low mutagenic activity [1]. In the case of TNT, DNT, and their products, consideration of dose ranges and the effects of modified methodologies led to a fairly consistent data set. The known products are predominantly base-pair mutagens. For the nitroaromatics, mutagenicity decreases as nitro groups are converted to amino groups, but the intermediate hydroxylamino derivatives and azoxy products are direct mutagens, which can be as potent or more potent than the parent compounds. With the exception of 2′,4-AZT, mutagenicity of TNT products is not significantly increased in the presence of mammalian metabolic enzymes present in the S9.

In contrast to TNT, DNTs and their products tend to be more mutagenic in the presence of S9 homogenate, and the presence of other enzymes such as nitroreductases and O-acetyltranferases dramatically increases mutagenicity. Although short-term testing has the advantage of rapid assessment, often unrealistically high doses are required to elicit effects, and in some cases (e.g., TNB), cytotoxic effects complicated the detection of mutagenicity. The wide variety of species tested and the use of modified *Salmonella* strains and methodologies generate difficulties in directly comparing results. However, the different species/tester strains help to highlight the different mechanisms of genotoxicity, and the use of modified methodologies often increases the sensitivity of the assays. The testing of environmental samples exemplified the complexity of the issue; the potential genotoxicity of a compound is not a simple story because it involves many breakdown products, both known and unknown, that may result from sequential metabolic pathways (e.g., oxidation followed by conjugation).

The evidence of the in vitro mammalian genotoxicity of explosives products is much less convincing than that found for microorganism-based assays. Although there is only partial agreement among the results obtained in the various mammalian assays, it can be concluded that these compounds are weak mutagens in mammalian cells. The relative mutagenic potency of the parent explosive compounds in the various mammalian systems is similar to what was described earlier in microorganism-based systems. Apparently, the cyclic nitramine explosives RDX and HMX are nonmutagenic, the DNTs have low mutagenicity (depending on the test conditions), and TNT is genotoxic in most assay systems. However, TNB is an exception to this generalization since only negative results were obtained so far. The reductive pathway seems to be pivotal in the production of active metabolites from nitroaromatics, and the lack of such activity could explain the nongenotoxicity of TNT in some in vitro systems. This pathway may also result in the production of the moderately potent mammalian genotoxicant MNX from inactive RDX. However, experimental results following the addition of S9 activating mixture were unclear. In some cases activity was abolished (TNT, DNT) [31], whereas in others it was left unaffected (TNT, MNX) [32,50]. In some assays, a strong enhancing effect was found (4-ADNT, MNX) [32,50]. Inactivation of reactive metabolites by reaction with proteins present in S9 mixture was a possible explanation for these results. Overall, it is apparent that S9 homogenates (primarily oxidative in nature) are not optimal for assessing nitro compounds due to the importance of reductive metabolism in their activation.

The availability of in vivo data is even more limited than that of in vitro assays. TNT, DNTs, and RDX have been classified as carcinogenic in mammalian model organisms, but in general, the ability of a compound to cause malignant tumors is highly dependent on many factors including the tissue examined, the gender of the organism, and the selected species; these differences make it difficult to predict potential impacts to wildlife populations.

8.8   DATA GAPS AND FUTURE DIRECTIONS

Many genotoxicity data gaps still exist, including mammalian cell-based and in vivo assays for tetryl and in vivo plant studies for compounds other than DNTs. The cyclic nitramines are not well characterized for any assay except *Salmonella*. Even in this case, few studies have been conducted with RDX products and no studies are available on products of HMX. As previously discussed, clastogenic effects have been linked to effects at the population level, yet assays detecting clastogenic effects have not been conducted for any compound except TNT.

Another major gap in toxicological studies, in general, is that the vast majority of studies are conducted with a single, pure compound. A single compound will rarely if ever dominate environmental situations, and the effects of mixtures on genotoxicity may be complex. Some interactions may be predictable, for example, the presence of compounds stimulates metabolic activation pathways resulting in increased genotoxicity. However, some interactions are unpredictable, for example, the copresence of BaP and TNT fails to result in a mutagenic response [128], probably due to a complexation between TNT and BaP that inhibits uptake of the compounds. Also, as previously discussed, both microbial and mammalian cell-based assays on environmental sample extracts have provided data that indicate potential synergistic effects of mixtures [20]. To assess the wide variety of mixtures typically present at contaminated sites, high throughput technologies such as those discussed in Section 8.6 will be required. Additionally, newer technologies, such as toxicogenomics, may begin to help us understand the mechanisms behind toxic interactions of mixtures.

The integration of genotoxicity into ecotoxicology is still a science in its infancy. Ecosystems include a wide variety of organisms, few of which are adequately characterized for factors that may affect genotoxicity (xenobiotic metabolism capability, DNA repair capability, etc.). Although data for rats, mice, and dogs serve to help regulatory classification and to develop cancer slope factors for risk assessment, they provide little input for ecological assessments. Some of these needs are addressed in new methodologies developed from advances in molecular biology. Toxicogenomics may offer in vivo methodologies by providing markers of exposure that can be linked mechanistically to genotoxicity; understanding the mechanisms behind genotoxicity in a test species may allow extrapolation to the various species of concern.

Additionally, genotoxicity assays should be capable of identifying significant and heritable genetic damages that can unfold at higher hierarchic levels. Although there are now several case studies linking genetic damage to effects on the population (reduced survival, fecundity, recruitment, and juvenile survival), no studies are available for explosives. Newer methodologies as well as expansion of traditional studies may eventually help fill these gaps. In the end, though, differential sensitivities between species for acute toxicity will probably be a bigger driving factor for ecological impacts, especially for vertebrate species typically included in ecological risk assessments.

REFERENCES

1. George SE, Huggins-Clark G, and Brooks LR, Use of *Salmonella* microsuspension bioassay to detect the mutagenicity of munitions compounds at low concentrations, *Mut. Res.*, 490, 45, 2001.
2. Lachance B et al., Cytotoxic and genotoxic effects of energetic compounds on bacterial and mammalian cells in vitro, *Mut. Res.*, 444, 25, 1999.

3. Kaplan DL and Kaplan AM, Mutagenicity of 2,4,6-trinitrotoluene surfactant complexes, *Bull. Environ. Contam. Toxicol.,* 28, 33, 1982.

4. Padda RS et al., Mutagenicity of trinitrotoluene and metabolites formed during anaerobic degradation by *Clostridium acetobutylicum* ATCC 824, *Environ. Toxicol. Chem.,* 19, 2871, 2000.

5. Spanggord RJ et al., Mutagenicity in *Salmonella typhimurium* and structure-activity relationships of wastewater components emanating from the manufacture of trinitrotoluene, *Environ. Mutagen.,* 4, 163, 1982.

6. Tan EL et al., Mutagenicity of trinitrotoluene and its metabolites formed during composting, *J. Toxicol. Environ. Health,* 36, 165, 1992.

7. Spanggord RJ, Stewart KR, and Riccio ES, Mutagenicity of tetranitroazoxytoluenes: A preliminary screening in *Salmonella typhimurium* strains TA100 and TA100NR, *Mut. Res.,* 335, 207, 1995.

8. Karamova NS, Il'inskaya ON, and Ivanchenko OB, Mutagenic activity of 2,4,6-trinitrotoluene: The role of metabolizing enzymes, *Russ. J. Genet.,* 30, 785, 1994.

9. Vaatanen AK et al., Spectrum of spontaneous and 2,4,6-trinitrotoluene (TNT)-induced mutations in *Salmonella typhimurium* strains with different nitroreductase and *O*-acetyltransferase activities, *Mut. Res.,* 379, 185, 1997.

10. Whong WZ and Edwards GS, Genotoxic activity of nitroaromatic explosives and related compounds in *Salmonella typhimurium,* *Mut. Res.,* 136, 209, 1984.

11. Won WD, DiSalvo LH, and Ng J, Toxicity and mutagenicity of 2,4,6-trinitrotoluene and its microbial metabolites, *Appl. Environ. Microbiol.,* 31, 576, 1976.

12. Honeycutt ME, Jarvis AS, and McFarland VA, Cytotoxicity and mutagenicity of 2,4,6-trinitrotoluene and its metabolites, *Ecotoxicol. Environ. Saf.,* 35, 282, 1996.

13. Einisto P, Role of bacterial nitroreductase and *O*-acetyltransferase in urine mutagenicity assay of rats exposed to 2,4,6-trinitrotoluene (TNT), *Mut. Res.,* 262, 167, 1991.

14. Assmann N et al., Genotoxic activity of important nitrobenzenes and nitroanilines in the Ames test and their structure-activity relationship, *Mut. Res.,* 395, 139, 1997.

15. Sayama M et al., Mutagenicities of 2,4- and 2,6-dinitrotoluenes and their reduced products in *Salmonella typhimurium* nitroreductase- and *O*-acetyltransferase-overproducing Ames test strains, *Mut. Res.,* 420, 27, 1998.

16. Maragos CM et al., Mutagenicity of glyceryl trinitrate (nitroglycerin) in *Salmonella typhimurium,* *Mut. Res.,* 298, 187, 1993.

17. Padda RS et al., Mutagenicity of nitroaromatic degradation compounds, *Environ. Toxicol. Chem.,* 22, 2293, 2003.

18. Griest WJ et al., Chemical and toxicological testing of composted explosives-contaminated soil, *Environ. Toxicol. Chem.,* 12, 1105, 1993.

19. Griest WH et al., Chemical characterization and toxicological testing of windrow composts from explosives-contaminated sediments, *Environ. Toxicol. Chem.,* 14, 51, 1995.

20. Berthe-Corti L et al., Cytotoxicity and mutagenicity of a 2,4,6-trinitrotoluene (TNT) and hexogen contaminated soil in *S. typhimurium* and mammalian cells, *Chemosphere,* 37, 209, 1998.

21. Brooks LR et al., Mutagenicity of HPLC-fractionated urinary metabolites from 2,4,6-trinitrotoluene-treated Fischer 344 rats, *Environ. Mol. Mutagen.,* 30, 298, 1997.

22. Arfsten DP, Davenport R, and Schaffer DJ, Reversion of bioluminescent bacteria (Mutatox™) to their luminescent state upon exposure to organic compounds, munitions, and metal salts, *Biomed. Environ. Sci.,* 7, 144, 1994.

23. Frische T, Screening for soil toxicity and mutagenicity using luminescent bacteria—A case study of the explosive 2,4,6-trinitrotoluene (TNT), *Ecotoxicol. Environ. Saf.,* 51, 133, 2002.

24. Whong WZ, Speciner ND, and Edwards GS, Mutagenic activity of tetryl, a nitroaromatic explosive, in three microbial test systems, *Toxicol. Lett.,* 5, 11, 1980.

25. McGregor DB et al., Genotoxic activity in microorganisms of tetryl, 1,3-dinitrobenzene and 1,3,5-trinitrobenzene, *Environ. Mutagen.*, 2, 531, 1980.
26. Simmon VF et al., Mutagenicity of Some Munitions Wastewater Chemicals and Chlorine Test Reagents, SRI International, Contract No. DAMD 17-76-C-6013, Report No. ADA057680, U.S. Army Medical Research and Development Command, Fort Detrick, Frederick, MD, 1977.
27. Karamova NS et al., Compounds 2,4,6-trinitrotoluene and 2,4-diamino-6-nitrotoluene: The absence of *rec*A-dependent mutagenesis, *Russ. J. Genet.*, 31, 528, 1995.
28. Bruning T, Thier R, and Bolt HM, Nephrotoxicity and nephrocarcinogenicity of dinitrotoluene: New aspects to be considered, *Rev. Environ. Health,* 17, 163, 2002.
29. Rosenkranz HS and Mermelstein R, Mutagenicity and genotoxicity of nitroarenes: All nitro-containing chemicals were not created equal, *Mut. Res.*, 114, 217, 1983.
30. Purohit V and Basu AK, Mutagenicity of nitroaromatic compounds, *Chem. Res. Toxicol.*, 18, 673, 2000.
31. Styles JA and Cross MF, Activity of 2,4,6-trinitrotoluene in an in vitro mammalian gene mutation assay, *Cancer Lett.*, 20, 103, 1983.
32. Kennel SJ et al., Mutation analyses of a series of TNT-related compounds using the CHO-HPRT assay, *J. Appl. Toxicol.*, 20, 441, 2000.
33. Kiefer F et al., Arylamines suppress their own activation and that of nitroarenes in V79 Chinese hamster cells by competing for acetyltransferases, *Environmental Health Perspectives,* 102 (Suppl. no. 6), 95, 1994.
34. Aaron CS et al., Mammalian cell gene mutation assays: Working group report, *Mut. Res.*, 312, 235, 1994.
35. Wang K et al., Single strand DNA breaks of lens epithelial cells induced by 2,4,6-trinitrotoluene and its metabolites, *Shengwu Huaxue Za*, 10, 81, 1994.
36. Lachance B and Sunahara GI, Effects of energetic substances on mammalian cells in vitro. 4.1–Use of TK6 human lymphoblastic cell culture to assess energetic compounds, in Characterization and Remediation of Soil Contaminated with Explosives: Development of Practical Technologies. Interim Report, Hawari J, Ed., NRCC No. 41825, Biotechnology Research Institute, National Research Council, Montréal, Canada, 1998, pp. 72–74.
37. Butterworth BE et al., Use of primary cultures of human hepatocytes in toxicology studies, *Cancer Res.*, 49, 1075, 1989.
38. Tchounwou PB et al., Transcriptional activation of stress genes and cytotoxicity in human liver carcinoma cells (hepG2) exposed to 2,4,6-trinitrotoluene, 2,4-dinitrotoluene, and 2,6-dinitrotoluene, *Environ. Toxicol.*, 16, 209, 2001.
39. Lachance B and Sunahara GI, Use of rat liver S9 in activated V79 mutagenicity tests of energetic compounds, in Characterization and Remediation of Soil Contaminated with Explosives: Development of Practical Technologies. Final Report, NRCC No. 43264, Environmental Biotechnology Sector, Biotechnology Research Institute, National Research Council, Montréal, Canada, 1999, pp. 150–156.
40. Bermudez E, Tillery D, and Butterworth BE, The effect of 2,4-diaminotoluene and isomers of dinitrotoluene on unscheduled DNA synthesis in primary rat hepatocytes, *Environ. Mutagen.*, 1, 391, 1979.
41. Couch DB et al., The influence of activation systems on the metabolism of 2,4-dinitrotoluene and its mutagenicity to CHO cells, in *Banbury Report. 2. Mammalian Cell Mutagenesis: The Maturation of Test Systems* , Hsie AW, O'Neil JP, and McElheny VK, Eds., Cold Spring Harbor Laboratory, NY, 1979, pp. 303–309.
42. Abernethy DJ and Couch DC, Cytotoxicity and mutagenicity of dinitrotoluenes in Chinese hamster ovary cells, *Mut. Res.*, 103, 53, 1982.

43. Lachance B and Sunahara GI, Use of Mammalian Cell Cultures to Assess the Cytotoxicity and Genotoxicity of Energetic Compounds. Final Report, Hawari, J. Editor, NRCC No. 41773, Environmental Biotechnology Sector, Biotechnology Research Institute, National Research Council, Montréal, Canada, 1998, pp. 99–103.

44. National Institute of Environmental Health Sciences (NIEHS), National Toxicology Program, http://ntp-server.niehs.nih.gov/ (accessed August 2008).

45. Holen I, Mikalsen S-O, and Sanner T, Effects of dinitrotoluenes on morphological cell transformation and intercellular communication in Syrian hamster embryo cells, *J. Toxicol. Environ. Health,* 29, 89, 1990.

46. Huang Q-G et al., Relationships between molecular structure and chromosomal aberrations in in vitro human lymphocytes induced by substituted nitrobenzenes, *Bull. Environ. Contam. Toxicol.,* 57, 349, 1996.

47. Dilley JV, Tyson CA, and Newell GW, Mammalian Toxicological Evaluation of TNT Waste Waters. Volume 11: Acute and Sub Acute Mammalian Toxicity of TNT and the LAP Mixture, SRI International Menlo Park, CA, Contract No. DAMD17-76-C-6050, Report No. ADA 080957, U.S. Army Medical Research and Development Command, Fort Detrick, Frederick, MD, 1978.

48. Yinon J, *Toxicity and Metabolism of Explosives*, CRC Press, Boca Raton, FL, 1990.

49. Cifone MA, L5178Y TK+/- Mouse Lymphoma Forward Mutation assay with a Confirmatory Assay, Covance Laboratories Inc., Vienna, VA, Covance No. 22636-0-431OECD, U.S. Army Center for Health Promotion and Preventive Medicine, Aberdeen Proving Ground, MD, 2002.

50. Snodgrass HL Jr, Preliminary Assessment of Relative Toxicity and Mutagenicity Potential of 1-Nitroso-3,5-Dinitro-1,3,5-Triazacyclohexane (Mononitroso-RDX), Final Report, Study No. 75-51-0345-85, Report No. ADA14935, Aberdeen Proving Ground, MD, U.S. Army Environmental Hygiene Agency, 1984.

51. Sandhu JK and Birnboim HC, Mutagenicity and cytotoxicity of reactive oxygen and nitrogen species in the MN-11 murine tumor cell lines, *Mut. Res.,* 379, 241, 1997.

52. Tamano S et al., Histogenesis and the role of p53 and K-ras mutations in hepatocarcinogenesis by glyceryl trinitrate (nitroglycerin) in male F344 rats, *Carcinogenesis,* 17, 2477, 1996.

53. Ellis HV et al., Subacute and chronic toxicity studies of trinitroglycerin in dogs, rats, and mice, *Fundam. Appl. Toxicol.,* 4, 248, 1984.

54. White PA and Claxton LD, Mutagens in contaminated soil: A review, *Mut. Res.,* 567, 227, 2004.

55. Mirsalis JC et al., Role of gut flora in the genotoxicity of dinitrotoluene, *Nature,* 295, 322, 1982.

56. Long RM and Rickert DE, Metabolism and excretion of 2,6-dinitro[^{14}C]toluene in vivo and in isolated perfused rat livers, *Drug Metab. Dispos.,* 10, 455, 1982.

57. Chadwick RW, George SE, and Kohan MJ, Potentiation of 2,6-dinitrotoluene genotoxicity in Fischer 344 rats by pretreatment with coal tar creosote, *J. Toxicol. Environ. Health,* 44, 319, 1995.

58. United States Environmental Protection Agency (EPA), IRIS database, http://www.epa.gov/iriswebp/iris/ (accessed July 2008).

59. Furedi EM et al., Determination of the Chronic Mammalian Toxicological Effects of TNT: (Twenty-Four Month Chronic Toxicity/Carcinogenicity Study of Trinitrotoluene (TNT) in the Fischer 344 Rat). Volume 1. Final Report on Phase 3, ITT Research Institute, Chicago, IL , Contract No. DAMD17-79-C-9120, Report No. ADA168637, U.S. Army Medical Research and Development Command, Fort Detrick, Frederick, MD, 1984.

60. Furedi EM et al., Determination of the Chronic Mammalian Toxicological Effects of TNT: (Twenty-Four Month Chronic Toxicity/Carcinogenicity Study of Trinitrotoluene (TNT) in the B6C3F1 Mouse). Volume 1. Final Report, Phase 4, ITT Research Institute, Chicago, IL, Contract No. DAMD17-79-C-9120, Report No. ADA168754, U.S. Army Medical Research and Development Command, Fort Detrick, Frederick, MD, 1984.
61. McConnell EE et al., Guidelines for combining neoplasms for evaluation of rodent carcinogenesis studies, *J. Natl. Cancer Inst.*, 76, 286, 1986.
62. Ashby J et al., Non-genotoxicity of 2,4,6-trinitrotoluene (TNT) to the mouse bone marrow and the rat liver: Implications for its carcinogenicity, *Arch. Toxicol.*, 58, 14, 1985.
63. Liu YY et al., In vivo covalent binding of [14C]trinitrotoluene to proteins in the rat, *Chem. Biol. Interact.*, 82, 1, 1992.
64. Homma-Takeda S et al., 2,4,6-Trinitrotoluene-induced reproductive toxicity via oxidative DNA damage by its metabolite, *Free Radic. Res.*, 36, 555, 2002.
65. Mirsalis JC and Butterworth BE, Induction of unscheduled DNA synthesis in rat hepatocytes following in vivo treatment with dinitrotoluene, *Carcinogenesis*, 3, 241, 1982.
66. National Cancer Institute (NCI), Bioassay of 2,4-Dinitrotoluene for Possible Carcinogenicity, Technical Report Series No. 54, U.S. Dept. Health, Education and Welfare, Washington, DC, 1978.
67. Ellis HV et al., Mammalian Toxicity of Munitions Compounds. Phase III. Effects of Lifetime Exposure. Part I. 2,4-Dinitrotoluene, Contract No. DAMD17-74-C-4073, Midwest Research Institute, Kansas City, MO, Report No. ADA077692, Army Medical Research and Development Command, Fort Detrick, Frederick, MD, 1979.
68. Chemical Industry Institute of Toxicology (CIIT), 104-Week Chronic Toxicity Study in Rats: Dinitrotoluene. Final Report, Vol. 1 and 2, Docket No. 12362, Research Triangle Park, NC, 1982.
69. Leonard TB, Graichen ME, and Popp JA, Dinitrotoluene isomer-specific hepatocarcinogenesis in F344 rats, *J. Natl. Cancer Inst.*, 79, 1313, 1987.
70. Popp JA and Leonard TB, Hepatocarcinogenicity of 2,6- dinitrotoluene (DNT), *Proc. Am. Assoc. Cancer Res.*, 24, 91, 1983.
71. Leonard TB, Lyght O, and Popp JA, Dinitrotoluene structure-dependent initiation of hepatocytes in vivo, *Carcinogenesis*, 4, 1059, 1983.
72. Leonard TB, Adams T, and Popp JA, Dinitrotoluene isomer-specific enhancement of the expression of diethylnitrosamine-initiated hepatocyte foci, *Carcinogenesis*, 7, 1797, 1986.
73. Lane RW, Simon GS, and Dougherty RW, Reproductive toxicity and lack of dominant lethal effects of 2,4-dinitrotoluene in the male rat, *Drug Chem. Toxicol.*, 8, 265, 1985.
74. Soares ER and Lock LF, Lack of indication of mutagenic effects of dinitrotoluenes and diaminotoluenes in mice, *Environ. Mutagen.*, 2, 111, 1980.
75. Levine BS et al., Determination of the Chronic Mammalian Toxicological Effects of RDX: (Twenty-Four Month Chronic Toxicity/Carcinogenicity Study of Hexahydro-1,3,5-Trinitro-1,3,5-Triazine (RDX) in the Fischer 344 Rat), Volume 1, Final Report on Phase 5, ITT Research Institute, Chicago, IL, Contract No. DAMD17-79-C-9161, Report No. ADA 160774, U.S. Army Medical Research and Development Command, Fort Detrick, Frederick, MD, 1983.
76. Hart ER, Two-Year Feeding Study in Rats, Litton Bionetics, Inc., Kensington, MD, Contract No. N00014-73-C-0162, Office of Naval Research, 1976.
77. Lish PM et al., Determination of the Chronic Mammalian Toxicological Effects of RDX: (Twenty-Four Month Chronic Toxicity/Carcinogenicity Study of Hexahydro-1,3,5-Trinitro-1,3,5-Triazine (RDX) in the B6C3F1 Hybrid Mouse (Final Report: Phase 6, Volume 1), ITT Research Institute, Chicago, IL, Contract No. DAMD17-79-C-9161 (IITRI Project No. L6121), Report No. ADA160774, U.S. Army Medical Research and Development Center, Fort Detrick, Frederick, MD, 1984.
78. Majors MA, Re-Evaluation of RDX, Draft Pathology Working Group (PWG), 2001.

79. Erexson GL, In Vivo Mouse Micronucleus Assay with RDX, Covance Study No. 22636-0-455 ECD, Covance Laboratories Inc., Vienna, VA, U.S. Army Center for Health Promotion and Preventive Medicine, Aberdeen Proving Ground, MD, 2002.

80. Cholakis JM et al., Mammalian Toxicological Evaluation of RDX. III, Midwest Research Institute, Contract No. DAMD 17-78-C-802, Report No. ADAO92531, U.S. Army Medical Research and Development Command, 1980.

81. Reddy G et al., Assessment of environmental hazards of 1,3,5-trinitrobenzene, *J. Toxicol. Environ. Health*, 52, 447, 1997.

82. Slaga TJ et al., Carcinogenesis of Nitrated Toluenes and Benzenes, Skin and Lung Tumor Assays in Mice, U.S. Army Report Order No. 1807, Dept of Energy Interagency Agreement 40-1016-79 Oak Ridge, TN, 1985.

83. Reddy TV et al., Chronic Toxicity Studies of 1,3,5-Trinitrobenzene in Fischer 344 Rats. Final report, U.S. Army Project No. 93MM358, Report No. ADA 315216, 1996.

84. Reddy T et al., Chronic toxicity of 1,3,5-trinitrobenzene in Fischer 344 rats, *Int. J. Toxicol.*, 20, 59, 2001.

85. Bingham E, Cohrssen B, and Powell CH, *Patty's Toxicology. Volumes 1-9*, 5th ed., John Wiley & Sons, New York, 2001, p. 606.

86. Grant WF, The present status of higher plant bioassays for the detection of environmental mutagens, *Mut. Res.*, 310, 175, 1994.

87. Grant WF, Higher plant assays for the detection of chromosomal aberrations and gene mutations, *Mut. Res.*, 426, 210, 1999.

88. Gong P, Kuperman RG, and Sunahara GI, Genotoxicity of 2,4- and 2,6-dinitrotoluene as measured by the *Tradescantia* micronucleus (Trad-MCN) bioassay, *Mut. Res.*, 538, 16, 2003.

89. Patlewicz G, Rodford R, and Walker JD, Quantitative structure activity relationships for predicting mutagenicity and carcinogenicity, *Environ. Toxicol. Chem.*, 22, 1885, 2003.

90. Debnath AK et al., Quantitative structure-activity relationship investigation of the role of hydrophobicity in regulating mutagenicity in the Ames test: 2. Mutagenicity of aromatic and heteroaromatic nitro compounds in *Salmonella typhimurium* TA100, *Environ. Mol. Mutagen.*, 19, 53, 1992.

91. Debnath AK et al., A QSAR investigation of the role of hydrophobicity in regulating mutagenicity in the Ames test: 1. Mutagenicity of aromatic and heteroaromatic amines in *Salmonella typhimurium* TA98 and TA100, *Environ. Mol. Mutagen.*, 19, 37, 1992.

92. Benigni R, Anreoli C, and Giuliani A, QSAR models for both mutagenic potency and activity: Application to nitroarenes and aromatic amines, *Environ. Mol. Mutagen.*, 24, 208, 1994.

93. Williams GM and Weisburger JH, Chemical carcinogens, in *Casarett and Doull's Toxicology: The Basic Science of Poisons*, Klaassen CD, Amdur MO, and Doull J, Eds., Macmillan Publishing Company, New York, 1986, pp. 117–120.

94. Talmage SS et al., Nitroaromatic munition compounds: Environmental effects and screening values, *Rev. Environ. Contam. Toxicol.*, 161, 1, 1999.

95. Tennant RW et al., Prediction of chemical carcinogenicity in rodents from in vitro genetic toxicity assays, *Science*, 236, 933, 1987.

96. Benigni R, Pellizzone G, and Giuliani A, Comparison of different computerized classification methods for predicting carcinogenicity from short-term test results, *J. Toxicol. Environ. Health*, 28, 427, 1989.

97. Special Issue on Genetic and Molecular Ecotoxicology, *Environmental Health Perspectives Supplement*, 102 (Suppl. no. 12), 1994.

98. Special Issue on Environmental Population Genetics, *Ecotoxicology*, 10, 2001.

99. Theodorakis CW, Integration of genotoxic and population genetic endpoints in biomonitoring and risk assessment, *Ecotoxicology*, 10, 245, 2001.

100. Vethaak AD and Wester PW, Diseases of flounder (*Platyichthys flesus*) in Dutch coastal waters, with particular reference to environmental stress factors. Part 2. Liver histopathology, *Dis. Aquat. Org.*, 26, 99, 2004.

101. Reichert WL et al., Molecular epizootiology of genotoxic events in marine fish: Linking contaminant exposure, DNA damage, and tissue level alterations, *Mut. Res.*, 441, 215, 1998.

102. Rao SS et al., Fish hepatic micronuclei as an indication of exposure to genotoxic environmental contaminants, *Environ. Toxicol. Water Qual.*, 12, 217, 1997.

103. Harrison FL and Anderson SL, Reproductive success as an indicator of genotoxicity in the polychaete worm, *Neanthes arenaceodentata*, *Mar. Environ. Res.*, 28, 313, 1989.

104. Anderson SL and Wild GC, Linking genotoxic responses and reproductive success in ecotoxicology, *Environ. Health Perspect.*, 102, 9, 1994.

105. Benigni R, Andreoli C, and Giuliani A, Relationships among in vitro mutagenicity assays: Quantitative vs. qualitative test results, *Environ. Mol. Mutagen.*, 26, 155, 1995.

106. Benigni R, Relationships between in vitro mutagenicity assays, *Mutagenesis*, 7, 335, 1992.

107. Theodorakis CW, Blaylock BG, and Shugart LR, Genetic ecotoxicology I: DNA integrity and reproduction in mosquitofish exposed *in situ* to radionuclides, *Ecotoxicology*, 5, 1, 1996.

108. Theodorakis CW and Shugart LR, Genetic ecotoxicology III: The relationship between DNA strand breaks and genotype in mosquitofish exposed to radiation, *Ecotoxicology*, 7, 227, 1998.

109. Theodorakis CW et al., Genetics of radionuclide-contaminated mosquitofish populations and homology between *Gambusia affinis* and *G. holbrooki*, *Environ. Toxicol. Chem.*, 10, 1992, 1998.

110. Theodorakis CW, Elbl T, and Shugart LR, Genetic ecotoxicology IV: Survival and DNA strand breakage is dependent on genotype in radionuclide-exposed mosquitofish, *Aq. Toxicol.*, 45, 279, 1999.

111. Theodorakis CW and Shugart LR, Genetic ecotoxicology II: Population genetic structure in radionuclide-contaminated mosquitofish (*Gambusia affinis*), *Ecotoxicology*, 6, 335, 1997.

112. Garcia-Villada L et al., Evolution of microalgae in highly stressing environments: An experimental model analyzing the rapid adaptation of *Dictyosphaerium chlorelloides* (Chlorophyceae) from sensitivity to resistance against 2,4,6-trinitrotoluene by rare preselective mutations, *J. Phycol.*, 38, 1074, 2002.

113. Hebert PDN and Luiker MM, Genetic effects of contaminant exposure—Towards an assessment of impacts on animal populations, *Sci. Total Environ.*, 191, 23, 1996.

114. Trabalka JR and Allen CP, Aspects of fitness of a mosquitofish *Gambusia affinis* exposed to chronic low-level environmental radiation, *Radiat. Res.*, 70, 198, 2004.

115. Heddle JA et al., In vivo transgenic mutation assays, *Environ. Mol. Mutagen.*, 35, 253, 2000.

116. Hayashi H et al., Novel transgenic rat for in vivo genotoxicity assays using 6-thioguanine and Spi selection, *Environ. Mol. Mutagen.*, 41, 259, 2003.

117. Kovalchuk I et al., Transgenic plants are sensitive bioindicators of nuclear pollution caused by the Chernobyl accident, *Nat. Biotechnol.*, 16, 1054, 1998.

118. Kovalchuk O et al., A sensitive transgenic plant system to detect toxic inorganic compounds in the environment, *Nat. Biotechnol.*, 19, 568, 2001.

119. Citterio S et al., Soil genotoxicity assessment: A new strategy based on biomolecular tools and plant bioindicators, *Environ. Sci. Technol.*, 36, 2748, 2002.

120. Casciano DA and Fuscoe JC, Preface to *Mutation Research* special issue on toxicogenomics, *Mut. Res.*, 549, 1, 2004.

121. Pennie W, Pettit SD, and Lord PG, Toxicogenomics in risk assessment: An overview of an HESI collaborative research program, *Environ. Health Perspect.*, 112, 417, 2004.

122. Steiner G et al., Discriminating different classes of toxicants by transcript profiling, *Environ. Health Perspect.*, 112, 1236, 2004.

123. Newton RK, Aardema M, and Aubrecht J, The utility of DNA microarrays for characterizing genotoxicity, *Environ. Health Perspect.,* 112, 420, 2004.
124. Hu T et al., Identification of a gene expression profile that discriminates indirect-acting genotoxins from direct-acting genotoxins, *Mut. Res.,* 549, 5, 2004.
125. Dickinson DA et al., Differentiation of DNA reactive and non-reactive genotoxic mechanisms using gene expression profile analysis, *Mut. Res.,* 549, 29, 2004.
126. Bickham JW and Smolen MJ, Somatic and heritable effects of environmental genotoxins and the emergence of evolutionary toxicology, *Environ. Health Perspect.,* 102, 201, 1994.
127. D'Surney SJ, Shugart LR, and Theodorakis CW, Genetic markers and genotyping methodologies: An overview, *Ecotoxicology,* 10, 201, 2001.
128. Washburn KS et al., A study of 2,4,6-trinitrotoluene inhibition of benzo[a]pyrene uptake and activation in a microbial mutagenicity assay, *Chemosphere,* 44, 1703, 2001.

9 Mechanisms of the Mammalian Cell Cytotoxicity of Explosives

Narimantas Čėnas, Aušra Nemeikaitė-Čėnienė,
Jonas Šarlauskas, Žilvinas Anusevičius, Henrikas
Nivinskas, Lina Misevičienė, and Audronė Marozienė

CONTENTS

9.1 Introduction 211
9.2 Enzymatic Reactions of Nitroaromatic Explosives 213
    9.2.1 Free Radical Reactions of Nitroaromatic Explosives 213
    9.2.2 Formation of Stable Metabolites of Nitroaromatic Explosives 218
9.3 Mammalian Cell Cytotoxicity of Nitroaromatic Explosives 221
9.4 Mammalian Cell Cytotoxiciy and Enzymatic Reactions of Other
    Explosives 222
9.5 Concluding Remarks 223
References... 223

9.1 INTRODUCTION

Explosives and their degradation products can be toxic and mutagenic to humans, and other mammalian species (see Chapters 7, 8, and 10). The toxic effects of 2,4,6-trinitrotoluene (TNT) were observed as early as the 1920s [1]. Oral administration of TNT, picric acid, and 2,4,6-trinitrophenyl-N-methylnitramine (tetryl) to rats and mice was accompanied by hemolytic anemia, methemoglobinemia (increased formation of the ferric state of hemoglobin unable to bind oxygen), reduced hematocrit (erythrocyte concentration in blood, v/v), liver damage, splenomegaly, hypercholesterolemia, and testicular atrophy [2–6]. The data on the mutagenic activity of TNT and its metabolites in mammalian cells were equivocal, depending on the cell line and conditions [7–10]. The exposure of humans to nitroaromatic explosives is accompanied by multiple toxic effects, which are influenced by the genetic or individual susceptibility, and by the workplace standards [1,11,12]. Increased exposure to TNT causes methemoglobinemia [11]; cataracts [12–14]; reproductive toxicity [15]; skin lesions and dermatitis [16]; and urinary tract, kidney, and liver tumors

[17,18]. An increased incidence of acute and chronic leukemia has also been reported following chronic exposure [19].

In spite of numerous clinical and ecotoxicological reports, the biochemical mechanisms underlying the toxicity of nitroaromatic explosives are not well understood. The data on the toxicity of other groups of explosives are scarce. However, one may expect that the enzymatic activation and mammalian cell culture cytotoxicity mechanisms of nitroaromatic explosives will be similar to other groups of nitroaromatic compounds used in pharmacy (nitrofurans and nitroimidazoles) or recognized as important environmental pollutants (polycyclic nitroarenes) [20]. In this chapter, we analyze the available data on the enzymatic reactivity and mammalian cell cytotoxicity of the classical explosives (TNT and its metabolites, tetryl, and 2,4,6-trinitrophenyl-N-nitramino-ethylnitrate [pentryl]), and the new generation of explosives, 1,3,6,8-tetranitrocarbazole (TNC) [21], 4,6-diamino-5,7-dinitrobenzofuroxan (CL14) [22], 5-nitro-1,2,4-triazol-3-one (NTO), 5-nitro-1,2,4-triazol-3-amine (ANTA) [23], and 4,5,6,7-tetranitrobenzimidazol-2-one (TNBO) [24] (Figure 9.1). The data on

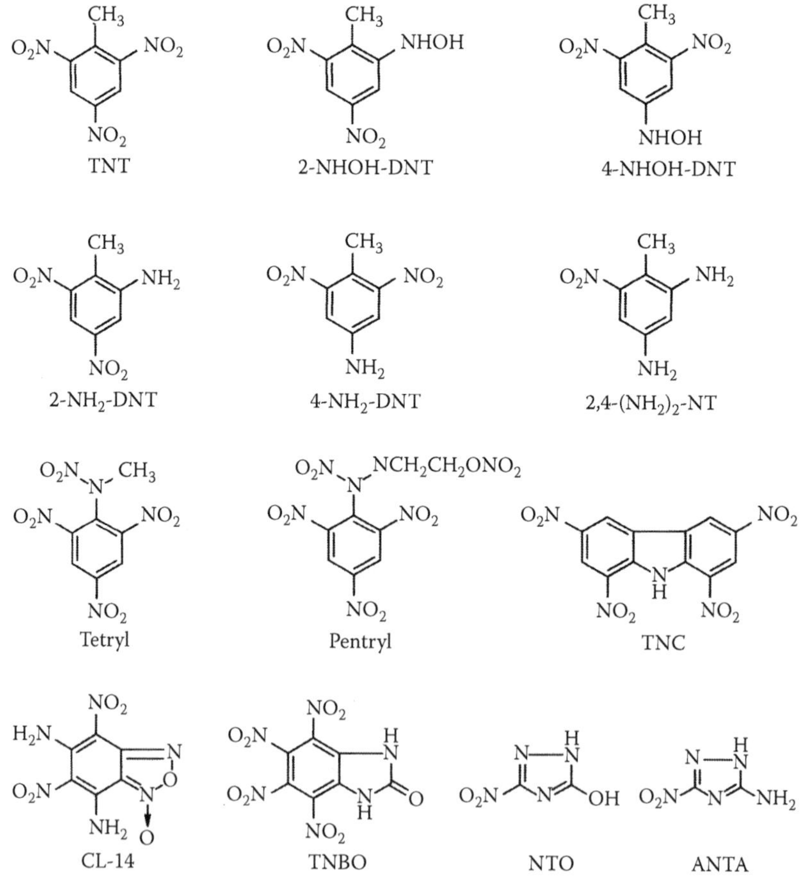

FIGURE 9.1 Structural formulae of nitroaromatic explosives and their metabolites.

the enzymatic reactions and cytotoxicity of other groups of explosives will be also presented.

9.2   ENZYMATIC REACTIONS OF NITROAROMATIC EXPLOSIVES

In this section, we present the data on the enzymatic reactions of explosives relevant to the general cytotoxicity mechanisms of nitroaromatics: (1) their single-electron enzymatic reduction to radicals accompanied by the formation of the reactive oxygen species (oxidative stress type of cytotoxicity); and (2) their two-electron reduction to nitroso and hydroxylamino metabolites causing the cytotoxicity by their covalent binding to proteins and DNA.

9.2.1   FREE RADICAL REACTIONS OF NITROAROMATIC EXPLOSIVES

TNT enhances the NAD(P)H-dependent formation of superoxide (O_2^{-}) and hydrogen peroxide (H_2O_2) in rat microsomes and mitochondria due to the single-electron enzymatic reduction of TNT and its subsequent redox cycling [25]. The single-electron enzymatic reduction of nitroaromatics ($ArNO_2$) to their anion-radicals ($ArNO_2^{-}$)

$$ArNO_2 + e^- \rightarrow ArNO_2^{-} \tag{9.1}$$

is followed by their reoxidation by O_2 with the formation of superoxide and hydrogen peroxide

$$ArNO_2^{-} + O_2 \rightarrow ArNO_2 + O_2^{-} \tag{9.2}$$

$$2O_2^{-} + 2H^+ \rightarrow H_2O_2 + O_2 \tag{9.3}$$

and, subsequently, of the cytotoxic hydroxyl radical (OH·) in transition metal-catalyzed Fenton reaction (Equations 9.4 and 9.5)

$$O_2^{-} + Fe^{3+} \rightarrow O_2 + Fe^{2+} \tag{9.4}$$

$$Fe^{2+} + H_2O_2 \rightarrow Fe^{3+} + OH^- + OH \tag{9.5}$$

The single-electron reduction of nitroaromatics is catalyzed mainly by flavin mononucleotide (FMN)- or flavin adenine dinucleotide (FAD)-containing dehydrogenases–electron transferases. These enzymes transfer electrons from NAD(P)H or other two-electron donors to single-electron accepting oxidants, for example, heme- or FeS-proteins. The nitroreductase reactions of microsomal NADPH:cytochrome P-450 reductase (P-450R, EC 1.6.4.2) and plant or algal ferredoxin:NADP+ reductase (FNR, EC 1.18.1.2) are most thoroughly studied [26–31]. P-450R is universally recognized as one of the most important generators of nitroradicals in mammalian cells [32].

The P-450R contains both FAD and FMN in the active center, and transfers electrons in the sequence NADPH → FAD → FMN → cytochrome P-450, or non-physiological oxidants (cytochrome c, quinones, nitroaromatics) [26,27]. The bimolecular rate constants (k_{cat}/K_m) of explosives reduction by P-450R vary depending on nitroaromatic from $\geq 10^7$ $M^{-1}s^{-1}$ (pentryl, tetryl, TNC), 10^6-10^5 $M^{-1}s^{-1}$ (TNT, TNBO, CL-14), and ~10^3 $M^{-1}s^{-1}$ (NTO, ANTA). FNR is responsible for the bioreductive activation of nitroaromatics in plants and is a useful model system for studying their reactivity. The rate-limiting step in the nitroreductase reaction of *Anabaena* FNR is the oxidation of the FAD semiquinone, whereas the oxidation of two-electron reduced FAD to semiquinone is 30 times faster [33]. The k_{cat}/K_m in FNR-catalyzed reactions vary from $\geq 10^5$ $M^{-1}s^{-1}$ (pentryl, tetryl, TNBO), $\geq 10^4$ $M^{-1}s^{-1}$ (CL-14, TNT), and ~10^2 $M^{-1}s^{-1}$ (NTO, ANTA). In general, the reactivity of nitroaromatics in P-450R- and FNR-catalyzed reactions is relatively insensitive to their molecular structure, but increases as their single-electron reduction potential ($E^1{}_7$, or standard potential of $ArNO_2/ArNO_2{}^{-}$ redox couple at pH 7.0) increases. The observed linear log k_{cat}/K_m vs. E^1 dependencies with the slope Δlog $k_{cat}/K_m /\Delta E^1{}_7$ ~ 10 V^{-1} [26–29] are consistent with an "outer-sphere electron transfer" model with a weak electronic coupling between the reactants [34].

Due to the instability of free radicals, the $E^1{}_7$ values of nitroaromatics (Table 9.1) are usually obtained from anaerobic pulse radiolysis experiments [35,36]. The $E^1{}_7$ values differ from half-wave potentials of electrochemical reduction, which reflect the four-electron transfer [37]. The use of Hammett constants or other thermodynamic

---

TABLE 9.1

Concentrations of Nitroaromatic Explosives and Their Metabolites for the 50% Cell Survival (LC$_{50}$) in Mammalian Cell Culture Cytotoxicity Studies

		LC$_{50}$ (µM)				
No.	Compound	FLK	KI	V79	H4IIE	TK-6
1	Pentryl	5.0 ± 1.0				
2	Tetryl	2.2 ± 0.3				
3	TNC	8.0 ± 2.0				
4	TNBO	30 ± 5.0				
5	N-methylpicramide	40 ± 7.0				
6	TNT	25 ± 5.0	106	197 ± 36	17.6	22 ± 5.0
7	CL-14	250 ± 40				
8	2-NHOH-DNT	40 ± 7.0				
9	4-NHOH-DNT	112 ± 10	18.8		28.2	
10	2-NH$_2$-DNT (2-ADNT)	440 ± 35	>1270	222 ± 76	91.3	168 ± 14
11	4-NH$_2$-DNT (4-ADNT)	316 ± 20	>1270	>328	335	248 ± 51
12	2,4-(NH$_2$)$_2$-NT (2,4-DANT)	350 ± 40	>1500	>600	>1500	>600
13	2,4,6-(NH$_2$)$_3$-T (TAT)			5.2 ± 0.8		0.4
14	NTO	>3500				
15	ANTA	3000 ± 400				

Note: Data taken from studies of FLK cells (24-h exposure) [29–31,33]; Chinese hamster ovary K1 (24 h) and rat hepatoma H4IIE cells (24 h) [72]; and Chinese hamster lung V79 (24 h) and human lymphoblast TK-6 cells (48 h) [9].

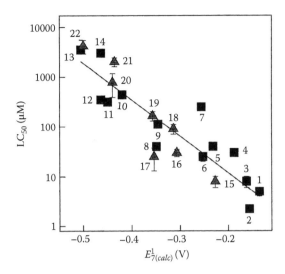

FIGURE 9.2 The dependence of LC_{50} of nitroaromatic explosives, their metabolites, and model nitroaromatic compounds in FLK cells [27–29,31] on their calculated single-electron reduction potentials ($E^l_{7(calc)}$) [33], the experimentally determined E^l_7 values [35,36] given in parentheses: 1. pentryl, −0.135 V; 2. tetryl, −0.156 V; 3. TNC, −0.162 V; 4. TNBO, −0.188 V; 5. N-methyl-picramide, −0.232 V; 6. TNT, −0.253 V (−0.253 V); 7. CL-14, −0.256 V; 8. 2-NHOH-DNT, −0.349 V; 9. 4-NHOH-DNT, −0.346 V; 10. 2-NH$_2$-DNT (2-ADNT), −0.420 V (−0.417 V); 11. 4-NH$_2$-DNT (4-ADNT), −0.450 V (−0.449 V); 12. 2,4-(NH$_2$)$_2$-NT (2,4-DANT), −0.464 V (−0.502 V); 13. NTO, −0.505 V; 14. ANTA, −0.464 V; 15. *p*-dinitrobenzene, −0.227 V (−0.257 V); 16. *o*-dinitrobenzene, −0.307 V (−0.287 V); 17. *p*-nitrobenzaldehyde, −0.354 V (−0.325 V); 18. *m*-dinitrobenzene, −0.314 V (−0.348 V); 19. *p*-nitroacetophenone, −0.357 V (−0.355 V); 20. *p*-nitrobenzoic acid, −0.440 V (−0.425 V); 21. *p*-nitrobenzyl alcohol, −0.436 V (−0.475 V); 22. nitrobenzene, −0.500 V (−0.485 V).

parameters to calculate the unavailable E^l_7 for explosives other than TNT (Figure 9.2 legend) using Hammett constants or other electronic parameters do not provide the reliable results due to the large data scattering ($r^2 = 0.65$–0.70) [33], whereas the use of the linear log k_{cat}/K_m vs. E^l_7 relationships in the nitroreductase reactions of P-450R and FNR [29–31] gives much better results ($r^2 = 0.9369$) [33]. The calculated reduction potentials ($E^l_{7(calc)}$, Figure 9.2 legend) deviate from the experimental ones by not more than 40 mV (standard deviation, ±18 mV), and thus should be considered as realistic.

The re-oxidation of nitroanion radicals by oxygen and their dismutation are most relevant to their cytotoxicity. The rate constants of nitroanion radical oxidation by O$_2$ (Equation 9.2) decrease with an increase in E^l_7 values, for example, 7.7×10^6 M^{-1}s^{-1} (nitrobenzene, $E^l_7 = -0.485$ V), 1.4×10^6 M^{-1}s^{-1} (*p*-nitroacetophenone, $E^l_7 = -0.355$ V), 1.5×10^6 M^{-1}s^{-1} (*p*-nitrobenzaldehyde, $E^l_7 = -0.325$ V), 2.5×10^5 M^{-1}s^{-1} (nitrofurantoin, $E^l_7 = -0.255$ V), 1.5×10^5 M^{-1}s^{-1} (nifuroxime, $E^l_7 = -0.255$ V) [38]. Thus, one may expect that the rate constant may be close to 10^5 M^{-1}s^{-1} for TNT$^-$ (Figure 9.2 legend) and even lower for the radicals of tetryl and TNC. The reduction of TNT and other nitroaromatics by P-450R and FNR is accompanied by redox cycling, that is, oxidation of a significant excess of NADPH over nitroaromatics compound and a stoichiometric to NADPH consumption of oxygen. During the FNR-catalyzed reduction of TNT and dinitrobenzenes, the superoxide dismutase-sensitive reduction of added cytochrome c takes place (180%–185% NADPH oxidation rate), thus demonstrating the formation of superoxide. However, the NADPH/cytochrome c stoichiometry does not exactly match the percentage of single-electron flux in the reduction of nitroaromatics, since the products of four-electron reduction, hydroxylamines, may reduce cytochrome c directly with $k = 90$ M^{-1}s^{-1} (2-NHOH-DNT) or $k = 39$ M^{-1}s^{-1} (4-NHOH-DNT) at pH 7.0 [31]. The single-electron reduction of tetryl and pentryl by FNR and P-450R is accompanied by its redox cycling, N-denitration, and nitrite formation [29,39]:

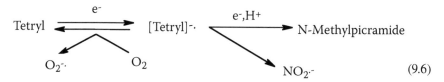

$$\text{(9.6)}$$

The pentryl radical undergoes N-denitration at a greater rate than tetryl, since pentryl forms a higher amount of nitrite during its redox cycling [29].

The dismutation of nitroanion radicals yields the nitroso compounds (ArNO):

$$2ArNO_2^- + 2H^+ \rightarrow ArNO_2 + ArNO + H_2O \qquad (9.7)$$

The dismutation rate constants ($2k_d$) for TNT$^-$ and other radicals of explosives are not reported; however, for the radicals of o-, m-, p-dinitrobenzenes, $2k_d$ are equal to 2.4×10^6 M^{-1}s^{-1}, 8.0×10^6 M^{-1}s^{-1}, and 3.3×10^8 M^{-1}s^{-1}, respectively, at pH 7.0 [40]. It is important to note that the fraction of stable reduction products of nitroaromatics may be formed under aerobic conditions due to the competition between the nitroradical oxidation by oxygen and its dismutation. The rate of dismutation of free radicals (V_{dism}) may be expressed as

$$V_{dism} = 2k_d ((k_{ox} \times [O_2]/4k_d)^2 + V/2k_d)^{1/2} - k_{ox} \times [O_2]/4k_d)^2 \qquad (9.8)$$

where V is the rate of free radical formation, k_{ox} is the rate constant of its oxidation by oxygen, and [O$_2$] is the oxygen concentration. Using V = 10^{-6} M/s, $k_{ox} = 10^5$ M^{-1}s^{-1}, [O$_2$] = 2.5×10^{-4} M, and $2k_d = 10^7$ M^{-1}s^{-1}, we obtain $V_{dism} = 1.6 \times 10^{-8}$ M/s, that is, 1.6% of the total nitroreduction rate. Using V = 10^{-5} M/s, we obtain $V_{dism} = 12.3 \times 10^{-7}$ M/s, that is, 12.3% of the total rate. It shows that the yield of the nitroreduction metabolites should exhibit a square dependence on a single-electron transferring enzyme concentration, which has been confirmed experimentally [41]. Explosives and other nitroaromatic compounds perform single-electron oxidation of oxyhemoglobin (Hb-Fe^{2+}O$_2$) into methemoglobin (Hb-Fe^{3+}):

$$Hb\text{-}Fe^{2+}O_2 + ArNO_2 \rightarrow Hb\text{-}Fe^{3+} + O_2^- + ArNO_2^- \qquad (9.9)$$

The rate constants of oxyhemoglobin oxidation vary from 9–10 M^{-1}s^{-1} (tetryl, pentryl, TNBO) to 3.3 M^{-1}s^{-1} (TNT) to 0.1 M^{-1}s^{-1} (NTO, ANTA), with an absence of the well-expressed reactivity dependence on $\dot{E}^l{}_7$ [30,42]. The interaction of hydroxylamino-DNTs with oxyhemoglobin may be more complex because hydroxylamine and phenylhydroxylamine convert $Hb\text{-}Fe^{2+}O_2$ into ferrylhemoglobin ($Hb\text{-}Fe^{4+}O_2$) [43,44]. However, the proposed reaction mechanisms are not well understood. Our preliminary observations show that 2-NHOH-DNT and 4-NHOH-DNT oxidize $Hb\text{-}Fe^{2+}O_2$ into $Hb\text{-}Fe^{3+}$ with the rate constants of 13 $M^{-1}s^{-1}$ and >100 $M^{-1}s^{-1}$, irrespectively, the reactions proceed further with the formation of $Hb\text{-}Fe^{4+}O_2$ [45].

Reactions of nitroaromatic explosives with a number of potentially important single-electron transferring flavoenzymes have not been studied. However, their reactivities may be predicted by the linear log k_{cat}/K_m vs. $E^l{}_7$ relationships that exist in nitroreductase reactions. The main source of nitroradicals in mammalian mitochondria is the outer membrane NAD(P)H-oxidizing nitroreductase(s), whose properties remain uncharacterized so far [46]. NADH:ubiquinone reductase (complex I of the inner mitochondrial membrane, EC 1.6.5.3) contains FMN, two Fe_2S_2, and a minimum three Fe_4S_4 clusters, and transfers electrons to ubiquinone-10. The nitroreductase reaction of complex I is not inhibited by rotenone, a competitive inhibitor to ubiquinone that binds at the N2 Fe_4S_4 cluster but is inhibited by NAD^+ and ADP-ribose [47]. It demonstrates that nitroaromatics are reduced via FMN. The stoichiometry of the coupled cytochrome c reduction implies that complex I may reduce nitroaromatics both by single- and two-electron transfer. FAD-dependent NADPH:adrenodoxin reductase (ADR, EC 1.18.1.2) transfers electrons to Fe_2S_2 protein adrenodoxin (ADX), which in turn reduces cytochromes P-450 in adrenal cortex mitochondria. Low nitroreductase activity of ADR is markedly stimulated by ADX acting as the redox mediator [48]. The nitroreductase reactions of complex I and ADX are characterized by the linear log k_{cat}/K_m vs. $E^l{}_7$ dependences. In terms of k_{cat}/K_m, the activity of complex I is close to that of FNR, and the activity of ADX is by 3 to 5 times higher [47,48]. FAD-dependent NADH:cytochrome b_5 reductase (EC 1.6.2.2), which performs the cytochrome b_5-mediated reduction of cytochromes P-450 in microsomes, and the reduction of $Hb\text{-}Fe^{3+}$ in erythrocytes, participates in microsomal nitroreduction as well [49]. Xanthine oxidase (EC 1.1.3.22) contains FAD, two Fe_2S_2 clusters, and molybdopterin cofactors, and catalyzes mixed single- and two-electron reduction of nitroaromatics [50]. This enzyme is supposed to be the main source for nitroreduction in cytosol [51]. However, the nitroreductase reactions of cytochrome b_5 reductase and xanthine oxidase have not been investigated in detail.

Remaining flavoenzymes are glutathione reductase (GR; EC 1.6.4.2) and thioredoxin reductase (TrxR; (EC 1.6.4.5). GR contains both FAD and a catalytic disulfide, the latter reducing oxidized glutathione (GSSG) in an obligatory two-electron way. Mammalian TrxRs contain an additional redox center, a terminal selenosulfide group, which transfers electrons to the 12 kD disulfide redox protein thioredoxin (Trx). Erythrocyte GR and rat TrxR perform single-electron reduction of tetryl (Equation 9.6) with the maximal rate (k_{cat}) of 5 s^{-1} and 1.8 s^{-1}, and $k_{cat}/K_m =$ 2.0×10^3 $M^{-1}s^{-1}$ and 1.4×10^4 $M^{-1}s^{-1}$, respectively, at pH 7.0 [52,53]. TNT is reduced 10 times slower. However, it is unclear whether these slow reactions may contribute to the oxidative stress-type cytotoxicity. On the other hand, GR and TrxR perform important antioxidant functions in the cell, and their inhibition by explosives may contribute to the cytotoxic effects. Tetryl and pentryl react with reduced glutathione (GSH) directly (k = 0.6 M^{-1}s^{-1}) forming trinitrophenyl-SG [42]. Tetryl, TNT, and trinitrophenyl-SG inhibit GR with the inhibition constants (K_i) of 4.5–14 μM [52]. Tetryl inhibits rat TrxR with K_I = 12 μM, and acts as the irreversible inactivator of the reduced enzyme with the maximal rate, 4 min^{-1}, and the bimolecular rate constant, 670 M^{-1}s^{-1} [53]. This reaction most probably proceeds via the covalent modification of reduced selenocysteine or cysteine. Finally, the flavin-independent ζ-crystallin from bovine lens also catalyzes the NADPH-dependent single-electron reduction and redox cycling of TNT [54]. This reaction is supposed to be responsible for TNT-induced cataracts.

9.2.2 FORMATION OF STABLE METABOLITES OF NITROAROMATIC EXPLOSIVES

In mammalian systems, nitroaromatic compounds are further reduced to amines and/or hydroxylamines, which may subsequently form DNA and protein adducts. These stable metabolites may be formed by the reduced oxygen tension and high local single-electron transferring enzyme concentration, for example, P-450R in microsomes may favor the free radical dismutation over their reoxidation by oxygen (Equation 9.8). Subsequently, the nitroso compounds formed (Equations 9.7 and 9.8) will be reduced to hydroxylamines (ArNHOH) and/or amines (ArNH$_2$). The stable metabolites may also be formed by the two-electron reduction of nitroaromatics by certain flavoenzymes. In fact, the enzymatic two-electron reduction may be considered as the four-electron reduction, since after the first two-electron (hydride) transfer, the reduction of an intermediate nitroso compound to hydroxylamine (ArNHOH) proceeds faster [55]:

$$ArNO_2 \xrightarrow[-H_2O]{+2e^- +2H^+} ArNO \xrightarrow{+2e^- +2H^+} ArNHOH \qquad (9.10)$$

Due to high redox potential (the polarographic reduction potentials of nitroso compounds are 0.2–0.8 V higher than those of parent nitroaromatics [56]), nitrosobenzenes may be also reduced by NAD(P)H, GSH, and other reductants nonenzymatically [55,57]. Alternatively, aromatic hydroxylamines may be formed during the N-hydroxylation of amines by cytochrome P-450 [58].

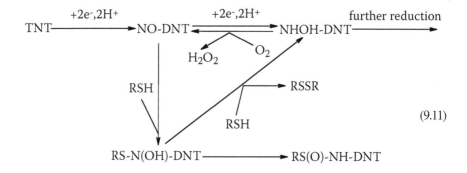

$$(9.11)$$

The administration of TNT to laboratory animals leads to the excretion of 4-NHOH-DNT, 2-NH$_2$-DNT, and 4-NH$_2$-DNT in the urine [59], and to the formation of covalent adducts with microsomal liver and kidney proteins, hemoglobin, and other blood proteins [60]. The acid hydrolysis of adducts yielded mainly 2-NH$_2$-DNT (2-ADNT) and 4-NH$_2$-DNT (4-ADNT). Incubation of rat liver microsomes with TNT and NADPH under aerobic conditions resulted in the formation of NH$_2$-DNTs and the transient metabolite 4-NHOH-DNT [57]. The formation of covalent protein adducts with TNT metabolites was enhanced by the presence of O$_2$ and decreased by GSH. This is consistent with the scheme of the TNT adduct formation with the central role of the nitroso metabolite (NO-DNT) reaction with protein or nonprotein thiols (RSH; Equation 9.11) [57]. The acid hydrolysis of the sulfinamide adduct (RS(O)-NH-DNT) formed after the rearrangement of the semimercaptal (RS-N(OH)-DNT; Equation 9.12) will yield NH$_2$-DNT. The mixture of NHOH-DNTs inhibits bacterial glyceraldehyde-3-phosphate dehydrogenase and glucose-6-phosphate dehydrogenase more efficiently than TNT [61]. This was attributed to the covalent modification of protein –SH groups.

Aromatic hydroxylamines can modify DNA either directly or via formation of an O-acetylated intermediate. The acetylated intermediate can be transformed to a strongly electrophilic nitrenium ion (ArNH$^+$) capable of modifying guanine bases with the formation of N-(deoxyguanosin-8-yl)-NHAr adducts. However, the formation of TNT adducts with DNA in mammalian cells has not been reported. On the other hand, NHOH-DNT may undergo the transition metal-catalyzed redox cycling, which may cause the DNA oxidative damage [62]:

$$NHOH\text{-}DNT + Cu^{2+} \rightarrow NHO^\cdot\text{-}DNT + Cu^+ + H^+ \tag{9.12}$$

$$Cu^+ + O_2 \rightarrow Cu^{2+} + O_2^{-\cdot} \tag{9.13}$$

$$NHO^\cdot\text{-}DNT + NADH \rightarrow NHOH\text{-}DNT + NAD \tag{9.14}$$

$$NAD^\cdot + O_2 \rightarrow NAD^+ + O_2^{-\cdot} \tag{9.15}$$

In liver microsomes, the P-450R-catalyzed reactions, according to Equations 9.1, 9.7, and 9.8 may be the main source of TNT metabolites, because the NADPH-dependent formation of nitrophenylamine and nitrophenylhydroxylamine from 1,3-dinitrobenzene possesses a lag period, evidently due to the reaction acceleration after partial oxygen exhaustion [63]. Since more thorough studies on the TNT metabolites in mammalian cells in vitro have not been performed, other formation pathways may be predicted by the analogous reactions of other polynitroaromatics. Rat liver cytosol catalyzed the NAD(P)H-dependent formation of the monoamino and mononitroso metabolites of 1,3- and 1,6-dinitropyrenes under both aerobic and anaerobic conditions [64]. The amine formation was inhibited by oxygen by 10% to 60%, whereas the nitroso metabolite formation was either not affected or even enhanced by oxygen. In contrast the microsomal formation of both amino and nitroso metabolites was inhibited by oxygen by 80% to 90% [64]. Xanthine oxidase and NAD(P)H:quinone oxidoreductase (DT-diaphorase, NQO1, EC 1.6.99.3) were responsible for the formation of 3-nitroaniline from 1,3-dinitrobenzene in the cytosolic fraction of rat small intestinal mucosa under aerobiosis [51], thus being the potential sources for the formation of the stable metabolites of TNT and other nitroaromatic explosives in cytosol.

The nitroreductase reactions of NQO1 have been studied more thoroughly [29,38,65,66]. Rat NQO1 contains FAD with the standard (two-electron reduction) potential at pH 7.0, $E^0{}_7$, of -0.159 V [67]. The ability of NQO1 to perform two-electron transfer is most probably determined by the instability of its anionic FAD semiquinone, since the $E^1{}_7$ of FAD/FAD$^-$ and FAD$^-$/FADH$^-$ couples are equal to -0.200 V and -0.118 V, respectively [67]. The majority of nitroaromatic compounds including TNT are very slow NQO1 substrates ($k_{cat}/K_m = 10^2$–10^4 M^{-1}s^{-1}, $k_{cat} = 0.1$–1.0 s^{-1}), with the exception of TNBO, tetryl, and pentryl, whose reactivities are intermediate ($k_{cat}/K_m \geq 10^5$ M^{-1}s^{-1}, $k_{cat} > 10$ s^{-1}). Possibly, NQO1 reduces TNT to NHOH-DNTs, which is further reduced to dihydroxylamino-NT at a similar rate. NQO1 performs reductive N-denitration of tetryl and pentryl (Equation 9.6) with the formation of picramides, other unidentified products, and O$_2{}^-$, which points to the involvement of single-electron transfer steps [29,66]. The relationship between the reactivity of nitroaromatics toward NQO1 and their reduction potential is absent; the structural parameters determining their selectivity are unclear.

TNT and other polynitrobenzenes may be reduced in erythrocytes with the formation of hydroxylamines and the covalent protein adducts [59,68], with possible involvement of an unidentified NADH-oxidizing flavoenzyme [69]. The formation of stable metabolites of TNT and other nitroaromatic explosives by intestinal microflora or their enzymes has not been investigated, except for the study of isolated *Enterobacter cloacae* NAD(P)H:nitroreductase (NR, EC 1.6.99.7) [70]. In the active center, *E. cloacae* NR contains FMN ($E^0{}_7 = -0.19$ V) whose semiquinone state is extremely unstable ($E^1{}_7$ of FMN$^-$/FMNH$^-$ couple, ~ -0.01 V, and $E^1{}_7$ of FMN/FMN$^-$ couple, ~ -0.37 V) [71]. The reactivity of nitroaromatics toward *E. cloacae* NR increases as their reduction potential increases, showing little specificity toward their molecular structure [70]. The nitroaromatic explosives with high $E^1{}_{7(calc)}$ values (tetryl, pentryl, TNC, TNBO) are reduced more rapidly ($k_{cat}/K_m = 10^6$–10^7 M^{-1}s^{-1}, $k_{cat} > 100$ s^{-1}) than NTO and ANTA ($k_{cat}/K_m \geq 10^4$ M^{-1}s^{-1}, $k_{cat} \geq 5$ s^{-1}). The reasons for the low enzymatic activity of CL-14 ($k_{cat}/K_m = 4 \times 10^4$ M^{-1}s^{-1}) are being investigated. NR reduces nitrobenzene to phenylhydroxylamine [55]; however, TNT, tetryl, and pentryl oxidize more than two equivalents of NADH [69]. The reduction of NHOH-DNTs by *E. cloacae* NR is accompanied by the close to stoichiometric O$_2$ consumption, as well as the oxidation of more than twofold excess of NADH by TNT [31]. Most probably, *E. cloacae* NR reduces TNT to dihydroxylamino-NT, which undergoes rapid autoxidation. The reduction mechanism of tetryl and pentryl by *E. cloacae* NR is different from that described by Equation 9.6, since the picramides are not formed and nitrite is formed not during the explosive reduction by two NADH equivalents, but in subsequent steps [70]. It is possible that other "oxygen-insensitive" nitroreductases of the intestinal microflora will reduce nitroaromatic explosives in a similar way, for example, the *Escherichia coli* nitroreductase NfsB which shares 80% sequence identity with *E. cloacae* NR and whose reactivity also increases with an increase in the reduction potential of nitroaromatics [72].

9.3 MAMMALIAN CELL CYTOTOXICITY OF NITROAROMATIC EXPLOSIVES

Mammalian cell culture cytotoxicity data of TNT, its metabolites, and other nitro-aromatic explosives, expressed as concentration causing 50% cell death (LC_{50}), are summarized in Table 9.1 [9,29–31,73]. Other less comprehensive studies show that LC_{50} of TNT and 2-NH_2-DNT (2-ADNT) are above 450 µM for human neuroblas-toma NG108 (7-h incubation) [74] and that LC_{50} of TNT for human hepatocarcinoma HepG2 is 460 ± 26 µM (48-h incubation) [10]. The explosives-contaminated soil extract containing 68% TNT, 12% NH_2-DNTs (ADNTs), and 12% RDX was toxic to human fibroblasts GM05757 with LC_{50} close to 2.1 mg L^{-1} (1-h incubation, 48-h growth in extract-free medium) [75]. The extracts of TNT-contaminated soil possess in vitro immunotoxicity [76]; however, the data were not expressed in quantitative terms.

The most thorough cytotoxicity studies were performed with bovine leukemia virus-transformed lamb kidney fibroblasts (line FLK) [29–31,65]. The cytotoxicity of TNT, tetryl, CL-14, NTO, ANTA, and tetranitrobenzimidazolone to this cell line was reduced by the antioxidant N,N'-diphenyl-p-phenylene diamine (DPPD), and desferrioxamine, an iron ion chelator that prevents the Fenton reaction (Equations 9.4–9.6). In contrast, the alkylating agent 1,3-bis-(2-chloroethyl)-1-nitrosourea (BCNU), which inactivates glutathione reductase and depletes intracellular GSH, potentiated the cellular toxicity. Cytotoxicity was accompanied by lipid peroxida-tion, suggesting the involvement of oxidative stress, resulting from the enzymatic redox cycling of anion-radicals of explosives (Equations 9.3–9.6). Enzymes respon-sible for free radical formation in FLK cells have not yet been identified. Dicumarol, an inhibitor of NQO1, partly reduced the cytotoxicity of TNT, TNBO, and CL-14, but surprisingly potentiated the cytotoxicity of tetryl and TNC. Both protective and potentiating effects of dicumarol may not be considered as significant, since it affected the LC_{50} values of the compounds by approximately 1.5 times. It shows that the cytotoxic consequences of reaction of NQO1 with all explosives in FLK cells are equivocal, evidently depending on the further metabolic fate of reaction prod-ucts. For a number of explosives and model nitroaromatics, log LC_{50} in FLK cells decreased with an increase in their E^1_7 or the geometric mean of their reactivity in P-450R and FNR-catalyzed reactions, (log k_{cat}/K_m (P-450R) + (log k_{cat}/K_m (FNR))/2) [29,31]. Analogous dependence may be obtained using their $E_1^7{}_{(calc)}$ as the correlation parameter (Figure 9.2). This type of dependence shows that the main mechanism of cytotoxicity of nitroaromatics is the oxidative stress, initiated by flavoenzyme-catalyzed single-electron reduction [77]. Because the amino metabolites of TNT are less cytotoxic than TNT in several cell lines (Table 9.1), it is possible that the flavoenzyme-catalyzed redox cycling is their universal cytotoxicity mechanism. One may note that FLK cell cytotoxicity of NHOH-DNTs, NH_2-DNTs (ADNTs), and 2,4-diamino-6-nitrotoluene ((NH_2)$_2$-NT; 2,4-DANT) is somewhat higher than expected from their redox cycling activities (Figure 9.2). Besides, in other cell lines, NHOH-DNTs are as toxic or even more toxic than TNT (Table 9.1). It points to the possibility of the parallel involvement of other cytotoxicity mechanisms, for example, redox cycling of hydroxylamines (Equations 9.12–9.15), their binding to macromole-cules, and the possibility of the formation of toxic hydroxylamines from amines [58]. The possibility of this mechanism is indirectly evidenced by high V79 and TK-6 cell cytotoxicity of 2,4,6-triaminotoluene ($(NH_2)_3$-T; TAT) (Table 9.1), which lacks the reducible nitro group. It is also possible that high TAT toxicity is caused by the action of its autoxidation products formed during its long incubation in solution [78].

Recent studies disclosed some important events in TNT cytotoxicity, demonstrat-ing the transcriptional activation of stress genes in HepG2 cells under the action of TNT [10] and the enhanced expression of *p53* tumor suppressor in MCF-7 human breast cancer cells under the action of 2-NH_2-DNT [79]. However, both the redox cycling of nitroaromatics and the formation of their DNA-alkylating products may be responsible for these phenomena [20].

In view of methemoglobinemia formation resulting from the exposure to explo-sives and other nitroaromatics [1–5,11], we briefly address the Hb-Fe^{3+} formation in human erythrocytes in vitro. The extent of Hb-Fe^{3+} formation in vitro shows strong variability among individuals and also depends on the storage conditions of erythro-cytes. The Hb-Fe^{3+} inducing potency is shown to be TNT, NHOH-DNTs, tetryl, pen-tryl, CL-14 > TNBO >>NH_2-DNT (ADNT), $(NH_2)_2$-NT > NTO, ANTA [30,42,45]. One may note that there does not exist a relationship between the rate of isolated Hb-$Fe^{2+}O_2$ and the extent of Hb-Fe^{3+} formation in erythrocytes. Although NHOH-DNTs can oxidize oxyhemoglobin into ferrylhemoglobin, only Hb-Fe^{3+} was formed during 24-h incubation of erythrocytes with NHOH-DNTs. Evidently, Hb-Fe^{3+} was formed in a subsequent reaction between Hb-$Fe^{2+}O_2$ and ·Hb-$Fe^{4+}O_2$.

9.4   MAMMALIAN CELL CYTOTOXICIY AND ENZYMATIC REACTIONS OF OTHER EXPLOSIVES

One may mention the relative lack of information on the possible toxicity mechanisms of other groups of explosives. The administration of hexahydro-1,3,5-trinitro-1,3,5-triazine (RDX; 30–300 mg kg^{-1} daily for 13 weeks) to rats caused hypotriglycidire-mia, convulsions, and death [4]. In contrast, pentaerythritol tetranitrate (PETN; 0.5%–1.0% in standard diet for 13 weeks) was nontoxic to rats [80]. RDX was much less cytotoxic to V79 and TK-6 mammalian cell cultures than TNT [9]. There also are very few data on their reactions with mammalian enzymes. Rabbit liver cyto-chrome P-450 2B4 (EC 1.14.14.1) converted RDX into 4-nitro-2,4-diazabutanal, two nitrite ions, ammonium, and formaldehyde, consuming one equivalent NADPH [81]. However, it is unclear whether this slow reaction ($k_{cat} < 0.01$ s^{-1}) may contribute to the toxicity of RDX. Xanthine oxidase transformed octohydro-1,3,5,7-tetranitro-1,3,5,7-tetrazocine (HMX) at a lower rate, 10.5 nmol h^{-1} mg^{-1} protein under anaero-bic conditions, into nitrite, formaldehyde, nitrous oxide, formic acid, and ammonium [82]. Our preliminary observations show that RDX was much less reactive substrate for P-450R and *E. cloacae* NR than NTO or ANTA [53]. Thus, the mechanisms underlying toxicity of RDX remain undisclosed.

9.5 CONCLUDING REMARKS

This chapter summarizes the present state of knowledge on the cytotoxicity mechanisms of nitroaromatic and other explosives. At present, the involvement of oxidative stress in the cytotoxicity of nitroaromatic explosives is unequivocally demonstrated. The data on the enzymatic reactivity, mammalian cell cytotoxicity, and Hb-Fe^{3+} in vitro formation of explosives provide some background for the understanding of their overall toxicity. For example, NTO is almost nontoxic in rats and mice ($LD_{50} > 5$ g kg^{-1} when given orally [83]), whereas LD_{50} for TNT and tetryl are equal to 660–1320 mg kg^{-1} and above 300 mg kg^{-1}, respectively [1,5]. This is consistent with much lower NTO enzymatic reduction rates and cell culture cytotoxicity (Table 9.1) [30]. This chapter also outlines some future directions for research: (a) the identification and characterization of both the enzyme and metabolites of nitroaromatic explosives in mammalian cell mitochondria, cytosol, erythrocytes, and gastrointestinal microflora; (b) elucidation of the mechanisms of explosive-induced methemoglobinemia; (c) the characterization of metabolic pathways and cytotoxicity mechanisms of nitroaliphatic explosives; (d) enzymatic and cytotoxicity studies of the nitroaromatic explosives of the new generation; and (e) identification of protein targets of the explosive toxic action in the organism including the proteomic approach, as well as the characterization of the mode of their binding to DNA.

REFERENCES

1. Voegtlin C, Hooper CW, and Johnson JM, Trinitrotoluene poisoning, *US Public Health Rep.*, 34, 1307, 1919.
2. Dilley JV et al., Short-term oral toxicity of 2,4,6-trinitrotoluene in mice, rats, and dogs, *J. Toxicol. Environ. Health*, 9, 565, 1982.
3. Wyman JF et al., Acute toxicity, distribution, and metabolism of 2,4,6-trinitrophenol (picric acid) in Fischer 344 rats, *J. Toxicol. Environ. Health*, 37, 313, 1992.
4. Levine BS et al., Subchronic toxicity of trinitrotoluene in Fischer 344 rats, *Toxicology*, 32, 253, 1984.
5. Levine BS et al., Toxic interactions of the munition compounds TNT and RDX in F344 rats, *Fundam. Appl. Toxicol.*, 15, 373, 1990.
6. Reddy TV et al., Toxicity of tetryl (N-methyl-N,2,4,6-tetranitroaniline) in Fischer F344 rats, *Int. J. Toxicol.*, 18, 97, 1999.
7. Styles JA and Cross MF, Activity of 2,4,6-trinitrotoluene in an in vitro mammalian gene mutation assay, *Cancer Lett.*, 20, 103, 1983.
8. Kennel SJ et al., Mutation analyses of TNT-related compounds using the CHO-HPRT assay, *J. Appl. Toxicol.*, 20, 441, 2000.
9. Lachance B et al., Cytotoxic and genotoxic effects of energetic compounds on bacterial and mammalian cells in vitro, *Mutation Res.*, 444, 25, 1999.
10. Tchounwou PB et al., Transcriptional activation of stress genes in human liver carcinoma cells (HepG2) exposed to 2,4,6-trinitrotoluene, 2,4-dinitrotoluene, and 2,6-dinitrotoluene, *Environ. Toxicol.*, 16, 209, 2001,
11. Djerassi L, Hemolytic crisis in G6PD-deficient individuals in the occupational setting, *Int. Arch. Occup. Environ. Health*, 71, S26, 1998.
12. Letzel S et al., Exposure to nitroaromatic explosives and health effects during disposal of military waste, *Occup. Environ. Med.*, 60, 483, 2003.

13. Harkonen H et al., Early equatorial cataracts in workers exposed to trinitrotoluene, *Am. J. Ophthalmol.*, 95, 807, 1983.

14. Zhou AS, A clinical study of trinitrotoluene cataract, *Pol. J. Occup. Med.*, 3, 171, 1990.

15. Yi L et al., Effects of exposure to trinitrotoluene on male reproduction, *Biomed. Environ. Sci.*, 6, 154, 1993.

16. Cherkasskaia RG et al., Toxic melanoderma in chronic intoxication with trinitrotoluene [in Russian], *Med. Tr. Prom. Ekol.*, 1, 39, 1993.

17. Bruning T et al., Occurrence of urinary tract tumours in miners highly exposed to dinitrotoluene, *J. Occupat. Environ. Med.*, 41, 144, 1999.

18. Yan C et al., The retrospective survey of malignant tumor in weapon workers exposed to 2,4,6-trinitrotoluene [in Chinese], *Zhonghua Lao Dong Wei Sheng Zhi Ye Bing Za Zhi* 20, 184, 2002.

19. Kolb G et al., Increased risk of acute myelogenous leukemia (AML) and chronic myelogenous leukemia (CML) in a county of Hesse, Germany, *Soz. Praeventivmed.*, 38, 190, 1993.

20. Purohit V and Basu AK, Mutagenicity of nitroaromatic compounds, *Chem. Res. Toxicol.*, 18, 63, 2000.

21. Meyer R, *Explosives,* VCH Verlagsgesellschaft mbH, Weinheim, 1987, 337.

22. Sikder HK, Sinha RK, and Gandhe BR, Cost-effective synthesis of 5,7-diamino-4,6-dinitrobenzofuroxan (CL-14) and its evaluation in plastic bonded explosives, *J. Hazard. Mat.*, 102, 137, 2003.

23. Sikder AK et al., Studies on characterization and thermal behaviour of 3-amino-5-nitro-1,2,4-triazole and its derivatives, *J. Hazard. Mat.*, 82, 1, 2001.

24. Tselinskii IV, Applications of energetic materials in engineering, technology and national economy [in Russian], *Soros Educat. J.*, 11, 46, 1997.

25. Kong LY, Jiang QC, and Qu QS, Formation of superoxide radical and hydrogen peroxide enhanced by trinitrotoluene in rat liver, brain, kidney, and testicle in vitro and monkey liver in vivo, *Biomed. Environ. Sci.*, 2, 72, 1989.

26. Orna MV and Mason RP, Correlation of kinetic parameters of nitroreductase enzymes with redox properties of nitroaromatic compounds, *J. Biol. Chem.*, 264, 12379, 1989.

27. Čénas N et al., The electron transfer reactions of NADPH:cytochrome P-450 reductase with nonphysiological oxidants, *Arch. Biochem. Biophys.*, 315, 400, 1994.

28. Anusevičius Ž et al., Electron transfer reactions of *Anabaena* PCC 7119 ferredoxin:NADP+ reductase with nonphysiological antioxidants, *Biochim. Biophys. Acta*, 1320, 247, 1997.

29. Čénas N et al., Quantitative structure-activity relationships in enzymatic single-electron reduction of nitroaromatic explosives: Implications for their cytotoxicity, *Biochim. Biophys. Acta*, 1528, 31, 2001.

30. Šarlauskas J et al., Enzymatic redox properties of novel nitrotriazole explosives: Implications for their toxicity, *Z. Naturforsch.*, 59c, 399, 2004.

31. Šarlauskas J et al., Flavoenzyme-catalyzed redox cycling of hydroxyl-amino and amino metabolites of 2,4,6-trinitrotoluene: Implications for their cytotoxicity, *Arch. Biochem. Biophys.*, 425, 184, 2004.

32. Rossi I et al., N-[5-nitro-2-furfurylidene]-3-amino-2-oxazolidinone activation by the human intestinal cell line Caco-2 monitored through noninvasible electron spin resonance spectroscopy, *Mol. Pharmacol.*, 49, 547, 1996.

33. Anusevičius Ž, unpublished data, 2004.

34. Marcus RA and Sutin N, Electron transfers in chemistry and biology, *Biochim. Biophys. Acta*, 811, 265, 1985.

35. Wardman P, Reduction potentials of one-electron couples involving free radicals in aqueous solutions, *J. Phys. Chem. Ref. Data*, 18, 1637, 1989.

36. Riefler RG and Smets BF, Enzymatic reduction of 2,4,6-trinitrotoluene and related nitroarenes: Kinetics linked to one-electron redox potentials, *Environ. Sci. Technol.*, 34, 3900, 2000.

37. Darchen A and Moinet C, Mecanisme E.C.E. de reduction du *para*-dinitrobenzene en *para*-nitrophenylhydroxylamine, *J. Electroanal. Chem.*, 78, 81, 1977.

38. Wardman P and Clarke ED, Oxygen inhibition of nitroreductase: Electron transfer from nitro radical-anions to oxygen, *Biochem. Biophys. Res. Commun.*, 69, 942, 1976.

39. Shah MM and Spain JC, Elimination of nitrite from the explosive 2,4,6-trinitrophenyl-methylnitramine (tetryl) catalyzed by ferredoxin NADP oxidoreductase from spinach, *Biochem. Biophys. Res. Commun.*, 220, 563, 1996.

40. Neta P, Simic MG, and Hoffman MZ, Pulse radiolysis and electron spin resonance studies of nitroaromatic radical anions. Optical absorption spectra, kinetics, and one-electron redox potentials, *J. Phys. Chem.*, 80, 2018, 1976.

41. Holtzman JL et al., The kinetics of the aerobic reduction of nitrofurantoin by NADPH-cytochrome P-450 (c) reductase, *Mol. Pharmacol.*, 20, 669, 1981.

42. Marozienė A et al., Methemoglobin formation in human erythrocytes by nitroaromatic explosives, *Z. Naturforsch.*, 56c, 1157, 2001.

43. Maples KR, Eyer P, and Mason RP, Aniline-, phenylhydroxylamine-, nitrosobenzene- and nitrobenzene-induced hemoglobin thiyl free radical formation in vivo and in vitro, *Mol. Pharmacol.*, 37, 311, 1990.

44. Stolze K et al., Hydroxylamine and phenol-induced formation of methemoglobin and free radical intermediates in erythrocytes, *Biochem. Pharmacol.*, 52, 1821, 1996.

45. Marozienė A, unpublished data, 2004.

46. Moreno SNJ, Mason RP, and Docampo R, Reduction of nifurtimox and nitrofurantoin to free radical metabolites by rat liver mitochondria, *J. Biol. Chem.*, 259, 6296, 1984.

47. Bironaitė DA, Čėnas NK, and Kulys JJ, The rotenone-insensitive reduction of quinones and nitrocompounds by mitochondrial NADH:ubiquinone reductase, *Biochim. Biophys. Acta*, 1060, 203, 1991.

48. Marcinkevičienė J et al., Nitroreductase reactions of the NADPH:adrenodoxin reductase and adrenodoxin complex, *Biomed. Biochim. Acta*, 49, 167, 1990.

49. Papadopoulos MV et al., Reductive metabolism of the nitroimidazole-based hypoxia-selective cytotoxin NLCQ-1 (NSC 709257), *Oncol. Res.*, 14, 21, 2003.

50. Tatsumi K, Inoue A, and Yoshimura H, Mode of reactions between xanthine oxidase and aromatic nitro compounds, *J. Pharmacobiodyn.*, 4, 101, 1981.

51. Adams PC and Rickert DE, Metabolism of [^{14}C]1,3-dinitrobenzene by small intestinal mucosa in vitro, *Drug Metab. Dispos.*, 23, 982, 1995.

52. Miškinienė V et al., Tetryl as inhibitor and "subversive substrate" for human erythrocyte glutathione reductase, *Flavins and Flavoproteins*, 13, 703, 1999.

53. Nivinskas H, unpublished data, 2004.

54. Kumagai Y et al., Zeta-crystallin catalyzes the reductive activation of 2,4,6-trinitrotoluene to generate reactive oxygen species: A proposed mechanism for the induction of cataracts, *FEBS Lett.*, 478, 295, 2000.

55. Koder RL and Miller A-F, Steady-state kinetic mechanism, stereospecificity, substrate and inhibitor specificity of *Enterobacter cloacae* nitroreductase, *Biochim. Biophys. Acta*, 1387, 395, 1998.

56. Kovacic P et al., Reduction potentials in relation to physiological activities of benzenoid and heterocyclic nitroso compounds: Comparison with nitro precursors, *Bioorg. Chem.*, 18, 265, 1990.

57. Leung KH et al., Mechanism of bioactivation and covalent binding of 2,4,6-trinitrotoluene, *Chem.-Biol. Interact.*, 97, 37, 1995.

58. Yamazaki H et al., Bioactivation of diesel exhaust particle extracts and their major nitrated polycyclic aromatic hydrocarbon components, 1-nitropyrene and dinitropyrenes, by human cytochromes P450 1A1, 1A2, and 1B1, *Mutat. Res.*, 472, 129, 2000.

59. Rickert DE, Metabolism of nitroaromatic compounds, *Drug Metab. Reviews*, 18, 25, 1987.

60. Liu YY et al., In vivo covalent binding of [^{14}C]trinitrotoluene to proteins in the rat, *Chem.-Biol. Interact.*, 82, 1, 1992.
61. Naumov AV et al., Transformation of 2,4,6-trinitrotoluene into toxic hydroxylamino derivatives by *Lactobacilii*, *Microbiology*, 68, 46, 1999.
62. Homma-Takeda S et al., 2,4,6-Trinitrotoluene-induced reproductive toxicity via oxidative DNA damage by its metabolite, *Free Rad. Res.*, 36, 555, 2002.
63. Reeve H and Miller MG, 1,3-Dinitrobenzene metabolism and protein binding, *Chem. Res. Toxicol.*, 15, 352, 2002.
64. Djuric Z et al., Aerobic and anaerobic reduction of nitrated pyrenes in vitro, *Chem.-Biol. Interact.*, 59, 309, 1986.
65. Šarlauskas J et al., Nitrobenzimidazoles as substrates for DT-diaphorase and redox cycling compounds: Their enzymatic reactions and cytotoxicity, *Arch. Biochem. Biophys.*, 346, 219, 1997.
66. Anusevičius Ž et al., DT-diaphorase catalyzes N-denitration and redox cycling of tetryl, *FEBS Lett.*, 436, 144, 1998.
67. Tedeschi G, Chen S, and Massey V, DT-diaphorase. Redox potential, steady-state, and rapid reaction studies, *J. Biol. Chem.*, 270, 1198, 1995.
68. Abbioni G and Jones CR, Biomonitoring of arylamines and nitroarenes, *Biomarkers*, 7, 347, 2002.
69. Belisario MA et al., Erythrocyte enzymes catalyze 1-nitropyrene and 3-nitrofluoranthrene reduction, *Toxicology*, 108, 101, 1996.
70. Nivinskas H et al., Quantitative structure-activity relationships in two-electron reduction of nitroaromatic compounds by *Enterobacter cloacae* NAD(P)H:nitroreductase, *Arch. Biochem. Biophys.*, 385, 170, 2001.
71. Koder RL et al., Flavin thermodynamics explain the oxygen insensitivity of enterobacteric nitroreductases, *Biochemistry*, 41, 14197, 2002.
72. Rau J and Stolz A, Oxygen-insensitive nitroreductases NfsA and NfsB of *Escherichia coli*. Functions under anaerobic conditions as lawsone-dependent azo reductases, *Appl. Environ. Microbiol.*, 69, 3448, 2003.
73. Honeycutt ME, Jarvis AS, and McFarland VA, Cytotoxicity and mutagenicity of 2,4,6-trinitrotoluene and its metabolites, *Ecotoxicol. Environ. Saf.*, 35, 282, 1996.
74. Banerjee HN et al., Cytotoxicity of TNT and its metabolites, *Yale J. Biol. Med.*, 72, 1, 1999.
75. Berthe-Corti L et al., Cytotoxicity and mutagenicity of 2,4,6-trinitrotoluene (TNT) and hexogen contaminated soil in *S. typhimurium* and mammalian cells, *Chemosphere*, 37, 209, 1998.
76. Beltz LA et al., Immunotoxicity of explosive-contaminated soil before and after bioremediation, *Arch. Environ. Contam. Toxicol.*, 40, 311, 2001.
77. O'Brien PJ et al., Toxicity of nitrobenzene compounds towards isolated hepatocytes: Dependence on reduction potential, *Xenobiotica*, 20, 945, 1990.
78. Lachance B, unpublished data, 1999.
79. Banerjee H et al., Effects of 2-amino-4,6-dinitrotoluene on *p53* tumor suppressor gene expression, *Mol. Cell. Biochem.*, 252, 387, 2003.
80. Bucher JR et al., No evidence of toxicity or carcinogenicity of pentaerythritol tetranitrate given in the diet to F344 rats and B6C3F1 mice for up to two years, *J. Appl. Toxicol.*, 10, 353, 1990.
81. Bhushan B et al., Biotransformation of hexahydro-1,3,5-trinitro-1,3,5-triazine (RDX) by a rabbit liver cytochrome P-450: Insight into the mechanism of RDX biodegradation by *Rhodococcus* sp. Strain DN22, *Appl. Environ. Microbiol.*, 69, 1347, 2003.
82. Bhushan B et al., Mechanism of xanthine oxidase catalyzed biotransformation of HMX under anaerobic conditions, *Biochem. Biophys. Res. Comun.*, 306, 509, 2003.
83. London JO and Smith DM, A Toxicological Study of NTO, Los Alamos National Laboratory Report LA-10533-MS, 1985.

10 Bioconcentration, Bioaccumulation, and Biomagnification of Nitroaromatic and Nitramine Explosives in Terrestrial Systems

Mark S. Johnson, Christopher J. Salice,
Bradley E. Sample, and Pierre Yves Robidoux

CONTENTS

10.1 Introduction 228
10.2 Environmental Degradation and Transformation of Energetic
       Compounds 230
       10.2.1 Nitroaromatics 230
       10.2.2 Nitramines 232
10.3 Bioaccumulation and Biotransformation of Energetic Compounds in
       Soil Biota 233
       10.3.1 Effects of Weathering and Aging on Energetic Compounds
                  Bioaccumulation 233
       10.3.2 Effects of Soil Properties on Bioaccumulation of Energetic
                  Compounds 233
       10.3.3 Development and Application of BAF Values for Ecological
                  Risk Assessment Process..234
10.4 Accumulation of Energetic Compounds in Plants............................235
       10.4.1 Nitroaromatics 235
       10.4.2 RDX 239
       10.4.3 HMX..243
10.5 Accumulation of Energetic Compounds in Soil Invertebrates 244
10.6 Accumulation of Nitroaromatics and Nitramines in Terrestrial Animals....245
10.7 Summary 247
Acknowledgments..247
References...247

10.1   INTRODUCTION

Ordnance manufacturing facilities and other sites that have been heavily contaminated with munitions compounds are currently in the process of evaluation and remediation by the military and other entities. A key component of this process is the estimation of trophic transfer of munitions for assessment of risks to wildlife and development of suitable remedial goals. This chapter outlines the current knowledge of bioaccumulation of munitions compounds in soil. General bioaccumulation principles, environmental degradation, and transformation of energetics (important to understanding of molecular properties; see Chapter 2), followed by summaries of bioaccumulation data for various classes of energetics in different biota are presented. The reader is referred to Chapter 6 for a review of these processes in aquatic environments.

Traditional food web models for organic compounds assume that uptake into biological tissues is based solely on measures of fat solubility. This logic is based on reports that have investigated those relationships for a wide range of organic compounds [1–4]. However, particularly in terrestrial systems, other factors such as sorption, molecular mass, hydrogen bond formation and other physicochemical aspects of a dynamic soil medium are also important, and lend significant variability to bioaccumulation models that use measures of fat solubility alone [5–8]. Among organics, energetic compounds are quite unique, because by design they tend to be strong oxidizing agents. This design yields substances that are not strongly water or fat soluble and with a tendency to form crystals. It is therefore important that laboratory studies focus on factors unique to energetics that influence their availability and uptake so that more predictive models can be developed for these relatively unique substances.

In addition to compound differences, factors that influence biotic uptake vary with the system and route of entry. Measures of water solubility are frequently important for estimating plant and oral uptake in animals. Chemicals that are neither water soluble or fat soluble are more likely to be absorbed through the dermal route in terrestrial animals (usually fat solubility alone or water solubility alone are exclusive of each other). Although these properties are important in estimating uptake, other processes internal to the organism are important in retention and accumulation.

Bioaccumulation results when uptake of chemicals by dietary and nondietary pathways exceeds metabolism and excretion. Rates at which substances are absorbed, altered, and then excreted are relatively important. Bioaccumulation may involve sequestration mechanisms, such as the deposition of polychlorinated biphenyls (PCBs) in fat, or the incorporation of lead in the mineral portion of bone. Incorporation into fat is dependent on the lipophilicity of the compound. The most commonly performed test of lipophilicity involves experimental determination of the equilibrium partitioning of a test compound between octanol, a nonmiscible organic solvent, and water, often expressed as the $\log_{10}$ of the ratio or the octanol/water partition coefficient ($\log K_{ow}$). Organic compounds in which the $\log K_{ow}$ value is less than 3.5 do not appreciably accumulate in the lipids of mammals [5]. Because energetic compounds have relatively low $\log K_{ow}$ values (Table 10.1), bioaccumulation cannot be explained solely by lipophilicity.

TABLE 10.1
Log K_{ow} and Water Solubility of Nitramine and Nitroaromatic Explosives

Compound	Molecular Weight	log K_{ow}[a]	Water Solubility (mg L^{-1})[b]
TNT	227	1.60	137
2,4-DNT	182	1.98	161
2,6-DNT	182	2.10	160
2-amino-4,6-DNT	197	1.95	38
4-amino-2,6-DNT	197	1.90	43
TNB	213	1.18	335
Tetryl	287	1.55	51
RDX	222	0.87	41
HMX	296	0.29	11

[a] The log K_{ow} values listed for 2,4,6-trinitrotoluene (TNT), 1,3,5-trinitrobenzene (TNB), cyclotrimethelenetrinitramine (RDX), 2,4-dinitrotoluene (2,4-DNT), and 2,6-dinitrotoluene (2,6-DNT) are experimental values taken from Hansch et al. [90]. Other K_{ow} values were determined experimentally in our laboratory by the high performance liquid chromatography (HPLC) retention time method of Harnisch et al. [91].

[b] All water solubility values were determined in our laboratory using EPA method OPPTS 830.7840. Although hydroxylamine and nitroso reduction products are generated, these are very short-lived intermediates and the majority of soil-borne contamination is in the form of TNT, TNB, DNT, or amino reduction products.

The term bioaccumulation factor (BAF) is a coefficient used to quantify the net body burden of chemicals (given uptake, distribution, metabolism, and excretion rates) and is expressed as the ratio of the concentration of a compound in the organism (or normalized to the lipid weight of the organism) to the concentration in its food or, for benthic or soil invertebrates, in the sediment or soil [9] (also see Chapter 6). Although by definition, BAF is measured at steady state, due to the complex nature of multiple-pathway exposure, steady-state measures of BAF are often difficult to obtain.

Bioconcentration describes the nondietary uptake of chemicals and is defined as an equilibrium partition of a compound between an organism and a contaminated medium [9–11]. The bioconcentration factor (BCF) is often expressed as the ratio of the concentration of a compound in an aquatic or semiaquatic organism (e.g., a fish, amphibian, or a submerged plant) to the concentration of the compound in the surrounding water.

Bioconcentration is related to bioaccumulation in that the lipophilic properties that induce a compound to bioaccumulate also tend to increase bioconcentration. In addition, the factors that increase the affinity of a molecule to other molecules in a soil phase (solid phase sorption rates) can reduce water solubility. The principal difference is that bioconcentration involves nondietary uptake from the surrounding medium (water or soil solution) and bioaccumulation is the net result of all uptake pathways. It follows that bioaccumulation may be a continuing process, whereby steady state may or may not be reached, that is, bioaccumulation is dependent on uptake, distribution, metabolism, and excretion rates. Compounds that bioconcentrate, however, normally reach a steady-state concentration that is related to concentration in the environment. Once steady state is achieved, further increases in concentration do not occur. The potential of a compound to bioconcentrate from water may often be estimated from its water solubility and its K_{ow} [12].

Biomagnification specifically describes trophic-level increases in body burdens resulting from transfer of bioaccumulating contaminants via the diet. Biomagnification characteristically results in disproportionately higher levels of contaminants occurring in higher trophic-level organisms compared to concentrations found in environmental media or in organisms at lower trophic levels. This effect is most dramatic in top carnivores that rely on prey species that have bioaccumulated residues of these compounds [3–11]. Biomagnification occurs when the rate of exposure exceeds the organism's ability to metabolize and excrete the compound, and then to effectively serve as food to a higher trophic-level organism. It is important to note that numerous studies on the natural fate/degradation of nitroaromatic, nitramine, and nitrate explosives have shown that these compounds are not resistant to chemical transformation (see Talmage et al. [13] for a review). Indeed, the fact that these compounds are explosives indicates that they contain elements in their molecular structure that are powerful oxidizers. The nitro groups of the nitroaromatics and nitramine explosives characteristically undergo biochemical reduction in living systems [14,15].

It is clear that any assessment of the bioaccumulation, bioconcentration, and biomagnification of military explosives must consider the behavior of the parent compound and all potential metabolites, whether they are formed within the organism or in the soil/water environment. This chapter will therefore begin discussions with descriptions of the chemical transformations that commonly occur with these compounds.

10.2   ENVIRONMENTAL DEGRADATION AND TRANSFORMATION OF ENERGETIC COMPOUNDS

10.2.1   NITROAROMATICS

Trinitrotoluene (TNT) is the most commonly used energetic compound and found in the soils at U.S. Army installations [16,17]. Military grades of TNT contain up to 8% DNTs (2,4-dinitrotoluene and 2,6-dinitrotoluene) as manufacturing impurities, and TNT is often degraded to DNT in hydric soils (i.e., wet anaerobic environments) [18,19]. Other compounds are also generated when TNT is degraded (e.g., 2-amino-4,6-dinitrotoluene [2-ADNT], 4-amino-2,6-dinitrotoluene [4-ADNT], 1,3,5-trinitrobenzene [TNB], and others). Contamination from TNT, DNTs, and their environmental breakdown products is known to persist in soil for years [20].

Natural processes in many environments degrade TNT, but the rates of decay can be slow [19–25]. Hydroxylation and subsequent ring cleavage is normally required for the biological catabolism of aromatic compounds. However, hydroxylation of TNT is difficult due to the electron withdrawing effect of the nitro groups on the aromatic ring. Rieger and Knackmuss [26] reported that oxygenase reactions (yielding hydroxylations) are "unknown" for the trinitro compounds due to the combined electron withdrawing power of the three-nitro groups. The electron withdrawing effect of the nitro groups makes the ring electron deficient, making electrophilic attack by hydroxylase enzymes difficult.

Although the electron withdrawing effect of the three-nitro groups may be sufficient in itself to protect TNT from biological attack, it does not explain the resistance amino metabolites of TNT have to mineralization by biological systems. Amino metabolites of TNT no longer have the electron withdrawing power of the three-nitro groups (amino groups are electron donors) but have retained the meta substitution pattern of the functional groups. The meta spacing of the three groups leaves only two unsubstituted carbons remaining on the ring, and these are also located meta to each other. However, biological cleavage of aromatic rings normally requires emplacement of hydroxyl groups that are ortho or para to each other [27,28]. Thus, even if hydroxylations of the aromatic ring were to occur, the required ortho or para spacing of the hydroxyl groups could not be achieved without subsequent isomerization. It is likely that the meta orientation of TNT and its amino metabolites results in inhibition of biological cleavage of the aromatic ring, and, hence, the persistence of these compounds in living organisms and in natural environments. TNT can be transformed either by oxidation of the methyl group or reduction of nitro groups [21,23–25].

TNT that resides in aerobic environments at the surface of the soil is often degraded by photocatalyzed oxidation of the methyl carbon. This oxidation is probably a multistep process by which the methyl group is initially oxidized to an alcohol, then to an aldehyde, and finally to a carboxylic acid. Decarboxylation of the resultant product yields trinitrobenzene. Evidence for this pathway was supplied by Spanggord et al. [29], who reported formation of trinitrobenzaldehyde and trinitrobenzoic acid during the degradation of TNT to trinitrobenzene. It has been reported that oxidation of the methyl group of TNT is mediated by surface catalysis on soil minerals [30], by ozonation [31], and by the action of sunlight [29]. At sites where the TNT contamination is localized to the soil surface, the concentration of trinitrobenzene may often exceed that of TNT [32].

Nitroaromatic compounds commonly encounter electrochemically reducing environments in the large bowel of vertebrates and in anaerobic sediment or wet soil environments. Reduction of one of the nitro substituents of TNT may occur biotically [33] or abiotically [34], in both aerobic and anaerobic environments [33]. Nitroaromatics are commonly reduced to aminonitro compounds in a multistep process involving nitroso (N=O) and hydroxylamino (HNOH) intermediates [24]. Although reduction of nitro groups occurs under aerobic conditions, the rate of the reaction and the equilibrium ratio of amino reduction products to the more oxidized nitro, nitroso, and hydroxylamino forms increase under reducing conditions [23,24,33,35]. The triamino reduction product of TNT is not formed in aerobic soil environments [23,33].

Reduction of the nitro group that resides para to the methyl group (at the #4 position) is thermodynamically preferable to reduction of those at the #2 and #6 positions, as predicted by the work of McCormick et al. [19]. It is known that the 4-amino reduction product is the predominant reduced form of TNT in aerobic environments, and 2,4-diamino-6-nitrotoluene predominates under mildly acidic (pH 6) anaerobic conditions [33]. It also has been reported that 2,4-diamino-6-nitrotoluene (2,4-DANT) can be generated under aerobic conditions by reaction with certain minerals associated with clays, and is stable in the surface soil environment [34].

Reduction of the nitro groups of TNT has pronounced toxicological impact because the amino reduction products of TNT differ in toxicity and bioavailability from the parent compound (see examples in Chapter 3). It is important to note that apart from electrochemical reduction, TNT is largely nonreactive. However, conversion of the nitro groups of TNT to less oxidized species in the body or in the environment yields much more reactive compounds. This reactivity is important from the standpoint of bioaccumulation. These reduction products do not differ greatly from the parent compounds in K_{ow} and are therefore insufficiently hydrophobic to bioaccumulate in lipids. However, they differ from their parent compounds in their ability to generate covalent bonds with biological macromolecules and this activity provides alternate routes for bioaccumulation.

10.2.2 NITRAMINES

Nitramines are a class of compounds in which the nitro groups are attached to the nitrogen components of the heterocyclic ring. The nitramine explosives of environmental interest are octahydro-1,3,5,7-tetranitro-1,3,5,7-tetrazocine (HMX) and hexahydro-1,3,5-trinitro-1,3,5-traiazine (RDX). The cyclic nitramines RDX and HMX are widely used as military explosives in applications that utilize their great explosive power (1.5 to 2 times that of TNT) and rapid detonating velocity (on the order of 1.3 times that of TNT). Most weapon grades of RDX contain 6%–10% HMX as an impurity [36].

It is known that RDX is biologically degraded by reductive processes under anaerobic conditions and recalcitrant in aerobic conditions [37–39]. Under these conditions, one or more of the nitro substituents are reduced to nitroso groups and then to hydroxylamines. According to McCormick et al. [37,38], reduction of the nitroso to the hydroxylamine is followed by an unusual reaction or reactions that result in the bilateral cleavage of the ring structure of the molecule. This cleavage yields two groups of compounds: one group retaining four of the original nitrogen atoms of RDX and another retaining two. In subsequent reactions, all products of the initial cleavage are converted to hydrazine, 1,1-dimethylhydrazine, 1,2-dimethylhydrazine, or formaldehyde. The degradation pathway described is problematic because of the existence of the proposed nitrosamine intermediates, some of which are known to be genotoxic carcinogens [40]. Moreover, dimethylnitrosamine is a demonstrated human carcinogen, whereas the carcinogenicity of RDX is under review. Therefore, accumulation of nitrosamine compounds would likely require more stringent remedial action than would the parent compound. Published studies have yielded results that support those of McCormick et al. [37–39]. In other work, the biochemical reduction potential was apparently maintained at a level that was sufficient to reduce nitro groups to nitroso groups but not sufficient to transform the nitroso groups to hydroxylamines [39,41]. In agreement with McCormick et al. [37,38], Hawari et al. [39] and Davis [41] observed formation of metabolites in which the ring structure remains intact even after all three nitro groups were reduced to nitroso functions. HMX is more recalcitrant than TNT or RDX, however, it is known to degrade under reductive conditions [23]. Unlike RDX, however, reduction of HMX proceeds only to the formation of reduction products. The intact ring is normally not degraded [23].

10.3 BIOACCUMULATION AND BIOTRANSFORMATION OF ENERGETIC COMPOUNDS IN SOIL BIOTA

Energetic compounds can be transformed by soil indigenous microorganisms, photo-degraded by sunlight, and can migrate through the soil subsurface [42]. TNT can be rapidly transformed to aminonitrotoluene intermediates [43] such as 2-ADNT, 4-ADNT, and 2,4-DANT. The cyclic nitramines RDX and HMX are transformed to nitroso derivatives, formed by the sequential reduction of $-NO_2$ functional groups, ring cleavage products, and ultimately in nitrous oxide (NO_2) and carbon dioxide (CO_2) [39,44]. Thus, in a soil environment, metabolites of RDX and HMX include hexahydro-1-nitroso-3,5-dinitro-1,3,5-triazine (MNX), hexahydro-1,3-dinitroso-5-nitro-1,3,5-triazine (DNX), hexahydro-1,3,5-trinitroso-1,3,5-triazine (TNX), bis-(hydroxymethyl)-nitramine, and methylenedinitramine [45]. Details on energetic compounds transformation processes in soil are presented in Chapter 2. TNT is also metabolized by plants and soil invertebrates and is transformed to aminonitrotoluene intermediates [46–48]. Nitroso compounds can also be found in soil invertebrates exposed to RDX [49].

10.3.1 Effects of Weathering and Aging on Energetic Compounds Bioaccumulation

There are a few studies that show the effect of weathering and aging of energetic compounds in soil on bioaccumulation. Experiments carried out by Lachance et al. [50] showed that weathering and aging of energetic material in soil enhanced the availability of RDX to the plants. The highest net concentration found in plant tissue was 6321 mg kg^{-1} in alfalfa exposed to RDX weathered and aged in Sassafras sandy loam (SSL) soil. BAFs ranged from 0.39 to 0.66. However, data obtained from the same studies in freshly amended soil showed that weathering and aging of HMX in soils had no significant effect on accumulation by plants.

10.3.2 Effects of Soil Properties on Bioaccumulation of Energetic Compounds

Few studies showed the effects of soil properties on bioaccumulation of energetic compounds. However, it was observed that toxicity of energetic compounds may be reduced in the presence of higher organic matter content or a higher sorption (or Kd

value) of explosive in soil, suggesting a lower bioavailability of the energetic compounds. Indeed, previous studies showed that toxicity of TNT was greater in natural
soil having low organic matter content (3.8%–4.2%) compared to Organisation of
Economic Cooperation and Development (OECD) artificial soil having 10% peat
[51–53]. Toxicity of CL-20 was also greater in soil having lower Kd and higher organic
carbon (OC) content [54]. The LC_{50} as well as the EC_{50} values (growth, cocoon, and
juvenile production) were lower in a soil having low organic carbon content (SSL;
0.33% OC, Kd = 2.4 L/kg^{-1}) compared to a forest soil having high organic carbon
content (20% OC, Kd > 300). Moreover, RDX and HMX tissue concentrations may
be also reduced in the presence of higher organic matter content [49]. It appears that
soil concentration of energetic compound can be an imprecise indicator of toxicity
to the earthworms. In this context, concentrations of energetic compounds and their
metabolites in tissues may be better indicators of exposure.

10.3.3 DEVELOPMENT AND APPLICATION OF BAF VALUES FOR ECOLOGICAL RISK ASSESSMENT PROCESS

Soil toxicity-based concentrations of explosive can be used as criteria for risk [55] or
soil screening assessment [56]. However, as stated earlier, soil concentrations may
be a poor reflection of exposure and environmental risks because the bioavailability
of energetics can be affected by soil characteristics and environmental conditions
[52]. Body residues can be a better indicator of exposure to explosives, especially for
the earthworms that are actively moving in soil organisms compared to plants. For
example, tissue concentrations of TNT metabolites in earthworms might be used as
a biomarker of exposure to TNT [57,58].

Direct measurements are preferable to estimated concentrations of contaminants
in biota [59]. When direct measurement of contaminants is not possible, estimation from an existing model is the only alternative [56,59,60]. The BAFs are used
to evaluate the bioaccumulation of contaminants. This information is essential for
adequately estimating the exposures by mammals and birds.

Different methods are used to estimate the contaminant concentrations in biota.
The regression model constitutes the preferred approach compared to the BAF or
models estimating BAF [56]. When regression models are not available, new ones
can be established if paired data (contaminant concentrations in soil organisms versus soil) are sufficient. However, few regression models are available for organic
contaminants [56,61], in particular for energetic compounds.

Because few regression models are available for energetic compounds and paired
data are limited, soil to invertebrate bioaccumulation models used to estimate uptake
in food items for energetic compounds are estimated from the K_{ow} [56]. Thus, concentrations of energetic compounds in earthworms are assumed to be a function of
partitioning between the soil water and the earthworm tissues. This approach was
used to derive the earthworm median BAF for RDX (9.91) and TNT (19.57) using
the RDX (0.87) and TNT (1.6) K_{ow} values, and the following bioaccumulation model:
$BAF = 10(\log K_{ow} - 0.6)/[\text{foc} \times 10(0.983 \log K_{ow} + 0.00028)]$.

Recently, Robidoux et al. [49] reported tissue concentrations accumulated in the
earthworm after exposure to different concentrations of TNT, RDX, and HMX in

OECD artificial soil and in a natural forest soil. However, these values were associated with specific experiments and were limited by the range of concentrations used. Lachance et al. [50] reported the BAF for earthworms and plants using radiolabeled RDX and HMX. These data can be used to estimate contaminant concentrations or calculate BAF.

Uptake of energetic compound by plants and earthworms can result in exposure and toxicity for other organisms such as bird and terrestrial mammals via ingestion, thus making bioaccumulation an important endpoint in ecological risk assessment (see Chapter 12). Additional experimental data will be required to validate the existing BAF values established for energetic compounds, and to develop regression models and BAF values for energetic compounds with insufficient data (e.g., TNT metabolites, HMX, and CL-20).

10.4   ACCUMULATION OF ENERGETIC COMPOUNDS IN PLANTS

10.4.1   NITROAROMATICS

Plants form the basis of the food web and are an important part of the ecosystem's biomass and energy transfer between trophic levels. Study of the bioaccumulation of TNT and other nitroaromatics in plant systems, both in soil and in hydroponic solutions, has been the focus of substantial effort in the past two decades. These studies show a tendency for these compounds to bioaccumulate in the roots and, less often, in the aboveground tissues of plants. Bioaccumulation of nitroaromatic compounds in plants is likely caused by covalent bonding of reactive TNT-reduction products to macromolecules within living tissue. One of the earliest studies to report this observation was Palazzo and Leggett [62]. In this study, the toxicity and uptake of TNT was evaluated in hydroponic culture with yellow nutsedge (*Cyperus esculentus*). The plants were grown for 42 d exposed to solutions of 0, 5, 10, or 20 (initial study) or of 0, 0.5, 2.0, or 5.0 mg L^{-1} TNT mixed in distilled water. Samples of leaves, rhizomes, tubers, and roots were collected and analyzed. TNT concentrations of 0.5 to 5 mg kg^{-1} produced increasing toxic effects, particularly in the roots and rhizomes. Plants exposed to the higher concentrations of TNT accumulated greater concentrations of TNT and TNT metabolites, compared to plants at lower concentrations (Table 10.2). The highest concentrations of TNT and metabolites were found in the roots and the lowest concentrations were found in the leaves. However, measurable accumulations of both TNT and its metabolites were detected in all plant tissues examined. It is important to note that 95% of the material found in the root tissue was in the form of monoamino metabolites (only 5% was TNT). It is revealing that the amino metabolites were bound to the plant tissue and could be detected only after digestion with strong acid, whereas TNT could be easily extracted with benzene. The bioavailability of these tissue-bound residues when the plant material is consumed is therefore questionable. This issue, however, was not addressed in these studies.

Checkai and Simini [63] further investigated the uptake of TNT in edible garden plants. TNT was dissolved in irrigation water that was added to the soil on a regular basis to maintain the plant/soil system at maximum water holding capacity. TNT was again found primarily in the roots and was not significantly absorbed by any

TABLE 10.2

Analyte Concentrations (mg kg⁻¹ dry weight) of Yellow Nutsedge Grown in Hydroponic Media

Treatment Level (mg L⁻¹)	Tissue	Extraction Type	N	TNT	4-ADNT	2-ADNT
10	Roots	Free	6	370 ± 64.0[a]	493 ±55.4	193 ± 19.5
5	Roots	Free	7	195 ± 17.1	390 ± 92.4	105 ± 11.7
0	Roots	Free	7	0.85	0.64	0.06
10	Leaves	Free	7	37.3 ± 22.7	47.6 ± 25.9	39.6 ± 10.6
5	Leaves	Free	7	8.9 ± 5.4	35.8 ± 6.4	22.0 ± 2.6
0	Leaves	Free	7	0.82	0.43	0.59
10	Roots	Bound	7	9.27 ± 5.15	276 ± 91.2	102 ± 31.0
5	Roots	Bound	7	8.31 ± 4.36	201 ± 45.1	58.5 ± 17.1
0	Roots	Bound	7	1.64	2.82	0.12
10	Leaves	Bound	7	2.36 ± 0.94	96.8 ± 35.9	76.0 ± 17.2
5	Leaves	Bound	7	2.17 ± 0.40	73.4 ± 11.1	39.8 ± 3.84
0	Leaves	Bound	7	0.11	0.32	0.15

[a] Results presented as the mean ±SD.
Source: Data taken from Ref. 62.

parts of plants commonly harvested for human consumption. Concentrations in the aerial portion of the plants were at or below the detection limits of the analytical method (1–40 μg TNT kg⁻¹ tissue wet weight). The mass balance of these studies ranged between 55% and 79%. Low mass balance results for TNT are expected because a hydrolysis step was not used, and covalent binding of metabolites to soils and plant tissue reduced the amount of extractable material.

In a series of studies by Cataldo et al. [64] and Harvey et al. [65], the uptake and fate of TNT was evaluated for plants grown in three types of soils (Table 10.3) amended with unlabeled and radiolabeled TNT. Plant tissues and soil were evaluated by chemical extraction and analysis methods that permitted resolution of parent TNT and more polar components. Bush bean, blando brome, and wheat shoot tissues accumulated an average of 27%, 37%, and 17% of the plant accumulated radiolabel of [¹⁴C]TNT, respectively. Plant uptake was inversely proportional to soil organic matter. Analysis of TNT and TNT-derived residues show that >80% of all accumulated and transported residues were either polar or conjugated metabolites of TNT-derived compounds [66]. Table 10.4 presents the range of detectable TNT concentrations in the root and shoot sections of wheat and blando brome in the three various soil types.

It is important to note that the reduced metabolites of TNT bind to the humic fraction of soil in a manner similar to that seen in plants. After 60 d in a growth chamber, up to half of the applied dose of TNT was converted to aminodinitrotoluenes and was "bound" to soil [64]. It is generally recognized that the amino reduction products can be retained by soil in two ways: by covalent bonding to reactive centers and by

TABLE 10.3
Properties of Tested Soils

Soil Property	Burbank Sandy Loam	Palouse Silt Loam	Cinebar Silt Loam
% Sand	45.1	1.1	35.2
% Silt	51.4	77.5	51.4
% Clay	4.0	21.4	13.4
% Ash	98.0	93.8	nd
pH (100% field capacity)	7.4	5.4	5.6
Organic carbon (%)	0.5	1.7	7.2
Sulfur (%)	0.053	0.043	nd
Nitrogen (%)	0.061	0.16	0.44
Total P (μg g^{-1})	2400	3770	3400
Phosphate-P (μg g^{-1})	4.8	5.8	26
Carbonate/bicarbonate (%)	<0.1	<0.1	<0.1
Ammonium-N (μg g^{-1})	6.1	18.3	15
CEC (meq per 100 g)	5.5	23.8	38.2

Note: nd = Not determined.
Source: Data taken from Ref. 66.

sequestration within the deep pores of soil particles [67]. Both mechanisms greatly diminish uptake by plants and, hence, reduce the toxicity of nitroaromatic compounds in the soil environment. Supporting this contention are the data presented by Zellmer et al. [68]. This work showed that only low concentrations of TNT and its monoamino reduction products could be detected in the roots of plants grown on soils historically contaminated with TNT. The highest TNT concentration (exceeding 39 g TNT kg^{-1} soil) produced bioaccumulation of TNT and the monoamino metabolites that averaged about 4 mg kg^{-1}. This represents a bioaccumulation factor of 10^{-4}.

Thompson et al. [69] evaluated TNT uptake in hybrid poplar trees using spiked and aged soils. As before (following a 2-d exposure period), 78% of the radiolabel absorbed by the plants remained in the roots, whereas 13% and 9% were translocated to the stem and leaves, respectively. After 7 d, slightly more radiolabel appeared in the stems and leaves, yet levels decreased thereafter [69]. Approximately 78% of the label remained in the soil after 20 d. An experiment with aged TNT soils (280 d after spiking with TNT) showed a greatly reduced propensity for TNT uptake (<10% of TNT in spiked soils). An analysis of TNT-derived compounds suggested that stems and leaves consisted primarily of polar heterocyclic TNT conjugates. The authors calculated root concentration factors (where RCF = (mL g^{-1}) × root mass (g) × water TNT concentration (mg L^{-1})) of 49 ± 30 mL g^{-1}.

Hughes et al. [70] evaluated TNT uptake in two aquatic plants (*Myriophyllum aquiticum, Myriophyllum spicaticum*) and plant tissue cultures (*Catharanthus roseus*) using the radiolabeled compound. Significant uptake occurred, with [^{14}C]-recoveries

TABLE 10.4

Concentration of Tissue TNT-Derived Radiocarbon and Distribution of Radiocarbon in Wheat and Blando Brome Grown in TNT-Amended Soil for 60 Days[a]

Soil	Shoot μg (as TNT) g^{-1} fresh wt $\pm$ SD	Root	Shoot % Total [^{14}C] in Plant	Root
	Plant Segment		**Plant Segment**	
	Shoot	Root	Shoot	Root
	μg (as TNT) g^{-1} fresh wt $\pm$ SD		% Total [^{14}C] in Plant	
		Wheat		
TNT-Amended				
Cinebar	0.62 ± 04	4.97 ± 1.48	11.1 ± 0.7	88.9 ± 26.5
Palouse	9.15 ± 1.27	40.9 ± 9.0	18.3 ± 2.5	81.7 ± 6.4
Burbank	54.0 ± 9.5	217 ± 59	19.9 ± 3.5	80.1 ± 21.8
Control				
Cinebar	0.08 ± 0.01	0.02 ± 0.001	80.0 ± 10.0	20.0 ± 1.0
Palouse	0.96 ± 0.01	0.37 ± 0.003	72.2 ± 7.5	27.8 ± 2.3
Burbank	0.28 ± 0.15	0.05 ± 0.01	84.8 ± 5.5	15.2 ± 3.3
		Blando Brome		
TNT-Amended				
Cinebar	0.77 ± 0.11	2.49 ± 0.52	23.6 ± 3.4	76.4 ± 15.9
Palouse	9.52 ± 1.84	19.98 ± 10.34	32.3 ± 6.2	67.7 ± 35.1
Burbank	65.28 ± 12.46	59.41 ± 11.35	52.4 ± 10.0	47.6 ± 9.1
Control				
Cinebar	0.07 ± 0.005	0.01 ± 0.004	87.5 ± 6.2	12.5 ± 5.0
Palouse	0.10 ± 0.01	0.02 ± 0.005	83.3 ± 8.3	16.7 ± 4.2
Burbank	1.01 ± 0.02	0.02 ± 0.005	98.1 ± 1.9	1.9 ± 0.5

[a] Plants grown in soils amended with 10 mg [^{14}C]TNT kg^{-1} (0.74 MBq 400 g^{-1} pot), and plants grown in nonamended control soils grown in the same chamber. Data are averages ±SD ($n = 9$ for TNT-amended plants, $n = 6$ for controls).
Source: Data taken from Ref. 66.

ranging from 93% to 99%. However, they were only able to extract 27% of the radiolabel from plant tissues, and only 6% could be identified (as 2-ADNT and 4-ADNT). Because TNT was rarely identified in plant tissues, it was suggested that the roots were primarily responsible for the initial reduction of the nitro groups to the primary amines. Further investigations into the TNT-derived metabolites beyond the primary amines were conducted by Bhadra et al. [71] using *C. roseus* root culture. Four TNT-derived metabolites were isolated and identified using a combination of extraction and characterization methods. These were identified as plant conjugates formed by the reactions of the amine groups of 2-ADNT and 4-ADNT. These conjugates comprised 22% of the bound residues. These results suggest that TNT is quickly reduced by nonspecific reductases to the primary amines and can be further conjugated or bound to facilitate excretion and reduce bioavailability.

Although there has been little work related to plant absorption/accumulation of tetryl, chemical residues in bush bean tissues have been characterized [72]. Employing hydroponics and radiolabeled tetryl, parent tetryl concentrations were less than 3% in the roots after only a 24-h exposure. There was a rapid and major shift to polar tetryl-derived residues in all tissues. Studies conducted to determine bioaccumulation factors in plants for TNB, 2,4-dinitrotoluene (2,4-DNT) and, 2,6-dinitrotoluene (2,6-DNT) were conducted at environmentally relevant concentrations using alfalfa, lettuce, and ryegrass using freshly amended and weathered/aged sandy loam soils [50]. Only alfalfa had measurable TNB concentrations when grown in freshly amended soils (BAF = 0.3); no TNB was detected in plants after weathering/aging TNB in soils. Unlike TNB, concentrations of 2,4-DNT were only found in plants grown in weathered/aged soils treatments, resulting in BAFs of 0.15 and 0.44 for alfalfa and ryegrass, respectively. Residues of 2,6-DNT were found at low concentrations in all three plant species with BAFs ranging from 0.54 to 1.7 for alfalfa, from 0.3 to 1.8 for Japanese millet, and from 0.26 to 0.74 for ryegrass [50]. Together, these results suggest a low potential for bioaccumulation and biomagnification for these compounds.

10.4.2   RDX

Plants have demonstrated high absorption of RDX both from soil and from irrigation water in hydroponic studies. Studies with radiolabeled RDX in plants conducted by Cataldo et al. [73] and Fellows et al. [74] have demonstrated that RDX was readily absorbed and migrated as an intact compound and as metabolites throughout the plant. In bush bean grown to maturity in [^{14}C]RDX amended soil, the relative order of tissue concentration was seed > leaves ≥ stem > root > pod. Concentrations of RDX were as high as 200 and 600 µg g^{-1} fresh weight for leaves and seeds, respectively. Roots contained less than 75 µg g^{-1}. Leaf concentrations of RDX in wheat and blando brome were as high as 75 µg g^{-1} while roots contained less than 45 µg g^{-1} (Table 10.5). This suggests an efficient uptake and distribution system for RDX compared with TNT. Chemical analysis of bush bean tissues grown in [^{14}C] RDX amended (5 mg L^{-1}) hydroponic solution showed that approximately 23% and 50% of the radiolabel present in the root and leaves, respectively, was parent RDX after a 7-d exposure. The remaining label was evenly divided between polar metabolites and insoluble conjugates. As with TNT, the availability and uptake of RDX in plants was inversely proportional to soil organic matter content. The chemical structures of these metabolites were not determined, but evidence was provided that these compounds were not mineralized (metabolized to $^{14}CO_2$). The majority (84%) of the radioactivity was recovered in the plant and in the soil at the end of the study. This and subsequent studies [74,75] demonstrated that RDX was absorbed at about 2 to 16 times the rate of TNT for different plant species. The absorption of RDX was greatest in the carrot and bush bean followed in order by blando brome, spinach,

TABLE 10.5

Tissue Concentrations[a] of RDX from Plants Grown in RDX-Amended Soil of 10 mg kg^{-1} for 60 Days

Soil Type	Concentration RDX in Plant Tissue (µg g^{-1} fresh weight ± SD)				
	Leaf or Shoot	Stem	Pod	Seed	Root
			Bush Bean		
Cinebar	22.43 ± 7.50	19.44 ± 5.61	10.52 ± 2.63	39.46 ± 15.78	8.13 ± 1.09
Palouse	119.37 ± 34.74	98.10 ± 32.19	33.37 ± 9.76	300.77 ± 169.78	49.12 ± 3.92
Burbank	216.74 ± 40.89	186.92 ± 38.09	44.54 ± 17.51	602.57 ± 256.64	75.01 ± 11.43
			Wheat		
Cinebar	75.72 ± 23.22	nd	nd	nd	17.92 ± 3.70
Palouse	422.70 ± 126.67	nd	nd	nd	45.10 ± 8.94
Burbank	549.88 ± 140.93	nd	nd	nd	40.11 ± 9.62
			Blando Brome		
Burbank	43.76 ± 9.95	nd	nd	40.00 ± 5.42	7.64 ± 2.03
Palouse	545.45 ± 155.36	nd	nd	257.96 ± 47.72	21.84 ± 3.78
Burbank	564.46 ± 128.15	nd	nd	317.06 ± 122.19	27.95 ± 5.51

Note: nd = not determined.

[a] Derived from specific activity relationships.

Source: Data taken from Ref. 74.

corn, alfalfa, and wheat [75]. Unlike TNT, the majority of the radiolabel remained as RDX within the plant, and the rate of conversion of RDX to metabolites in soil was insignificant. Less than 2% of the radiolabel became bound to the soil during the two months of the study. Also interesting was that some plant species were able to absorb as much as 33% of the RDX added to their soil over the course of a 60-d incubation period. The highest specific activity of RDX (µg [^{14}C]RDX equivalents g^{-1} fresh weight) was found in the seeds. Bush bean seeds were found to contain as much as 600 µg [^{14}C]RDX equivalents g^{-1} fresh weight at the conclusion of the study and the concentration in the leaves and stem were both greater than the concentration in the root. This concentration in the seeds could represent a 60-fold bioaccumulation from the 10 mg kg^{-1} concentration in the soil. The efficiency of absorption in these studies varied between plant species and was inversely proportional to the concentration of organic matter in the soil.

Winfield [76] evaluated RDX uptake as well as toxicity in sunflower (*Helianthus annuus*) for exposures of 2, 4, and 6 weeks at soil concentrations of 5.8, 50, and 100 µg RDX g^{-1} dry weight of Grenada soils. Two life stages, seeds/embryos and two-week-old seedlings, were used. The highest concentrations of RDX were found in the leaves and flowers for both life stages. Bioaccumulation factors were calculated for both life stages during each exposure duration from soil RDX concentrations. The BAF values with the highest associated coefficient of determination were

TABLE 10.6

Bioaccumulation Factors and Associated Coefficients of Determination (R^2) for Sunflower

Life Stage	Exposure Duration	Total RDX Concentration (g g^{-1} dry weight)	R^2
2-Week-old seedlings	2 weeks	49.3	0.74
2-Week-old seedlings	4 weeks	247.8	0.76
2-Week-old seedlings	6 weeks	310.8	0.82
Embryos	2 weeks	804.0	0.85
Embryos	6 weeks	381.5	0.71

Note: n = 9.
Source: Data taken from Ref. 76.

310.8 ($r^2 = 0.82$) for the two-week-old seedling life stage (six-week exposure) and 804.0 ($r^2 = 0.85$) for the embryo life stage (six-week exposure; Table 10.6).

Studies by Checkai and Simini [63] produced markedly different results (Table 10.7). In this work, the uptake of RDX was evaluated in tomato, bush bean, corn, soybean, alfalfa, lettuce, and radish. The compound was introduced by dissolution in the irrigation water. Exposures consisted of nominal concentrations of 100, 20, 2, and 0 μg kg^{-1} RDX solutions in irrigation water until fruit was harvestable (two weeks to four months). Unlike previous studies, RDX did not appreciably bioaccumulate. Uptake was not significant in tomato fruit, bush bean seeds, radishes, or soybean seeds. Plant concentrations in these studies were less than that of the irrigation water. The results of this study are applicable to estimations of exposures caused by irrigation of garden vegetables with contaminated groundwater but may not be useful for estimation of exposures caused by growing crops in contaminated soil. However, earlier studies with soils indicated that plant accumulation may increase

TABLE 10.7

RDX Concentrations in Plant Tissue (in mg kg^{-1})

Concentration RDX in Treatment	Lettuce Leaves	Tomato Fruit	Radish Root	Bush Bean Fruit	Soybean Seed	Corn Stover[a]	Alfalfa Shoot[a]
100	77	16	14	7	<56	126	186
20	18	6	6	4	<56	107	25
2	<8	11	9	<4	<56	5	<19
0	<8	<2	<1	<4	<56	<2	<19
Detection limit[b]	8	2	1	4	56	2	19

[a] Determined from dry weight; other tissues were determined from fresh weight.
[b] Determined from using two-times the mean standard deviation of background.
Source: Data taken from Ref. 63.

TABLE 10.8

Calculated Bioaccumulation Factors (BAFs) from Plant Tests Using Selected Nitramines

Measured Soil Concentration[a] (Exposure Periods in Days)	Plant Tissue Concentration (mg kg⁻¹ dry weight; calculated BAF in parentheses)		
	Alfalfa	Japanese Millet	Ryegrass
RDX 87 ± 2 (F, 42 d)	6871 ± 307 (79)	5802 ± 339 (70)	4115 ± 503 (45)
RDX 985 ± 20 (F, 42 d)	6820 (6.8)	5190 ± 86 (5.1)	3288 ± 570 (3.5)
RDX 9740 ± 154 (F, 16 d)	2610 ± 90 (0.27)	1658 ± 201 (0.17)	1330 ± 126 (0.14)
RDX 9537 ± 214 (W/A, 16 d)	6321 (0.66)	5047 ± 251 (0.53)	3748 (0.39)
HMX 101 ± 2 (F, 28 d)	218 ± 16 (2.2)	2.16 ± 3 (2.2)	148 ± 4 (1.5)
HMX 1126 ± 44 (F, 28 d)	253 ± 19 (0.22)	400 ± 37 (0.36)	141 ± 2 (0.13)
HMX 10,411 ± 807 (F, 16 d)	288 ± 50 (0.03)	133 ± 6 (0.01)	182 (0.02)
HMX 9341 ± 804 (W/A, 16 d)	349 (0.04)	241 ± 11 (0.001)	166 ± 11 (0.001)

Note: F = freshly amended soils; W/A = material weathered and aged in soils.

[a] Value units are mg kg⁻¹.

Source: Data taken from Ref. 50.

after continuous, long-term irrigation of soils containing RDX, after soil sorption/reaction sites are saturated [65,74].

Recent studies by Lachance et al. [50] showed that [^{14}C]RDX is accumulated by different plant species with maximum BAF of 79 (Table 10.8). These plant bioaccumulation tests were performed according to modified protocols of the ASTM standard guide for conducting terrestrial plant toxicity tests [77] and U.S. Environmental Protection Agency (USEPA) early seedling growth test [78]. Tests were performed using alfalfa (*Medicago sativa*), Japanese millet (*Echinochloa crusgalli*), perennial ryegrass (*Lolium perenne*), corn (*Zea mays*) and lettuce (*Lactuva sativa*). For their bioaccumulation tests, Lachance et al. [50] used a modified clear polycarbonate vacuum desiccator to construct an enclosed system (microcosm) that can house the plants in pots or earthworms in jars. Figure 10.1 shows the setup for the earthworm microcosm. The microcosm was made air tight by using two metal rings and associated polytetrafluoroethylene-rubber O-rings, fastened by bolted nuts. An access hole was drilled in the top to allow watering of plants, and to allow filling and emptying the KOH traps. One of the ports was used to pump air while the second was connected as an outlet to a series of tubes containing KOH to trap CO_2. A catalytic conversion unit made of potassium permanganate mixed with activated charcoal was used to convert putative volatile organic compounds into CO_2.

Lachance et al. [50] conducted bioaccumulation studies on corn for 19 days, and studies on alfalfa, Japanese millet, and perennial ryegrass for 42 d in sandy loam soil using [^{14}C]RDX or nonlabeled RDX. The corn study was conducted using a soil concentration of 27 mg kg⁻¹ and evaluated various tissue types using analytical

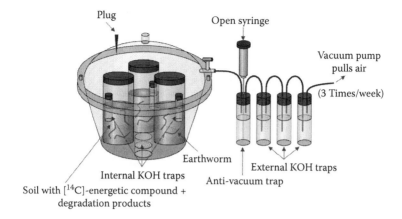

FIGURE 10.1   Microcosm design for assessing the accumulation of [^{14}C]-RDX or [^{14}C]-HMX in earthworms.

chemistry as well as radiological methods. The 42-d study evaluated species differences and was conducted using target soil concentrations of 100 and 1000 mg kg^{-1}. Leaves consistently contained the greatest concentrations of RDX followed by the roots and stems. Calculated bioaccumulation factors decreased at higher soil concentrations (Table 10.8), suggesting that bioaccumulation potential (i.e., saturation) was independent of soil concentration.

10.4.3   HMX

Studies conducted with aged sandy loam soils from a firing range in Alberta (Canada) were used to determine HMX uptake in native and agricultural plants under laboratory conditions [79]. When adjusting for total plant mass, concentrations of HMX in leaf tissue approximated that found in the soil (~20–50 mg kg^{-1} dry weight), and accumulating to over 500 mg kg^{-1} in senescent leaf tissue. Only minute concentrations were found in the root tissue. Generally, there were no marked differences in HMX uptake between agricultural and native plants [79]. Concentrations of HMX in plant tissue collected from the range were slightly lower than those detected in laboratory investigations (as was the moisture content), suggesting that uptake was dependent upon transpirational flux. No evidence of phytodegradation was found in these studies [58].

HMX uptake was evaluated from 21-d hydroponic exposures using hybrid poplar trees (*Populus deltoids × nigra*) at concentrations of 4.70, 2.84, and 1.37 mg L^{-1} [80]. After 21 d, 44.83% of the HMX in solution was sequestered by the trees. No toxic effects were evident at any of the exposures. The authors also calculated a transpirational stream concentration factor of 0.18 ± 0.03 suggesting that HMX was moderately available for uptake by plant roots. Root concentration factors were also calculated at 5.55 ± 1.78 [80]. As with other studies, most HMX was found in the leaves (42.9% using radiolabeled HMX). No HMX metabolites were found.

Lachance et al. [50] also conducted a suite of experiments to evaluate the potential of HMX to accumulate in several plant species. Consistent with other reports, the highest concentrations were found in the leaves, followed by roots and stems. The bioaccumulation potential for HMX was lower than that for RDX, and was independent of soil concentration. Indeed, Lachance et al. [50] showed that [^{14}C]HMX was accumulated by different plant species with maximum BAF of 2.2. No distinction could be made regarding clear differences between monocotyledonous and dicotyledonous species in bioaccumulation potential under these conditions (Table 10.8).

10.5 ACCUMULATION OF ENERGETIC COMPOUNDS IN SOIL INVERTEBRATES

Several studies demonstrated that TNT can be rapidly transformed by the earthworm *Eisenia andrei* in a forest soil [48], OECD artificial soil, or filter paper [52]. However, TNT metabolites can be accumulated in the earthworm [50,52]. Using standard toxicity exposure methods and different substrates, Robidoux et al. [49] determined that RDX or HMX accumulate in the earthworms. In these studies, earthworms were exposed for 28 d to different concentrations of TNT (ranging from 10 to 881 µg g^{-1} dry soil), RDX (ranging from 12.5 to 1671 µg g^{-1} dry soil), and HMX ranging from 12.5 to 987 µg g^{-1} dry tissues in OECD artificial soil or a forest soil. Concentrations of RDX and HMX measured in tissues were <682 mg RDX kg^{-1} and <484 mg HMX kg^{-1} (dry weight), respectively. No TNT was detected in earthworms exposed to TNT-amended soils, but the mean maximum concentrations of TNT metabolites in tissues reached 52.7 and 98.9 µg g^{-1} for 2-ADNT and 4-ADNT, respectively. In this study, the maximum BAF measured for RDX and HMX were 3.3 (in artificial soil; <1.0 in forest soil) and 0.3, respectively.

Lachance et al. [50] also showed that accumulation in the earthworm was low for RDX with a BAF of 2.9–13 and was negligible for HMX with a BAF of 0.3–1.0. The earthworm bioaccumulation tests were performed according to modified protocols of the ASTM [77] standard guide for conducting laboratory soil toxicity or bioaccumulation tests with the lumbricid earthworm *E. fetida*. Modifications included reducing the exposure from the recommended 28 d to 14 d to eliminate the need for feeding the earthworms, and using glass instead of plastic pots to avoid adsorption of energetic compounds. For their bioaccumulation tests, Lachance et al. [50] used a microcosm design described in the previous section (Figure 10.1). In these studies, earthworms (*E. andrei*) were exposed to nominal concentrations of RDX in soil of 10 and 100 mg kg^{-1} for 14 d. Tissue concentrations were 125 ± 11 and 283 ± 28 mg kg^{-1} at the 10 and 100 mg kg^{-1} treatments (11 and 99 mg kg^{-1} measured), respectively, and resulted in BAFs of 13 ± 1 and 2.9 ± 0.2. Bioaccumulation rates of RDX in earthworms were inversely proportional to soil concentrations. The same experiment was repeated with HMX using the same nominal concentrations of 10 and 100 mg kg^{-1} [53] resulting in tissue burdens of 9 ± 1 and 26 ± 3 mg kg^{-1} for each treatment, respectively. These data suggest that HMX does not accumulate in earthworms (BAFs = 1.0 ± 0.2 and 0.32 ± 0.02, for the 10 and 100 mg kg^{-1} treatments, respectively).

10.6   ACCUMULATION OF NITROAROMATICS AND NITRAMINES IN TERRESTRIAL ANIMALS

The evidence for significant bioaccumulation of explosive compounds in terrestrial animal species is lacking. One of the most comprehensive studies of explosive residues in wildlife was Shugart et al. [81]. This study evaluated body and tissue burdens of TNT and nine probable metabolites in game animals whose ranges overlapped areas of TNT contamination at the Alabama Army Ammunition Plant. The study included analysis of muscle and liver tissues from deer ($N = 12$), rabbit ($N = 5$), and quail ($N = 5$). The detection limit for each compound was 0.2 mg kg^{-1}. Chemical analyses revealed that no TNT-related compounds were found in tissues of any of the test species. Because of the lack of a positive result, a subacute study was conducted with mice to determine if the target metabolites could be detected in a controlled experiment. Laboratory mice were dosed via gavage with four equal doses of 100 mg kg^{-1} body weight of radiolabeled TNT over a 9-d period. At the end of the treatment, blood, liver, muscle, and excreta (i.e., urine and feces) were analyzed. A small amount of the administered radioisotope was found in blood (0.01%), muscle (0.07%), and liver (0.12%). It was estimated that only 11% of the recovered dose was actually free product and that TNT would be a minor component (<10%) in a diverse mixture of metabolites. It is important to note that the test animals received a very large dosage of [^{14}C]TNT in order to find detectable levels of radioisotope in the blood and organs. This dosage would be equivalent to about 28 g in a 70 kg human over the 9-d period. Animals residing in proximity to areas contaminated with TNT would receive only a small fraction of this dosage, and the concentration of TNT and metabolites would probably be too low to detect.

Shugart [82] evaluated deer tissue from Badger Army Ammunition Plant, Baraboo, Wisconsin, for TNT metabolites (2,4- and 2,6-dinitrotoluene), and, again, these compounds were not seen above the analytical detection limit of 0.1 mg kg^{-1}. The U.S. Army Environmental Hygiene Agency [83] conducted an extensive study of whitetail deer (*Odocoileus virginiana*) and white-footed mice (*Peromyscus leucopus*) from Joliet Army Ammunition Plant (Joliet, Illinois). Various tissues were analyzed for RDX, HMX, 2,4- and 2,6-dinitrotoluene, and TNT and its metabolites. The detection limits ranged from 0.05 to 0.2 mg kg^{-1}. Although levels in the soil were high especially around the explosive manufacturing areas, contamination with nitroaromatic compounds was found in only two mice from contaminated areas and one from a noncontaminated reference location. Due to the extremely low concentration in the tissues of these two mice, it was assumed that these concentrations in tissues were not the result of uptake but arose from cross-contamination from the skin/fur during necropsy. The Army also sampled deer and white-footed mice from Aberdeen Proving Ground (Maryland) in 1994 and 1995, and white-footed mice from Volunteer Army Ammunition Plant (Chattanooga, Tennessee) in 1996, and again found no explosives [84].

Schneider et al. [85] completed a study of RDX accumulation in the tissue of laboratory rats following a 90-d exposure in drinking water (50–70 mg L^{-1}) and by oral gavage (20 mg kg^{-1} day). About one-third of the total radiolabel was excreted in the urine, another third of the total label was excreted as $^{14}CO_2$, and only 5% excreted in the feces. RDX did not accumulate in any tissue examined.

Studies to determine the relative significance of two routes of exposure of TNT and a lipophilic compound (PCB mixture) in soil were conducted by Johnson et al. [86]. In these studies, salamanders were exposed to TNT from soil alone, by eating worms acclimated to contaminated soil, or by both routes of exposure (contaminated soil and food). It was determined that the principal route of uptake for TNT was dermal absorption from contaminated soil. In the first 10 d of this study, the concentration of TNT in the soil inhabited by salamanders decreased 10-fold. This is probably due to conversion to amino compounds that were subsequently bound to the soil. It was determined that the animals that received dual exposure (both feeding and residence on contaminated soil) accumulated the most TNT and metabolites. Accumulations of TNT and the two monoaminodinitrotoluenes (taken as a sum from all routes) were 1.3 mg kg^{-1} with an initial soil concentration of about 740 mg kg^{-1}.

Another study evaluated organ-specific concentrations of TNT and metabolites in tiger salamanders (*Ambystoma tigrinum*) exposed via contaminated soil and food for 14 d [87]. Although TNT concentrations in the soil were approximately an order of magnitude above the monoamino metabolites (280, 39, and 62 μg g^{-1} for TNT, 2-ADNT, and 4-ADNT, respectively), only a single individual had trace levels of TNT in skin and liver samples. Earthworms and salamander tissue samples had similar levels of the monoamino metabolites (approximately 2.3–3.5 μg g^{-1}). These studies demonstrate that significant exposures can occur through the dermal pathway in salamanders, and, once absorbed, TNT is readily reduced to the monoamino reduction products.

A study of the movement of RDX through trophic levels was conducted by Fellows et al. [88]. In these studies, corn and alfalfa were grown on soil amended with radiolabeled RDX. The leaves from the corn (containing RDX at 19 μg g^{-1}) were mixed with the alfalfa, which had accumulated RDX at a concentration of 187 μg g^{-1}. The leaves containing naturally incorporated RDX were added to commercial rabbit feed, and the mixture was ground and made into pellets. The radiolabeled pelleted chow (10 μg g^{-1} RDX) was fed to prairie voles (*Microtus ochrogaster*) for 4 d or 7 d with a 4-d posttest "chase" period, and the uptake and excretion of the label were measured. In contrast to the extensive metabolism of the waterborne RDX in the study by Schneider et al. [6], only about 7% and 11% of the plant incorporated radioactivity was exhaled or excreted in the urine, respectively. It was found that about 60% of the total radioactivity was excreted in the feces, suggesting that the RDX in the plant tissue was bound to the less digestible (fiber) component of the diet. The RDX was not mineralized, but less than 20% was excreted after 5 d of feeding and a 5-d "chase" period. Analysis of tissues indicated that the label was distributed throughout the animal. Retention of nearly 20% of the radioactivity in the tissues of the voles indicated that herbivorous animals such as the prairie vole, which are able to retain and ferment less digestible dietary components [89], can bioaccumulate RDX or RDX-radiolabeled fragments. Accumulation of RDX in an herbivore that was fed plants grown in RDX-contaminated soil suggests further trophic transfer of the compound is possible.

10.7 SUMMARY

Nitroaromatic and nitramine explosives have been found in environmental media at considerable concentrations at numerous military sites. Some of these locations provide valuable habitat to plants and animals. Therefore, it is important to understand the dynamics of how these compounds can move through the food web to determine if exposure is sufficient to cause harm.

Evidence from controlled laboratory studies and field evaluations suggest that nitroaromatics do not bioaccumulate in the aboveground portion of plants or in the body burdens of terrestrial wildlife species. Chemical reduction of the nitro groups of the nitroaromatic TNT occurs relatively quickly in anaerobic soils that yield compounds that may then form chemical linkages to structural macromolecules, thus binding these reduction products to the subsurface portion of plants and soils. These primary reduction products are formed quickly in vivo, but can bioaccumulate to a small extent if exposures are high and continuous. In the earthworm, TNT can also be transformed, and TNT metabolites (2-ADNT, 4-ADNT and 2,4-DANT) are accumulated in the tissues. These compounds have not been found in tissue from animal species whose range includes areas of highly contaminated soil, but have been predominantly found in the root portions of plants. Because of the rapid transformation, excretion, and lack of potential for high, continuous exposure, it is unlikely that nitroaromatic compounds could be transferred between predator and prey animal species.

Current information suggests that plants can absorb RDX from soil or irrigation water and accumulate this compound unchanged in both subsurface and aboveground tissues. The BAF values for RDX can be quite high, especially in the seeds and other edible portions of the plant, and are roughly equivalent to the evapotranspiration rates in plants. Because the compound is largely unchanged by its absorption into plants, transfer from soil to plants and ultimately to herbivores may occur. The magnitude of such transfer remains uncertain. RDX and HMX can also be transferred to soil invertebrates but current information suggests that their BAF would be relatively low.

ACKNOWLEDGMENTS

We would like to thank Mick Major for initiating our first manuscript and the helpful contributions of two anonymous reviewers.

REFERENCES

1. Kenaga EE, Correlation of bioconcentration factors of chemicals in aquatic and terrestrial organisms with their physical and chemical properties, *Environ. Sci. Technol.*, 14, 553, 1980.
2. Travis CC and Arms AD, Bioconcentration of organics in beef, milk, and vegetation, *Environ. Sci. Technol.*, 22, 271, 1988.

3. Macay D and Clark KE, Predicting the environmental partitioning of organic contaminants and their transfer to biota, in *Organic Contaminants in the Environment*, Jones KC, Ed., Elsevier Science Publications, New York, 1991, 159.

4. Fries GF, Organic contaminants in terrestrial food chains, in *Organic Contaminants in the Environment*, Jones KC, Ed., Elsevier Science Publications, New York, 1991, 207.

5. Garten CT Jr. and Trabalka JR, Evaluation of models for predicting terrestrial food chain behavior of xenobiotics, *Environ. Sci. Technol.*, 17, 590, 1983.

6. Esser EO, A review of the correlation between physicochemical properties and bioaccumulation, *Pestic. Sci.*, 17, 265, 1986.

7. Hansen LG and Lambert RJ, Transfer of toxic trace substances by way of food animals: Selected examples, *J. Environ. Qual.*, 16, 200, 1987.

8. Connell DW and Markwell RD, Bioaccumulation in the soil to earthworm system, *Chemosphere*, 20, 91, 1990.

9. Barron MG, Bioaccumulation and bioconcentration in aquatic organisms, in *Handbook of Ecotoxicology*, Hoffman DJ et al., Eds., Lewis Publishers, Boca Raton, FL, 1995, 652.

10. Thomann RV, Bioaccumulation model of organic chemical distribution in aquatic food chains, *Environ. Sci. Technol.*, 23, 699, 1989.

11. Hoffman DJ et al., *Handbook of Ecotoxicology*, Lewis Publishers, Boca Raton, FL, 1995.

12. Lyman WG, Reehl WF, and Rosenblatt DH, *Handbook of Chemical Property Estimation Methods*, McGraw-Hill, New York, 1982.

13. Talmage SS et al., Nitroaromatic munition compounds: Environmental effects and screening values, *Rev. Environ. Contam. Toxicol.*, 161, 1, 1999.

14. Major MA, Biological degradation of explosives, in *Bioremediation of Contaminated Soils*, Adriano DC et al., Eds., American Society of Agronomy, the Crop Science Society of America and the Soil Science Society of America, Madison, WI, 1999, 111.

15. Yinon J, *Toxicity and Metabolism of Explosives*, CRC Press, Boca Raton, FL, 1990.

16. Hathaway JA, Trinitrotoluene: A review of reported dose-related effects providing documentation for a workplace standard, *J. Occup. Med.,* 19, 341.

17. Hovatter PS et al., Ecotoxicity of nitroaromatics to aquatic and terrestrial species at army superfund sites, in *Environmental Toxicology and Risk Assessment: Modeling and Risk Assessment* (Vol. 6), ASTM STP 1317, Dwey FJ, Doane TR, and Hinman ML, Eds., American Society for Testing and Materials, West Conshohocken, PA, 1997.

18. Gong P et al., Effects and bioavailability of 2,4,6-trinitrotoluene in spiked and field contaminated soils to indigenous microorganisms, *Environ. Toxicol. Chem.*, 18, 2681, 1999.

19. McCormick NG, Feeherry FE, and Levinson HS, Microbial transformation of 2,4,6-trinitrotoluene and other nitroaromatic compounds, *Appl. Environ. Microbiol.*, 31, 949, 1976.

20. Carpenter DF et al., Microbial transformation of ^{14}C-labeled 2,4,6-trinitrotoluene in an activated sludge system, *Appl. Environ. Microbiol*, 35, 949, 1978.

21. Kaplan DL and Kaplan AM, Thermophilic biotransformations of 2,4,6-trinitrotoluene under simulated composting conditions, *Appl. Environ. Microbiol*, 44, 757, 1982.

22. Kaplan DL and Kaplan AM, 2,4,6-trinitrotoluene-surfactant complexes: Decomposition, mutagenicity and soil leaching studies, *Environ. Sci. Technol.*, 19, 566, 1982.

23. Kaplan DL, Biotransformation pathways of hazardous energetic organonitro compounds, in *Advances in Applied Biotechnology 4: Biotechnology and Biodegradation*, Kamely D, Chakrabarty A, and Omenn GS, Eds., Portfolio Publishing, Dallas, TX, 1990, 155.

24. Walsh ME, Environmental Transformation Products of Nitroaromatics and Nitramines, U.S. Army Corps of Engineers Cold Regions Research & Engineering Laboratory, Special Report 90-2, Hanover, NH, 1990.

25. Isbister JD et al., Composting for decontamination of soils containing explosives, *Microbiologica*, 7, 47, 1984.

26. Rieger PG and Knackmuss HJ, Basic knowledge and perspectives on biodegradation of 2,4,6-trinitrotoluene and related nitroaromatic compounds in contaminated soil, in *Bioremediation of Nitroaromatic Compounds*, Spain JC, Ed., Plenum Press, New York, 1995.

27. Simpson JR and Evans WC, The metabolism of nitrophenols by certain bacteria, *Biochem. J.*, 55, 24, 1953.

28. Dagley S, Microbial degradation of organic compounds in the biosphere, *American Scientist*, 63, 681, 1975.

29. Spanggord RJ et al., Environmental Fate Studies on Certain Munition Wastewater Constituents Lab Studies, Menlo Park, California, SRI International, Menlo Park, CA, ADA099256, U.S. Army Medical Research and Development Command, Fort Detrick, MD, 1980.

30. Checkai RT et al., Transport and Fate of Nitroaromatic and Nitramine Explosives in Soils from Open Burning Open Detonation Operations at Anniston Army Depot, ERDEC-TR-135, Edgewood Research Development and Engineering Center, Aberdeen Proving Ground, MD, 1993.

31. Kearney PC, Zeng Q, and Ruth JM, Oxidative pretreatment accelerates TNT metabolism in soils, *Chemosphere*, 12, 1583, 1983.

32. Checkai RT et al., Transport and Fate of Nitroaromatic and Nitramine Explosives in Soils from Open Burning Open Detonation Operations at Radford Army Ammunition Plant, ERDEC-TR-133, Edgewood Research Development and Engineering Center, Aberdeen Proving Ground, MD, 1993.

33. Funk SB et al., Initial Phase Optimization for the Bioremediation of Munition-Contaminated Soils, Idaho Agricultural Experiment Station, Moscow, ID, 1993.

34. Ainsworth CC et al., Relationship Between the Leachability Characteristics of Unique Energetic Compounds and Soil Properties, Final Report on Project Order No. 91PP1800, U.S. Army Biomedical Research and Development Laboratory, Fort Detrick, MD, 1993.

35. Gilcrease PC and Murphy VG, Bioconversion of 2,4-diamino-6-nitrotoluene to a novel metabolite under anoxic and aerobic conditions, *Appl. Environ. Microbiol.*, 61, 4209, 1995.

36. U.S. Army, Military Explosives, TM 9-1300-214, Department of the Army Technical Manual. Headquarters, Department of the Army, Washington, D.C., 1994.

37. McCormick NG, Cornell JH, and Kaplan AM, Biodegradation of hexahydro-1,3,5-trinitro-1,3,5-triazine, *Appl. Environ. Microbiol.*, 42, 817, 1981.

38. McCormick NG, Cornell JH, and Kaplan AM. The Anaerobic Biotransformation of RDX, HMX and Their Acetylated Derivatives, U.S. Army Natick Research, Development, and Engineering Center, Technical Report No.85-007, Natick, MA, 1985.

39. Hawari J et al., Characterization of metabolites during biodegradation of hexahydro-1,3,5-trinitro-1,3,5-triazine (RDX) with municipal anaerobic sludge, *Appl. Environ. Microbiol.*, 2652, 2000.

40. Pitot HC III and Dragan YP, Chemical carcinogens, in *Casarett and Doull's Toxicology*, Klaassen CD, Ed., McGraw-Hill, New York, 2001, 241.

41. Davis J, personal communication, 1999.

42. Halasz A et al., Detection of explosives and their degradation products in soil environments, *J. Chromatogr. A*, 963, 411, 2002.

43. Hawari J et al., Characterization of metabolites in the biotransformation of 2,4,6-trinitrotoluene with anaerobic sludge: Role of triaminotoluene, *Appl. Environ. Microbiol.*, 64, 2200, 1998.

44. Fournier D et al., Biodegradation of octahydro-1,3,5,7-tetranitro-1,3,5,7-tetrazine (HMX) by *Phanerochaete chrysosporium*: New insight into the degradation pathway, *Environ. Sci. Technol.*, 38, 4130, 2004.

45. Groom CA et al., Detection of the cyclic nitramine explosives hexahydro-1,3,5-trinitro-1,3,5-triazine (RDX) and octahydro-1,3,5,7-tetranitro-1,3,5,7-tetrazine (HMX) and their degradation products in soil environments, *J. Chromatogr. A*, 909, 53, 2001.
46. Talmage SS et al., Nitroaromatic munition compounds: Environmental effects and screening values, *Rev. Environ. Contam. Toxicol.* 161, 1, 1999.
47. Lachance B et al., Toxicity and bioaccumulation of reduced TNT metabolites in the earthworm *Eisenia andrei* exposed to amended forest soil, *Chemosphere*, 55, 1339, 2004.
48. Renoux AY et al., Transformation of 2,4,6-trinitrotoluene (TNT) in soil in the presence of the earthworm *Eisenia andrei*, *Environ. Toxicol. Chem.*, 19, 1473, 2000.
49. Robidoux PY, Hawari J, and Sunahara GI, Uptake and transformation of RDX, HMX, and TNT in the earthworm *(Eisenia Andrei)* using different substrates (submitted).
50. Lachance B et al., Bioaccumulation of Nitro-Heterocyclic and Nitroaromatic Energetic Materials in Terrestrial Receptors in a Natural Sandy Loam Soil, Biotechnology Research Institute, Applied Ecotoxicology Group, National Research Council Canada, Montreal, Quebec, Canada, 2004.
51. Robidoux PY et al., Acute toxicity of 2,4,6-trinitotoluene (TNT) using the earthworm *(Eisenia andrei)*, *Ecotoxicol. Environ. Saf.*, 44, 311, 1999.
52. Robidoux PY et al., TNT, RDX and HMX decrease earthworm *(Eisenia andrei)* life-cycle responses in a spiked natural forest soil, *Arch. Environ. Contam. Toxicol.*, 43, 379, 2002.
53. Robidoux PY et al., Chronic toxicity of energetic compounds in soil using the earthworm *(Eisenia andrei)* reproduction test, *Environ. Toxicol. Chem.*, 19, 1764, 2000.
54. Robidoux PY et al., Acute and chronic toxicity of a new explosive CL20 in the earthworm *(Eisenia andrei)* exposed to amended natural soil, *Environ. Toxicol. Chem.*, 23, 1026, 2004.
55. Renoux AY et al., Derivation of environmental soil quality guidelines for 2,4,6-trinitrotoluene: CCME approach, *Human Ecol. Risk Assess.*, 7, 1715, 2001.
56. USEPA, Guidance for Developing Ecological Soil Screening Level, OSWER Directive 92857-55, Office of Solid Waste and Emergency Response, U.S. Environmental Protection Agency, Washington, D.C., November 2003, http://rais.ornl.gov/homepage/ecossl.pdf (accessed August 2008).
57. Robidoux PY et al., Evaluation of tissue and cellular biomarkers to assess 2,4,6-trinitrotoluene (TNT) exposure in earthworms: Effects-based assessment in laboratory studies using *Eisenia andrei*, *Biomarkers*, 7, 306, 2002.
58. Robidoux PY et al., Assessment of a 2,4,6-trinitrotoluene (TNT) contaminated site using *Aporrectodea rosea* and *Eisenia andrei* in mesocosms, *Arch. Environ. Contam. Toxicol.*, 48, 56, 2004.
59. Sample BE et al., Development and Validation of Bioaccumulation Models for Earthworms, ES/ER/TM-220, Oak Ridge National Laboratory, Oak Ridge, TN, 1998.
60. Sample BE et al., Literature-derived bioaccumulation models for earthworms: Development and validation, *Environ. Toxicol. Chem.*, 18, 2110, 1999.
61. Jager T et al., Elucidating the routes of exposure for organic chemical in the earthworm, *Eisenia andrei* (Oligochaeta), *Environ. Sci. Technol.*, 37, 3399, 2003.
62. Palazzo AJ and Leggett DC, Effect and Disposition of TNT in a Terrestrial Plant and Validation of Analytical Methods, CRREL Report No. 86-15, U.S. Army Corps of Engineers Cold Regions Research and Engineering Laboratory, Hanover, NH, 1986.
63. Checkai RT and Simini M, Plant Uptake of RDX and TNT Utilizing Site Specific Criteria for the Cornhusker Army Ammunition Plant (CAAP), Nebraska, U.S. Army ERDEC Technical Report, Project Order No 560786M8AA, Edgewood Research Development and Engineering Center, Aberdeen Proving Ground, MD, 1996.
64. Cataldo DA et al., An Evaluation of the Environmental Fate and Behavior of Munitions Materiel (TNT, RDX) in Soil and Plant Systems, Environmental Fate and Behavior of TNT, AD A223546, Prepared for the U.S. Army Biomedical Research and Development Laboratory, Fort Detrick, Frederick, MD, 1989.

65. Harvey SD et al., Fate of the explosive hexahydro-1,3,5-trinitro-1,3,5-triazine (RDX) in soil and bioaccumulation in bush bean hydrophonic plants, *Environ. Toxicol. Chem.*, 10, 845, 1991.

66. Fellows RL, Harvey SD, and Cataldo DA, An Evaluation of the Environmental Fate and Behavior of Munitions Materiel (Tetryl and Polar Metabolites of TNT) in Soil and Plant Systems, Preliminary Evaluation of TNT-Polar Metabolites in Plants, PNL-8823, prepared for the U.S. Army Biomedical Research and Development Laboratory, Fort Detrick, Frederick, MD, 1993.

67. Alexander M, Aging, bioavailability and overestimation of risk from environmental pollutants, *Environ. Sci. Technol.*, 34, 4259, 2000.

68. Zellmer SD et al., Plant Uptake of Explosives from Contaminated Soil at the Joliet Army Ammunition Plant, Report No. SFIM-AEC-ET-CR-95014, U.S. Army Environmental Center, Aberdeen Proving Ground, MD, 1995.

69. Thompson PL, Ramer LA, and Schnoor JL, Uptake and transformation of TNT by hybrid popular trees, *Environ. Sci. Technol.*, 32, 975, 1998.

70. Hughes JB et al., Transformation of TNT by aquatic plants and plant tissue cultures, *Environ. Sci. Technol.*, 31, 266, 1997.

71. Bhadra R et al., Confirmation of conjugation processes during TNT metabolism by axenic plant roots, *Environ. Sci. Technol.*, 33, 446, 1999.

72. Harvey SD et al., Analysis of the explosive 2,4,6-trinitrophenylmethylnitramine (tetryl) in bush bean plants, *J. Chromatogr.*, 630, 167, 1993.

73. Cataldo A, Harvey SD, and Fellows RJ, An Evaluation of the Environmental Fate and Behavior of Munitions Material (TNT, RDX) in Soil and Plant Systems, Environmental Fate and Behavior of RDX, PNL-7529; ADA223540, prepared for the U.S. Army Biomedical Research and Development Laboratory, Fort Detrick, Frederick, MD, 1990.

74. Fellows RJ, Harvey SD, and Cataldo DA, Evaluation of the Metabolic Fate of Munitions Material (TNT & RDX) in Plant Systems and Initial Assessment of Material Interaction with Plant Genetic Material, Validation of the Metabolic Fate of Munitions Material (TNT & RDX) in Mature Crops, PNL-10825, prepared for the U.S. Army Biomedical Research and Development Laboratory, Fort Detrick, Frederick, MD, 1995.

75. Fellows RJ et al., Biotic and Abiotic Transformations of Munitions Materials (TNT, RDX) by Plants and Soils, Potentials for Attenuation and Remediation of Contaminants, International Business Communications Symposium on Phytoremediation [Abstract], May 8–10, Arlington, VA, 1996.

76. Winfield LE, Development of a Plant-Based, Short-Term (<12 days) Screening Method for Determination of Hexahydro-1, 3, 5-trinitro-1, 3, 5-triazine (RDX) Bioavailability, Bioconcentration, and Phytotoxicity to Selected Terrestrial Plants from Soil Matrices, Ph.D. dissertation, University of Mississippi, University, MS, 2001.

77. ASTM, Standard Guide for Conducting Laboratory Soil Toxicity or Bioaccumulation Tests with the Lumbricid Earthworm *Eisenia Fetida*, E 1676-97, 1998.

78. USEPA, Early Seedling Growth and Vigor Toxicity Test, EPA 560/6-82-002, 1982.

79. Groom CA et al., Accumulation of HMX (octahydro-1,3,5,7-tetranitro-1,3,5,7-tetrazocine) in indigenous and agricultural plants grown in HMX-contaminated anti-tank firing range soil, *Environ. Sci. Technol.*, 36, 112, 2002.

80. Yoon JM et al., Uptake and leaching of octahydro-1,3,5,7-tetranitro-1,3,5-tetrazocine by hybrid poplar trees, *Environ. Sci. Technol.*, 36, 4649, 2002.

81. Shugart LR et al., TNT Metabolites in Animal Tissues, Final Report, ORNL/M-1336, Oak Ridge National Laboratory, Oak Ridge, TN, 1990.

82. Shugart LR, Dinitrotoluene in Deer Tissues, Oak Ridge National Laboratory Final Report, ORNL/M-1765, Oak Ridge National Laboratory, Oak Ridge, TN, 1991.

83. Whaley JE and Leach GJ, Health Risk Assessment for Consumption of Deer Muscle and Liver from Joliet Army Ammunition Plant, Joliet, Illinois, Final Report, Project # 75-51-YF23, U.S. Army Environmental Hygiene Agency, Aberdeen Proving Ground, MD, 1994.

84. Whaley JE, personal communication, 2001.

85. Schneider NR, Bradley SL, and Andersen ME, The distribution and metabolism of cyclotrimethylenetrinitramine (RDX) in the rat after subchronic administration, *Toxicol. Appl. Pharmacol.*, 46, 163, 1978.

86. Johnson MS et al., Bioaccumulation of 2,4,6-trinitrotoluene and polychlorinated biphenyls through two routes of exposure in a terrestrial amphibian: Is the dermal route significant? *Environ. Toxicol. Chem.*, 18, 873, 1999.

87. Johnson MS et al., Fate and biochemical effects of 2,4,6-trinitrotoluene exposure to tiger salamanders (*Ambystoma tigrinum*), *Ecotoxicol. Environ. Saf.*, 46, 186, 2000.

88. Fellows RJ et al., Bioavailability of hexahydro-1,3,5-trinitro-1,3,5-triazine (RDX) to the prairie vole (*Microtus ochrogaster*), *Environ. Toxicol. Chem.*, 25, 1881, 2006.

89. Steven CE and Hume ID, *Comparative Physiology of the Vertebrate Digestive System* (2nd ed.), Cambridge University Press, New York, 1995.

90. Hansch C, Leo A, and Hoekman D, *Exploring QSAR, Hydrophobic, Electronic, and Steric Constants*, ACS Professional Reference Book, American Chemical Society, Washington, D.C., 1995.

91. Harnisch M, Mockel HJ, and Schulze G, Relationship between log K_{ow} shake-flask values and capacity factors derived from reversed-phase high performance liquid chromatography for n-alkylbenzenes and some reference substances, *J. Chromatogr.*, 282, 315, 1983.

11 Habitat Disturbance at Explosives-Contaminated Ranges

Rebecca A. Efroymson, Valerie Morrill, Virginia H. Dale, Thomas F. Jenkins, and Neil R. Giffen

CONTENTS

11.1 Introduction 253
11.2 Munitions Ranges 254
11.3 Spatial Distribution of Contamination and Phytotoxicity 257
11.4 Spatial Scale of Physical Effects 260
11.5 Habitat Suitability and Connectivity 261
11.6 Quantifying Habitat Change 263
11.7 Responses of Species to Disturbance 265
11.8 Resiliency and Recovery 267
11.9 Confounding Effects of Multiple Stressors 269
11.10 Conclusion 271
Acknowledgments 271
References 272

11.1 INTRODUCTION

The sustainability of wildlife populations at explosives-contaminated ranges depends on the presence of adequate habitat as well as the absence of bioavailable concentrations of energetic chemicals in soil that would adversely affect these populations. The extent and importance of habitat disturbance is rarely investigated on ranges where explosives are used. Risk assessments for wildlife at contaminated sites occasionally consider habitat preferences in models of trophic uptake of chemicals [1,2] (see Chapter 10), but almost never consider the potential habitat loss associated with those contaminants or physical disturbance [3]. Ecological risk assessments for explosives-contaminated ranges should consider physical habitat disturbance in addition to exposure to explosives contaminants in order to distinguish habitat-based effects from putative toxicity observed in the field. Additionally, ecological risk assessments that are intended to include all ecological stressors from live-fire training, (e.g., those that may support installation Integrated Natural Resources

Management Plans) [4], should incorporate the effects of habitat loss, even if these are small in scale compared to the often large areas of military installations characterized by relatively intact vegetation communities. Therefore, a discussion of habitat disturbance on explosives-contaminated ranges is included in this volume on the ecotoxicology of explosives.

Little research has been performed to investigate ecological effects of munitions use within impact areas, including the possible spatial scales of these effects [5]. The phytotoxicity of explosives has been examined primarily in a laboratory context, which does not consider whether additional potential sources of habitat disturbance are present on ranges. Habitat disturbance can result from construction of ranges, range management practices (e.g., clearing of woody vegetation), phytotoxicity of explosives, cratering and associated localized disturbance of soil, unintentional wildfire from explosives detonation, and removal of ordnance scrap and associated chemicals. Those plant communities and wildlife populations that require a minimum contiguous area of soil having specific bulk densities or vegetation cover are potentially at risk from large-scale activities at explosives ranges. In contrast, impact areas and surrounding buffer zones sometimes provide beneficial habitats for species that thrive in disturbed areas and those that prefer open or edge habitats.

Few studies have summarized the scale and magnitude of potential habitat loss and species sensitivities due to detonations, contamination from explosives, and range construction and management practices. This chapter describes munitions ranges (mission, design, and construction practices), the spatial distribution of contamination and phytotoxicity, and the spatial scale of physical effects (e.g., detonation craters, fire, clearing of vegetation). Principles of habitat suitability and connectivity are summarized, as well as ways to quantify habitat change, responses of species to disturbance, resiliency and recovery, and confounding effects of multiple stressors, such as foot traffic, vehicle maneuvers, noise, fire, and invasive vegetation. Together, these topics form the basis for understanding the nature and scale of habitat disturbances at explosive-contaminated ranges. This discussion is not meant to imply that poor wildlife habitat is ubiquitous at military installations; many of these installations and surrounding buffer areas have large vegetation communities that serve as reservoirs for protected species [6]. This chapter is meant to argue that in ecological risk assessments and environmental management decisions, ecotoxicity should be considered in the context of habitat quality, which may be poor in isolated locations for particular species. Although issues of habitat quality are equally important in aquatic and terrestrial environments, this chapter has a terrestrial emphasis.

11.2 MUNITIONS RANGES

U.S. forces must practice live-fire training and thoroughly test weapons and munitions that will be relied upon in battle. The military manages its land, air, and water assets in ranges to assure that its arsenal has high quality and reliable weapons. The size, construction practices, uses, duration, and intensity of uses, target maintenance, and contamination of military ranges affect wildlife habitat suitability.

Ranges have multiple definitions in the military context. A recent U.S. Department of Defense Directive [7] attempts to clarify the definitions of *range* and *operating area* as: "specifically bounded geographic areas that may encompass a landmass, body of water (above or below the surface), and/or airspace used to conduct operations, training, research and development, and test and evaluation of military hardware, personnel, tactics, munitions, explosives, or electronic combat systems. Those areas shall be under strict control of the Armed Forces or may be shared by multiple Agencies." Terrestrial ranges within the U.S. Air Force, Navy, and military testing community typically are large expanses of land, sea, or air, and the term *range* sometimes denotes the entire installation, as with the White Sands Missile Range. In contrast, the Army and some Marine trainers often view ranges as a specialized configuration of land features conducive to or designed to safely and effectively train specific skills and missions. Examples of these are rifle, artillery, and tank gunnery ranges. Because this chapter emphasizes habitat issues, including the spatial distribution of a habitat, we adopt the latter, more specific definition of range, because it refers to physically disturbed areas.

As defined in the previous paragraph, different types of ranges may be designed and used for varied purposes, such as flight training, vehicle maneuvering, and urban warfare. This discussion includes only those ranges intended for weapons firing, which are more narrowly defined in the Military Munitions Rule as "designated land or water area set aside, managed, and used to conduct research on, develop, test, and evaluate military munitions and explosives, other ordnance, or weapon systems, or to train military personnel in their use and handling ..." [8]. In this chapter, these ranges are designated "explosives ranges," which is not a military term and is not limited to Explosives Ordnance Demolition Ranges.

The design, construction, and mission of firing ranges are highly variable, although efforts have been made to standardize these elements [9]. For example, a small arms range for soldier training, especially a hand grenade range, may be a considerably altered environment. The clearing of vegetation, grading of surfaces, earth moving to construct containment berms, and construction of support facilities (such as firing lanes, targets, observation towers, and staging areas) are often involved. However, for some uses, such as scattered artillery firing points, the range alterations may be limited to markers designating the firing points, and the impact area, which is often out of sight several kilometers away, may have no intentional alterations other than the posting of warning signs or fencing. Often, single impact areas serve multiple weapons systems. Clearly, air-to-air combat training and ground-to-air training, such as antiaircraft firing, require little more than secure land (or open water) for spent rounds to land. Air-to-ground combat training ranges typically require ground target placement and surface improvements, such as service roads, to maintain them.

Recent range design trends include investment in highly instrumented ranges capable of creating multiple scenarios of targets, obstacles, and objectives. Many of the existing landscape features and vegetation are often retained for tactical concealment [10]. On the other hand, such ranges are often heavily used and require intensive maintenance and support facilities, such as access roads, observation towers, and instrumented targets. These training ranges are similar to weapons testing ranges in their investment in instrumentation.

Test ranges conduct firing tests to certify that weapons and munitions meet performance specifications and to report failure analyses back to program managers. Often the test article is an experimental prototype or a limited lot tested against performance criteria under controlled conditions, in contrast to the synergistic approach of soldiers, weapons, tactical equipment, and systems common to training doctrine. Support services at test events add assets, such as sensitive optic, acoustic, and other sensors; observation bunkers and towers; meteorological stations and other monitors; or high-speed networking and communication systems. Fired rounds or their explosion fragments may be recovered from impact areas for further analysis. These activities nominally involve off-road vehicle operations and excavation to retrieve buried rounds. A further complication of munitions testing is the risk that prototype items in different stages of development may not perform as planned (which is why testing is important in the first place). In response, heightened safety precautions and conservatively estimated surface danger zones are in place to counter the elevated risk of misfire or malfunction. Additionally, developmental munitions testing is not typically located in permanent firing positions and impact areas, because each test is driven by the proponent's requirements and specifications.

Production or lot acceptance testing (PAT), whereby samples are routinely taken and fired from current munitions inventories for quality assurance, can generate a tremendous amount of expended ordnance downrange. Such a mission at the now-closed Jefferson Proving Ground in Indiana generated an estimated 23,000,000 rounds fired (and an estimated >1,000,000 duds, i.e., rounds that did not explode on impact) during its more than 40 years of operation [11]. In most instances, recovery of spent rounds is not a requirement of PAT.

In short, troop weapons training needs fixed, standardized ranges for use on a frequent and intense basis. PAT requires the same, but at an even higher, routine rate of use, depending on the current munitions inventory. Developmental testing requires highly variable, custom range configurations for short-term, low intensity use, with increased ancillary services and support areas. In the future, the differences between training and testing will be less apparent as the National Defense Strategy mandates large sophisticated ranges on multiple installations functioning synergistically in order to conduct exercises combining missions of both testing and training [12].

An additional factor that determines potential exposure to habitat disturbance is the spatial extent of munitions ranges. Ranges at Army training installations where energetic compounds are used vary in size from hand grenade ranges (400–500 m^2) and demolition ranges (a couple of hectares), to antitank rocket ranges (a hundred hectares), to mortar and artillery ranges (hundreds of square kilometers). It is important to note that range size is increasing dramatically as operational speed, range, and firepower increase. For example, the typical large range requirement for World War II era tactics was 10 km^2. That required area has tripled recently and is soon anticipated to expand to over 3000 km^2 to meet needs of future forces [12]. Additionally, the use of joint forces in complex, networked exercises implies more frequent training objective scenarios that cross multiple installations.

11.3 SPATIAL DISTRIBUTION OF CONTAMINATION AND PHYTOTOXICITY

Among other factors, an investigation of habitat loss requires consideration of the spatial extent of contamination from a field perspective. Potential habitat loss can be quantified if the sizes and shapes of contaminated areas and chemical concentrations available to plants are known, as well as the area of physical disturbance. Shapes of contaminated areas help to determine the connectivity of species-specific habitat (see discussion below). Chemical concentrations determine the likelihood and spatial extent of phytotoxicity and the time to recovery of vegetation. In addition, the depth of contamination, relative to that of plant roots, is a factor that determines the potential for plant uptake of explosives.

Activities that potentially contaminate portions of explosives ranges include small arms and artillery firing (training and testing), blow-in-place operations by ordnance recovery teams, the use of depleted uranium (DU) tank penetrator rounds, and the use of smokes and obscurants. Munitions-related energetic compounds may be deposited at detectable levels near firing points and impact areas. The spatial distribution of these compounds is related to the munition and targetry type. Artillery ranges increasingly use "shoot and scoot" procedures of rapid deployment and withdrawal, rather than fixed firing points, and this practice leads to dispersed contamination. At firing points, the major potential contaminants are related to the propellants. For artillery, mortars, and shoulder-fired rockets, these propellants are nitrocellulose based. For single-base propellant, 2,4-dinitrotoluene (2,4-DNT) is added; for double-base propellant, nitroglycerin is added; and for triple-base propellant, nitroglycerin and nitroguanidine are added. Many larger rocket and missile systems use ammonium perchlorate as the oxidant.

The major energetic compounds deposited at the impact areas are high explosives. These include trinitrotoluene (TNT), Composition B (60% royal demolition explosive [RDX], 39% TNT, 1% wax), octol (60%–70% high melting explosive [HMX], 30%–40% TNT), tritonol (TNT and aluminum), and Composition A5 (98.5% RDX). When duds are destroyed within active impact areas, the donor charge is generally C-4, which is 91% RDX and 9% oil. The military also fires smoke rounds downrange, and one of the major smoke-producing chemicals is white phosphorus (P_4), although restrictions on where these may be used have reduced their numbers somewhat, especially in situations where the impact area consists of marshes or wetlands.

The potential for phytotoxicity from explosives is discussed at length elsewhere in this volume (see Chapter 3). Few studies of the phytotoxicity of these chemicals have been undertaken, and most of these, which are reviewed in Rocheleau et al. [13], were tests of TNT. Concentrations of TNT that reduce shoot or root biomass in crop test plants in the laboratory range from about 0.1 mg kg^{-1} to >1600 mg kg^{-1} dry soil [14–17]. Phytotoxicity tests with amended soils produced lowest observed effect concentration (LOEC) values for TNT ranging from 0.1 to 64 mg kg^{-1} depending on plant species and exposure type used [15] (see Chapter 3). Plant growth was not affected by HMX at concentrations up to approximately 1900 mg kg^{-1} dry forest soil [16], and additional studies suggest that nitroheterocyclic compounds are not as toxic as nitroaromatic compounds [13–14]. Notably, concentrations of TNT below 50 mg kg^{-1} have sometimes been observed to stimulate seedling growth in the laboratory [15,17]. Clearly, more studies of phytotoxicity of these chemicals are needed, especially field investigations and laboratory studies using native species. In general, the concentrations of explosives in soil are not sufficiently high to denude the area of all vegetation [18].

The nature of constituent residues observed in surface soils on training ranges is determined by the characteristics of the munitions fired. The following discussion describes residues associated with hand grenade, antitank, and artillery/mortar ranges. Hand grenade ranges are relatively small, sparsely vegetated, and heavily cratered, and the soils contain high concentrations of metallic fragments. The energetic compounds detected at the highest concentrations at these ranges include RDX, TNT, HMX, and two environmental transformation products of TNT, 2-amino-4,6-dinitrotoluene (2-ADNT) and 4-amino-2,6-dinitrotoluene (4-ADNT) [19–20]. The energetic compounds are largely in the top 15 cm of soil with concentrations of RDX and TNT ranging from high parts per billion (µg kg^{-1}) to low parts per million (mg kg^{-1}) levels. The bulk of the residue appears to be due to low-order (partial) detonations and destruction (blow-in-place) of duds, which results in deposition of particles of the undetonated explosive [21].

The spatial distribution of contamination at several antitank rocket ranges has also been studied [19,22–27]. These relatively small ranges (e.g., 5 ha) generally consist of grasslands, because most current weapons systems require a line of sight from the firing point to the target. The weapon fired with the greatest frequency at these ranges was the M72 66-mm LAW rocket. The propellant used with this weapon is double base, and concentrations of nitroglycerine as high as a few percent have been detected behind the firing line where deposition occurs from the back blast [24–25]. The high explosive used in the warhead of this rocket is octol, and the major residue detected at the impact areas is HMX. Concentrations of HMX are inversely related to the distance from the target with concentrations as high as hundreds of parts per million adjacent to targets. Even though octol is 70/30 HMX/TNT, the concentration ratios of these compounds in surface soils are about 100/1 HMX/TNT due to the much lower solubility of HMX and its much greater half-life in the soil compared to TNT [24,28]. The major source of the deposition of octol at these ranges appears to be the rupture of these thin-skinned rockets when they do not directly impact a target, and the fuse fails to function normally. Such an occurrence can spread relatively high concentrations of explosive compounds at target locations.

Research on energetic compounds has also been conducted at artillery/mortar ranges at several installations [19–20,27,29–35]. These ranges are very large, and the terrain varies significantly, encompassing grasslands, forests, wetlands, and dry, sparsely vegetated areas of the desert Southwest. Typical shoot-and-scoot exercises do not utilize fixed positions, so the deposition that occurs near the firing point is widely dispersed at low concentrations. The energetic compounds that can sometimes be detected at firing points are 2,4-DNT from single-base propellants and nitroglycerin from double- and triple-base propellants. These chemicals are found at parts per billion to low parts per million concentrations in the top few centimeters of soil.

Only a small portion of the large tracts of land designated as impact areas have artillery targets; thus, the majority of the surface of these ranges is largely uncontaminated [19,27,29,30,32,34]. Live-fire testing has indicated that when artillery and mortar rounds detonate as designed, only microgram to milligram quantities of high explosives are deposited. The major contamination at artillery and mortar ranges results from low-order detonations of TNT and Composition B-containing munitions. Using the data presented by Ref. 21, we estimate that it can take as many as 100,000 normal detonations to deposit as much residue of RDX and TNT as would one low-order detonation in which half of the explosive fill remains undetonated. These "hot spots" of contamination are often (but not always) located in the general vicinity of targets, and they result in an extremely heterogeneous distribution of residues in surface soils at these ranges. A recent study indicated that concentrations of RDX varied over five orders of magnitude within a 10 m × 10 m area whose primary contamination came from a single Composition B-filled mortar round that had undergone a low-order detonation [36].

Distributional heterogeneity of residues of energetic compounds is observed at the impact (detonation) area of all ranges, but especially at artillery/mortar ranges. Compositional heterogeneity is also present, as residues of energetic compounds are often deposited as particles of energetic compounds of various sizes [35]. It is very difficult to collect truly representative soil samples under these conditions, and studies indicate that characterization based on discrete samples would result in huge uncertainties [35,37]. Multi-increment composite samples using 30 increments or more have been used to reduce sampling error [20,24,33,35], but uncertainties can still be significant.

Very little research has been conducted on residue deposition at Air Force or Naval terrestrial ranges. One study conducted at a Canadian Air Force range found that most training activities do not use high explosive-filled bombs and rockets [38]. Typically, no detonators are present and spotters rely on visual observation of impact, or there is a small smoke marker. Only one relatively small grassland range (a few hundred acres) was used for training with high explosives-filled bombs and rockets. The soil was tilled to reduce the potential for forest fires at this largely forested range, and TNT was found in concentrations up to several hundred ppm in surface soils near the target. RDX was not present; Air Force bombs are often filled with tritonol (TNT and aluminum).

White phosphorus-based smoke rounds are also used frequently at artillery/mortar training ranges. White phosphorus is a form of elemental phosphorus that is pyrophoric, spontaneously igniting in the presence of air. At one Army artillery range where the impact area was in a salt marsh, particles of white phosphorus present in shallow sediments resulted in mortality of a variety of waterfowl through ingestion of these particles while feeding [39]. As with other energetic contaminants, the deposited white phosphorus particles are spatially heterogeneous, complicating the characterization of these areas. White phosphorus is less of a problem in terrestrial environments because of its pyrophoric behavior, although it can sometimes crust over, protecting chunks of white phosphorus from oxidation and resulting in issues of safety when these particles are broken open at a later date.

Based on the limited field sampling that has been performed, only small fractions of the area of explosives ranges are contaminated with explosives, and often these are at concentrations that may not be bioavailable or phytotoxic. However, concentrations of TNT in hot spots of soil in most impact areas exceed the lowest levels that have been observed to be toxic to some test plant species (7 mg kg^{-1}) as discussed earlier [14–15].

The uncertainty associated with spatial predictions of contamination (and therefore phytotoxicity) is typically higher on testing grounds, where new ordnance items are tested, than on training installations. Targeting error may be unknown, including propulsion, ballistic, and detonation reliability; however, technological advances may allow impact locations to be calculated more precisely.

11.4   SPATIAL SCALE OF PHYSICAL EFFECTS

The exposure of wildlife to physical disturbance on explosives ranges may be as important (or unimportant) as exposure to chemical explosives. Habitat disturbance from range construction may occur at the scale of the entire range. This type and scale of disturbance may be intense and large enough to fragment habitat (see discussion of fragmentation in Section 11.5). However, other sources of physical disturbance generally occur at much smaller spatial scales. A single firing point may be used (and therefore disturbed) repeatedly by rotational training units in free-maneuvering combat vehicles that establish a hasty firing position and fire into a distant impact area. Detonation craters are a source of local disturbance in topography, soil organic matter, and vegetation. Although these craters are dense on hand grenade ranges [18], they may represent no more than 10% of typical impact areas [40]. If line of sight is required (e.g., for small arms and many tank gunnery and antitank weapons, but not digital ranges), then range management practices may include clearing of all woody vegetation between the firing points and targets, sometimes with mowing and herbicide use. The area cleared of vegetation around individual firing points of artillery and mortar ranges is typically no more than 0.2 ha, unless groupings of firing points occur in close proximity [40]. Vegetation at U.S. Air Force bombing ranges is often cleared of vegetation at the larger scale of 1 km^2 or more. The spatial scales of fires, including wildfires from detonations and tracer rounds, depend on meteorology, as well as the presence of adequate fuel. Range management practices often include controlled burns or disking (turning over vegetation attached to soil clods with a harrow) to control potential wildfires. The spatial scales of controlled burns used to protect against wildfires from detonations and those conducted for habitat management (e.g., for red-cockaded woodpecker at Fort Benning, Georgia) are prescribed at each installation. Blow-in-place of unexploded ordnance (UXO) on ranges and excavation to remove explosives at closed ranges (e.g., for open burning/open detonation treatment) are intense but small-scale disturbances. Safety buffer areas, often more than a thousand hectares around major ordnance impact areas, tend to be almost undisturbed habitat, and human safety restrictions within ranges can increase wildlife habitat quality in areas where firing points and targets are not located.

11.5   HABITAT SUITABILITY AND CONNECTIVITY

Habitat for a wildlife species is the sum of the locations from which it derives food, water, and cover, as well as locations where reproductive activities occur. Particular environments may be well suited, partially suited, or not suited to specific species. Winter and summer habitats differ for migrating species, and breeding habitats may differ in spatial extent and other characteristics from nonbreeding habitats. Some of the variables that determine wildlife habitat include: soil characteristics (particle size, moisture content, pH, nutrient content, etc.), topography (slope, aspect), temperature, precipitation, vegetation characteristics (type, height, basal area, cover), distance to a specified land feature, and edge length per unit area [41]. In addition, the connectivity of habitat is important. The exact nature as well as the relative importance of each of these factors depends on the species of concern. For example, soil particle size would be expected to be important to burrowing wildlife species and to species that require plant cover or forage vegetation that is associated with soil of a particular particle size, but not for other species. Moreover, some species have specific habitat requirements for breeding territories, such as grassland birds that form leks (male group displays) in open areas. These habitat suitability factors are incorporated into U.S. Fish and Wildlife Service habitat models. Numerical indices of habitat suitability are derived on a scale of 0.0 to 1.0 based on the assumption that key environmental variables are related to habitat carrying capacity [42].

Habitat suitability factors may be affected by the use of explosives, at least on the small scale (Table 11.1). Areas may be cleared by blading (removing surface soil) and removing vegetation [43], particularly during construction and excavation of UXO. Detonations remove surface soil (including organic matter), alter soil particle size, induce downslope erosion, decrease plant cover, and may kill or alter the spatial distribution of key invertebrate foods. Indirect changes to plant cover or food

TABLE 11.1
Effects of Explosives Ranges on Terrestrial Wildlife Habitat Variables

Stressor	Direct Effect	Habitat Variable Affected
Explosives contamination	Phytotoxicity	Forage, cover
	Invertebrate toxicity	Forage
Detonation craters	Alteration of surface soil	Soil moisture, particle size, composition Local slope, aspect
Range clearing	Plant removal	Forage, cover, habitat connectivity
	Alteration of vertebrate population abundance	Prey
	Alteration of surface soil	Soil composition, particle size
Remediation (excavation)	Removal of soil	Soil composition, particle size
	Removal of plant community	Forage, cover, habitat connectivity
	Removal of invertebrate community	Forage
	Alteration of vertebrate population abundance	Prey

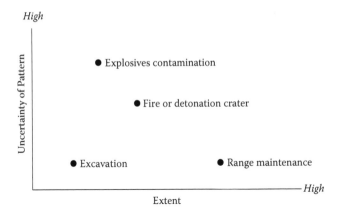

FIGURE 11.1 Vegetation removal due to various stressors during a single training, testing, or range maintenance event. Example relative extents of disturbance and uncertainties of spatial disturbance pattern of many of these stressors are depicted.

availability can also occur as a result of fires initiated by munitions. Detonation craters may locally alter slope and aspect, as well as surface soil characteristics, thus altering soil moisture. Contamination might be present at phytotoxic concentrations. The relative extent of disturbance and uncertainty of spatial disturbance patterns of many of these stressors is depicted in Figure 11.1. In this example, the spatial extent of range maintenance activities (e.g., vegetation clearing) is greater than the area affected by detonation craters, explosives contamination, or excavation. The locations of disturbance associated with excavation and range maintenance activities are planned with exact boundaries, unlike areas of contamination or fire, where affected locations are not entirely predictable (Figure 11.1). In addition, large-scale disturbances at explosives ranges can lead to the fragmentation of wildlife habitat and increases in disturbed and edge habitats (e.g., transitions from forest to grassland), which are not depicted on Figure 11.1.

Habitat Suitability Index models provide habitat suitabilities for specific locations [42] but do not consider the importance of landscape pattern. The connectivity of habitat is often just as important as soil or vegetation type in determining if habitat for a particular species is adequate [44]. Habitat fragmentation is the process of dividing an area into unassociated pieces that affect an organism's use of the area for all or part of its life cycle. Fragmentation can have three components: loss of area of the original habitat, reduction in habitat patch size, and increasing isolation of habitat patches. The major effect of fragmentation is change in biodiversity, including genetic, species, and landscape diversity [45]. Typically, fragmentation results in a decline of those species that have restricted habitat requirements. However, species that thrive in ecotones become more abundant [46]. Other effects include changes in predation, competition, pollination, seed dispersal, and mating behavior [47]. Fragmentation concepts have spawned the field of metapopulation dynamics (the idea that the regional persistence of a species depends on the maintenance, colonization, and extinction of subpopulations).

Fragmentation effects are most intensely observed for species that require large areas of habitat or those that avoid or will not move across unsuitable habitats. Fragmentation is of greatest concern in relatively large ecosystems that have a history of minor human disturbances but are now subjected to major human disturbances, as well as those where wildlife movement corridors (e.g., to water sources or breeding sites) are interrupted [48]. The intensity and frequency of disturbances are important predictors of effects. Several models that predict population response to habitat fragmentation on military installations have been developed. For example, a model that projects how the spatial distribution of nesting habitat affects the reproductive success of territorial migrant bird species breeding in fragmented, patchy landscapes was implemented for Henslow's sparrow (*Ammodramus henslowii*) at Fort Riley, Kansas, and at Fort Knox, Kentucky [49]. Results indicated that persistence of the sparrow at Fort Knox appears to require recruitment of individuals from other parts of the species' range, which in turn may reflect the marginal habitat at Fort Knox, for it is on the southern edge of the species' summer range.

Patches of habitat may sometimes be lost from a landscape without the isolation of patches that denotes fragmentation, producing holes in the landscape (Figure 11.2). The relative importance of habitat loss and fragmentation on military installations is likely to depend on the habitats and sensitivities of particular species. The areas of patches of suitable habitat may be compared to what Carlsen et al. [50] term "critical patch size," the contiguous habitat area needed to maintain a population. Estimates of critical patch sizes are available for 249 species [50]. If these critical thresholds in patch size are not met because of range construction or clearing practices, cratering, phytotoxicity due to explosives, a population may not persist. The effect of detonation craters on populations depends on the cumulative number and distribution of craters and rates of natural recovery. As stated earlier, phytotoxicity is observed in the field if soil concentrations exceed phytotoxic concentrations at rooting depths [51] (see Chapter 3).

Studies of habitat alteration on military installations have not focused on explosives ranges. Investigations of the responses of soil and vegetation communities to disturbance by tracked vehicles call attention to some of the habitat factors that could also be studied on explosives ranges. Tracked vehicles can cause direct plant mortality by crushing and can indirectly affect plant communities through the compaction of soil and species competition [52]. Prose [53] measured the extent of soil compaction laterally from and below track locations. Several researchers studied vegetation cover and succession patterns [52,54–56]. Ayers [57] related tracked vehicle operations such as turning radius and pad–load ratio to soil and vegetation disturbance.

11.6   QUANTIFYING HABITAT CHANGE

Numerous methods are available to measure or model the impact of explosives contamination and detonation craters or other physical stressors on vegetation. U.S. Army installations may collect data on the distribution of vegetation and soils to describe plant communities and the distribution of small mammals, birds, reptiles, and amphibians to characterize species-specific habitat as part of the Land Condition

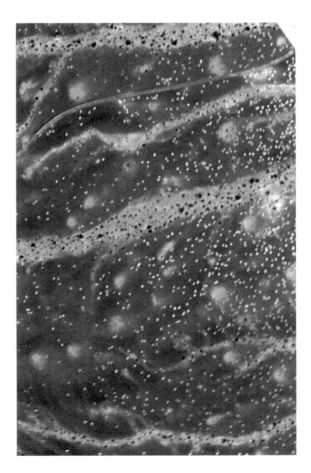

FIGURE 11.2 Aerial photograph of an impact area at Yuma Proving Ground. The small bright circles are ordnance impact scars on the dark desert pavement. Larger, circular light areas are scars in the pavement where plants are growing or have grown in the past. Four large, light gray desert washes cross the photo, with trees and shrubs seen as dark spots. (U.S. Army file photo, scale of photo unknown.)

Trend Analysis (LCTA) program [58,59]. Permanent inventory plots are located in a stratified random manner based on soil data and satellite imagery, but less frequently within demarked impact zones because of the dangers of sampling there. Because military testing and training typically result in intense, local, and broadly spaced impacts, the LCTA plots typically do not capture the full spatial distribution of the effects. For example, at Yuma Proving Ground (YPG), Arizona, 60% to 70% of the plots showed no use during 1991 to 1993 and again in 1998 and 2003 [60,61]. Thus, the LCTA approach should be supplemented by a method designed to focus on discerning impacts of use and integrating over broader spatial scales. Studies evaluating LCTA data at the Kansas Army National Guard Training Facility compared LCTA results with a modified methodology designed to place sampling transects in field-identified rather than satellite-identified land-cover types. The studies found that

LCTA sampling was too limited in the ecologically important riparian woodland habitat, with the result that bird species were not adequately sampled [62]. In 2004, the LCTA program was restructured into the Range and Training Land Assessment program [63] with the understanding that modifications of LCTA's standardized approach are needed to address these issues and to reflect local landscapes and land uses.

Remote sensing is useful to detect large-scale (and increasingly, small-scale) changes in vegetation cover and type where field sampling is impractical because of safety issues or cost. General methods are described by Lillesand et al. [64]. Underwood et al. [65] describe hyperspectral methods for detecting invasive plants, and that research group is currently mapping nonnative species at military installations. Moreover, the U.S. Army Corps of Engineers Engineer Research and Development Center is attempting to develop remote sensing methods for detecting explosives contamination [66].

Maps of potential habitat distribution can be produced when information from ground surveys is not available, such as occurs in impact areas. Mann et al. [67] developed a geographic information system (GIS) model that predicts potential location and successional status of threatened calcareous habitat at the Fort Knox Military Reservation, including heavily impacted ordnance and tank training areas that are unsafe for public access. Their model uses ecosystem information contained in the U.S. Department of Agriculture Natural Resources Conservation Service State Soil Geographic Database, as well as satellite imagery. These threatened calcareous habitats support several rare plant species. The combined soil–geology–slope GIS approach is useful in conservation management and restoration, especially where intensive ground surveys are impractical [67].

Models are also available to predict vegetation or soil changes in the future. A random optimization procedure (neural network model) has been used to estimate vegetation cover probabilities based on past disturbance pattern and vegetation coverage data collected according to LCTA methods at Fort Sill, Oklahoma [68]. Transition models can also be used to project potential vegetation and habitats [69]. Ecological dynamics simulation modeling (EDYS) is an ecosystem model that has been used in a wide range of applications, including the assessment of potential impacts of different training regimes on vegetation and endangered species habitats at the U.S. Air Force Academy, Colorado; Fort Bliss, New Mexico; and Fort Hood, Texas [70]. EDYS models soil water and nutrient dynamics, species-specific plant uptake and growth, herbivory, fire, contaminant dynamics, physical disturbance, and management actions. The Army Training and Testing Carrying Capacity model measures training load in terms of maneuver impact miles (MIMs) of an M1A2 tank driving one mile in an armor battalion field training exercise [71]. Impacts from firing and bombing ranges could be quantified in a similar manner.

11.7 RESPONSES OF SPECIES TO DISTURBANCE

Numerous threatened, endangered, and other valued wildlife species are present on U.S. Department of Defense (DoD) installations in greater abundance on a per area basis than on most other federal lands [6,72]. These and other federal lands often form refuges for species from land-use changes and activities on adjacent, privately held lands [73]. For example, habitat complexity in the prairie-forest ecotone of southwestern Oklahoma is reduced by agricultural development and enhanced by protection afforded by the Fort Sill Military Reservation [74]. Large tracts of intact habitat on military lands are important for the sustainability of these species; therefore, the potential habitat loss associated with explosives use is worthy of examination.

We are aware of no studies that attribute any species declines to the presence of explosives contamination or firing ranges. However, most vegetation and wildlife (other than disturbance-adapted species) are often removed by the initial construction or clearing to create an impact area. In addition, the cumulative habitat loss and fragmentation associated with explosives ranges and other military training and testing activities (see discussion of multiple stressors in Section 11.9) could adversely affect particular wildlife populations. The loss of shrub cover during training in the Mojave Desert was previously determined to lower the relative abundance of the little pocket mouse (*Perognathus longimembris*) and southern grasshopper mouse (*Onychomys torridus*) [75,76].

Several examples exist of species that have benefited from disturbance at explosives-contaminated ranges or similar sites. At Jefferson Proving Ground in Indiana, artillery impact craters were posited as the habitat enhancement factor supporting increased vegetation diversity [77]. In the Chocolate Mountain Aerial Gunnery Range in southeastern California, 250-lb bombs create detonation craters that break through and expose the sandy soil beneath the desert pavement. Kangaroo rats (*Dipodomys merriami*) colonize the soft walls of the crater, resulting in an apparent population increase [78]. Anecdotal evidence suggests that Sonoran pronghorns (*Antilocapra americana sonoriensis*) prefer the watering holes and young vegetation found in craters of bombing ranges, areas that also have fewer creosote bushes that impede the pronghorns' view of predators [79]. Tadpoles have been observed in water-filled impact craters at Eglin Air Force Base, Florida, [80] and it is likely that these ephemeral pools serve as beneficial habitat for amphibians at other military installations. At Fort Hood, the black-capped vireo (*Vireo atricapillus*) is an example of an endangered species that colonizes early successional scrub arising within several years after range fires caused by explosives, flares, or other military activities. Another endangered species, the Karner blue butterfly (*Lycaecides melissa samuelis*), is on a precipitous decline, probably due to habitat fragmentation, yet it occurs in a habitat that persists only with regular light fires or some other similar disturbance, such as military training [81]. At Fort McCoy, Wisconsin, populations of the Karner blue butterfly are found in oak and pine barren communities, which are habitat for wild lupine, a disturbance-dependent forb required by Karner blue butterfly larvae. The butterfly is more abundant in some impact zones, although obtaining data on these butterflies has proven perilous to biologists [82]. Severe disturbances may cause the demise of these butterfly habitats. At YPG and similar arid installations, impact craters collect surface runoff, allowing plants to colonize otherwise barren desert pavement (Figure 11.3). Although individual plants benefit, the impediment to surface runoff may adversely affect plant communities downstream [83].

In general, ranges where explosives are used (and their boundaries) may provide valuable habitats for endangered species highly specialized to open or edge landscapes. For example, disturbance by tracked vehicles promoted the lupine (*Lupinus*

FIGURE 11.3   Drought-stressed white bursage (*Ambrosia dumosa*) that has apparently colonized an impact crater at Yuma Proving Ground, Arizona. (U.S. Army file photo.)

perennis) vegetation required by the federally endangered Karner blue butterfly [84]. Imbeau et al. [85] argue that species that prefer edge habitats at agriculture–forest junctures are actually species that prefer early successional habitats wherever they are available. Early successional habitats are common and are regularly created on explosives ranges.

11.8   RESILIENCY AND RECOVERY

The habitat disturbance associated with explosives and explosives-contaminated ranges is temporary. The duration of the disturbance depends on the life expectancy of the range and the time periods during which: (1) explosives reside in surface soil; (2) plant communities that are adapted to the contamination or physical disturbance become dominant; (3) surface soil in detonation craters is replaced; and (4) the ecosystem recovers from unintentional wildfire from the detonation of explosives.

A detailed review of environmental fate of energetic materials is provided in Chapter 2. Houston et al. [5] provided metrics for assessing the resiliency of generic environmental settings to explosive residue contamination. These ecological characteristics influence the fate and transport of explosives. Resilient impact areas are characterized by: (1) low moisture values from Thornwaite Moisture Regions that indicate low dissolution of chemicals; (2) high cation exchange capacity of soils that indicates limited desorption; (3) long effective growing seasons that indicate high potential for biodegradation and biological uptake; and (4) high soil organic carbon that suggests high potential for chemical transformation. Houston et al. [5] also ranked the resilience of 11 installations based on these criteria, but other important predictors of resiliency were not included, such as plant regeneration, fragmentation, or sensitive populations.

Recovery typically refers to the colonization, growth, or succession of ecological entities, following the effective removal of the direct pressure of a stressor. The stressors that are most relevant here include those listed in Table 11.1: clearing of vegetation, explosives contamination, detonation craters, and remediation. Factors that influence the recovery of ecosystems from disturbance include current state, disturbance severity and frequency, successional history, history of disturbance, preferred state, management of the disturbance, and random factors such as weather [86]. Recolonization time is dependent on the size of the site and proximity to a recolonization source. Species that are characteristic of early successional communities recover relatively rapidly from disturbance to colonize disturbed areas, due to their high reproductive rates and rapid dispersal mechanisms [87]. Diersing et al. [88] defined recovery by estimating the average number of years for tracked vehicle-affected areas to regrow vegetation cover equivalent in C-value (for the universal soil loss equation) to untracked areas.

Steiger and Webb [43] studied the recovery of desert vegetation in military target sites in the Mohave and Cerbat Mountains of northwestern Arizona. The sites were cleared between 1942 and 1944 with up to 0.2 m of the surface material displaced by blading. The sites were used for strafing runs and did not use impact explosives. Steiger and Webb found greater variability in the extent of recovery for sites on older geomorphic surfaces than on younger surfaces and a weak inverse relationship between the degree of recovery and geomorphic age. Vegetation parameters in desert washes had generally recovered since the 1940s. In some cases, the survival of root crowns (e.g., of *Larrea* and *Ambrosia* species) facilitated revegetation of study sites. The recovery of vegetation in detonation craters might be expected to display similar dynamics to the recovery in blading locations. Steiger and Webb note that blading could enhance the moisture and perennial vegetation in disturbed sites on desert pavement, which is consistent with the effects observed in detonation craters at YPG, as described earlier. Recovery times in deserts likely represent a worst-case scenario for terrestrial ecosystems. Similarly, low moisture (ice-bound water, low precipitation) and low temperature conditions result in slow ecological recovery in subarctic training areas.

The time periods that are required for recovery from possible phytotoxicity and physical disturbance are uncertain [89]. In general, recovery of vegetation from physical disturbance in arid ecosystems can take hundreds of years [90]. An estimate of the minimum time to recovery could be provided by the average age of the lost vegetation [90]. It should be noted that the recovery of one ecological property can be impeded by restoration or reclamation of another. For example, the maintenance of impervious soil caps over a waste disposal site involves the removal (and therefore prevention of recovery) of deeply rooted vegetation and burrowing mammals [91].

The potential for a habitat to recover following fragmentation depends on the degree of fragmentation and rates of recovery of the vegetation. For example, the demise and fragmentation of longleaf pine (*Pinus palustris*) forest from timber harvesting, fire suppression, and land-use change [92,93] resulted in the decline of red-cockaded woodpeckers (RCW; *Picoides borealis*) in the southeastern United States, but active management for longleaf pine (as well as the continued presence of other pines) on military installations has contributed to the stabilization of remaining RCW populations [94].

11.9 CONFOUNDING EFFECTS OF MULTIPLE STRESSORS

Multiple stressors that are associated with explosives ranges and range operation and management include, but are not limited to: foot traffic from soldiers, maneuvering of tracked vehicles, maintenance of targets, noise, fire, heat, water- and wind-eroded soils, flooding, forest management, grazing, installation infrastructure development, petroleum spills and other contaminants such as metals, and encroachment of invasive vegetation associated with surface disturbance. Together, these stressors can result in changes to soil or vegetation components of wildlife habitat. For example, military training in longleaf pine ecosystems and in grasslands was associated with increased bare soil, reduced total plant cover, and compositional shifts in plant communities [95,96]. Military training resulted in reductions in both soil carbon and soil nitrogen levels, and greater surface soil bulk density at the Fort Benning Military Reservation [97]. Furthermore, soil microbial biomass and community composition were significantly altered by military training at Fort Benning [98]. In general, training has been shown to increase the abundance of early successional species that replace less common climax and other native species [76].

Range construction practices can involve the use of bulldozers, excavators, and other earth moving equipment for potentially extensive clearing and grubbing of large areas of the landscape to meet design requirements for long-range firing of both small and large arms. These practices can temporarily strip most existing vegetation and topsoil, with potentially significant impacts to wildlife habitats. Invasions by nonnative vegetation may reduce habitat quality. The potential interruption of site hydrology may impact wetland ecosystems and increase flooding of ranges.

Tracked vehicles, including tanks and Bradley armored fighting vehicles, may be maneuvered across range areas to established firing positions, leading to erosion along unpaved roads and trails. Aerial maneuvers using helicopters can disperse significant fugitive dust. Foot traffic by soldiers can also promote erosion and slow vegetation recovery in areas of exposed subsoils, though these impacts can be minimized if soldiers are directed to different areas on a rotational basis. Frequent foot traffic through long-leaf pine ecosystems at Fort Benning resulted in more trees in the understory, probably because small woody vegetation was better able to withstand the mechanical stressor than herbaceous species [95].

A common practice of planting or hydroseeding of fescue and other nonnative "turf" grasses to hold soils and to promote line of sight, or to stabilize disturbed ground may not reestablish adequate habitat for many species. Native, fire-climax, mixed grassland communities provide superior wildlife food and shelter, as well as better soil-holding capabilities, compared to those of the "turf" grass monocultures, but may be very costly to establish following construction of ranges.

A major source of habitat loss on closed explosives ranges can be remedial activities such as excavation. Remedial goals for explosives ranges are typically defined on the basis of human health and safety, but remedial technologies for contaminated sites are chosen based primarily on two engineering criteria, the ability to achieve those goals and cost-effectiveness, rather than ecological criteria [99]. Even deep UXO that poses little risk is sometimes removed along with the associated soil ecosystem and vegetation community. Whicker et al. [100] describe putatively unrealistic land-use assumptions that result in highly conservative risk assessments and habitat removal during contaminant remediations at U.S. Department of Energy sites. Similar conclusions may apply to explosives ranges at closed DoD installations. A net environmental benefit analysis is recommended prior to the remediation of explosives-contaminated soil, especially in arid or semiarid regions, where recovery from disturbance can take centuries [99].

Effects due to multiple stressors are observed if stressors overlap in space or if wide-ranging wildlife are exposed to stressors in different locations. The stressors discussed here may be found in close proximity to or at a distance from chemical contamination in soil. Noise is one of the stressors that always overlaps spatially with contamination from explosives and clearing of range vegetation. Noise affects habitat suitability of many species, such as the endangered Sonoran pronghorn. At the Barry M. Goldwater Range in southwestern Arizona, habitat areas with noise levels greater than or equal to 55 dB are used less by the Sonoran pronghorn than quieter, equally suitable areas [101]. In the Snake River Birds of Prey National Conservation Area in Idaho, firing of artillery, small arms, and main turret guns or machine guns on tanks reduced counts of raptors on ranges, whereas tank preparation (i.e., assembling and loading ammunition), driving, laser training, and convoy traffic had no effect [102]. To our knowledge, noise has never been included in habitat suitability indices, but future knowledge about the impacts of blast noise could be incorporated into habitat models.

In addition to on-base stressors, urban development encroaches near the boundaries of many military installations. Development may increase species abundance in relatively undisturbed areas of installations and buffers, but it decreases the available habitat for populations or metapopulations that extend beyond installation boundaries.

The combined effects of particular groups of stressors at specific ranges may be evaluated through field studies of wildlife populations if adequate control sites or predisturbance data are available, if natural variation in population abundance is not too high, and if sample sizes are adequate. A simulation model is currently under development that can integrate effects from multiple stressors if adequate exposure and effects data are available from military installations. The Regional Simulator (RSim) is designed to simulate: (1) land cover changes caused by urban development, road development, and changes in military training activities; (2) resulting changes in air quality, water quality, soil nutrients, and noise; and (3) changes in vertebrate populations and their habitats [95]. For example, RSim could be employed to assess the ecological benefits of establishing native grasslands on ranges, as discussed earlier. A framework that addresses a process for determining the additivity of effects or exposures associated with various stressors can be found in Ref. 103. This framework was developed in the context of military activities, and it addresses the importance of spatial and temporal overlap of multiple stressors.

An even greater challenge than integrating the effects of multiple stressors is determining the cause of observed effects (e.g., distinguishing the ecotoxicity of explosives from other effects of detonation craters, range clearing, or noise). A risk assessment for a single stressor is easy if multiple stressors are present that together produce no effect (e.g., the absence of effects of firing of small arms and artillery on red-cockaded woodpecker at Fort Benning) [104]. Principles for determining causation of ecological impairments in streams were developed by Suter et al. [105] and are applicable here. For example, they recommend an evaluation of the association of measurements of exposure and effects, including spatial co-occurrence, spatial gradients, temporal relationships, and temporal gradients, as well as the association of effects with mitigation or manipulation of causes. Thus, the effects of contamination by explosives could be distinguished from those of noise if a range is inactive or closed.

11.10   CONCLUSION

Changes in species-specific habitat suitability may arise from localized chemical contamination and physical stressors present on explosives ranges. However, existing studies do not attribute any species declines on or around military installations to the presence of explosives contamination or firing ranges. The relative and cumulative contributions of chemical contaminants, detonation craters, range clearing practices, and environmental management to the loss, gain, or fragmentation of habitats of various species have not been investigated. The characterization of munitions range variables, such as ordnance type, amount and delivery, and clearance rules, is a necessary precursor to studies of habitat change. Studies are in progress to describe the spatial distribution of contamination, and these will support investigations of potential phytotoxicity of nitroaromatic and other chemicals on explosives ranges. Preliminary data suggest that major contamination and detonation craters are typically limited to small areas of explosives ranges (i.e., firing points [contamination] and targets [contamination and craters]). The RSim model, which is under development, will be able to integrate risks to wildlife populations from multiple chemical and physical stressors if sufficient supporting data are available. Results from models of habitat loss should be verified by field studies on all military installations. Although ecotoxicity of explosives is important, an understanding of ecological effects of explosives and firing and bombing ranges is not complete without a thorough understanding of the potential and actual removal and fragmentation of wildlife habitat, as well as the beneficial effects of open and edge habitats and undisturbed range buffer areas. Demarais et al. [76] assert, "physical modification of habitat [from all training exercises] resulting in changed levels of available resources is the primary disturbance affecting vertebrate populations on military installations."

ACKNOWLEDGMENTS

This chapter was funded partly by the Strategic Environmental Research and Development Program (SERDP; Project CS-1259). Oak Ridge National Laboratory is managed by the UT-Battelle, LLC, for the U.S. Department of Energy under contract DE-AC05-00OR22725. Contributions of Thomas Jenkins were funded by SERDP Project CP1155. Earlier drafts of this manuscript were reviewed by Hal Balbach of the U.S. Army Construction Engineering Research Laboratory, Jim Loar of ORNL, Judy C. Pennington of the U.S. Army Engineer Research and Development Center, and three anonymous peer reviewers.

REFERENCES

1. Henriques WD and Dixon KR, Estimating spatial distribution of exposure by integrating radiotelemetry, computer simulation, and geographic information system (GIS) techniques, *Hum. Ecol. Risk Assess.*, 2, 527, 1996.
2. Hope BK, Generating probabilistic spatially-explicit individual and population exposure estimates for ecological risk assessments, *Risk Anal.*, 20, 573, 2000.
3. Efroymson RA et al., Toward a framework for assessing risk to vertebrate populations from brine and petroleum spills at exploration and production sites, in *Landscape Ecology and Wildlife Habitat Evaluation: Critical Information for Ecological Risk Assessment, Land-Use Management Activities, and Biodiversity Enhancement Practices*, ASTM STP 1458, Kapustka L et al., Eds., ASTM International, West Conshohocken, PA, 2004, chap. 15.
4. DoD and USFWS (U.S. Department of Defense and U.S. Fish and Wildlife Service), Integrated Natural Resources Management Plans. Fact Sheet, http://www.fws.gov/habitatconservation/INRMP%20Fact%20Sheet.pdf (accessed August 2008).
5. Houston ST, Doe WWIII, and Shaw RB, Environmental risk of army ranges and impact areas: An ecological framework for assessment, *Federal Facilities Environ. J.*, Spring, 93, 2001.
6. Tazik DJ and Martin CO, Threatened and endangered species on U.S. Department of Defense lands in the arid west, USA, *Arid Land Res. Manage.*, 16, 259, 2002.
7. DOD, U.S. Department of Defense Directive 3200.15 SUBJECT: Sustainment of Ranges and Operating Areas (OPAREAs), USD(P&R), January 10, 2003.
8. USEPA, Military Munitions Rule, *Federal Register*, Volume 62, Number 29, February 12, 1997, http://www.epa.gov/fedrgstr/EPA-WASTE/1997/February/Day-12/f3218.htm (accessed August 2008).
9. HQDA, Training Circular 25-8, Training Ranges, Headquarters Department of the Army, Washington, DC, April 2004.
10. Hale T et al., Tactical Concealment Area Planning and Design Guidance Document, SFIM-AEC-EQ-CR-99031, Champaign, IL, August 1999.
11. JPG, Jefferson Proving Ground Records, 1951-1993, from Indiana Historical Society Manuscripts and Archives, http://www.indianahistory.org/library/manuscripts/collection_guides/M0440.html, 1994 (accessed August 2008).
12. HQDA, Army Range and Training Land Strategy, Headquarters Department of the Army, Washington, DC, January 30, 2004.
13. Rocheleau S et al., Toxicity of Nitro-Heterocyclic and Nitroaromatic Energetic Materials to Terrestrial Plants in a Natural Sandy Loam Soil, Technical Report No. ECBC-TR-351, U.S. Army Edgewood Chemical Biological Center, Aberdeen Proving Ground, MD, 2005.
14. Simini M et al., Evaluation of soil toxicity at Joliet Army Ammunition Plant, *Environ. Toxicol. Chem.*, 14, 623, 1995.
15. Rocheleau S et al., Phytotoxicity of nitroaromatic energetic compounds in freshly amended or weathered/aged sandy loam soil, *Chemosphere*, 62, 545, 2006.
16. Robidoux PY et al., Phytotoxicity of 2,4,6-trinitrotoluene (TNT) and octahydro-1,3,5,7-tetranitro-1,3,5,7-tetrazocine (HMX) in spiked artificial and natural forest soils, *Arch. Environ. Contam. Toxicol.*, 44, 198, 2003.
17. Gong P, Wilke BM, and Fleischmann S, Soil-based phytotoxicity of 2,4,6-trinitrotoluene (TNT) to terrestrial plants, *Arch. Environ. Contam. Toxicol.*, 36, 152, 1999.
18. Pennington JC, personal communication, 2004.
19. Jenkins TF et al., Characterization of Explosives Contamination at Military Firing Ranges, Technical Report, ERDC TR-01-05, U.S. Army Engineer Research and Development Center, Hanover, NH, 2001.

20. Pennington JC et al., Distribution and Fate of Energetics on DoD Test and Training Ranges: Report 4, ERDC TR-04-4, U.S. Army Engineer Research and Development Center, Environmental Laboratory, Vicksburg, MS, 2004.

21. Hewitt AD et al., Estimates for Explosives Residues from the Detonation of Army Munitions. Technical Report, ERDC/CRREL TR-03-16, U.S. Army Engineering Research and Development Center, Hanover, NH, 2003.

22. Jenkins TF et al., Assessment of sampling error associated with the collection and analysis of soil samples at explosives contaminated sites, *Field Anal. Chem. Technol.*, 1, 151, 1997.

23. Jenkins TF et al., Site Characterization for Explosives at a Military Firing Range Impact Area, U.S. Army CRREL Special Report 98-9, U.S. Army Engineer Research and Development Center, Hanover, NH, 1998.

24. Jenkins TF et al., Representative Sampling for Energetic Compounds at an Antitank Firing Range, Technical Report, ERDC/CRREL TR-04-7, U.S. Army Engineer Research and Development Center, Hanover, NH, 2004.

25. Thiboutot S et al., Environmental Conditions of Surface Soils and Biomass Prevailing in the Training Area at CFB Gagetown, New Brunswick, Technical Report, TR-2003-152, Defence Research and Development Canada, Valcartier, 2003.

26. Thiboutot S et al., Characterization of Antitank Firing Ranges at CFB Valcartier, WATC Wainwright and CFAD Dundurn, Quebec Report DREV-R-9809, Defence Research Establishment Valcartier, 1998.

27. Pennington JC et al., Distribution and Fate of Energetics on DoD Test and Training Ranges: Interim Report 2, ERDC TR 02-8, U.S. Army Engineer Research and Development Center, Vicksburg, MS, 2002.

28. Grant CL, Jenkins TF, and Golden SM, Experimental Assessment of Analytical Holding Times for Nitroaromatic and Nitramine Explosives in Soil, Special Report 93-11, USA Cold Regions Research and Engineering Laboratory, Hanover, NH, 1993.

29. USACHPPM, Training Range Site Characterization and Risk Screening, Camp Shelby, Mississippi, 7–23 September 1999, Geohydrogeological Study No. 38-EH-8879-99, U.S. Army Center for Health Promotion and Preventative Medicine, Aberdeen Proving Ground, MD, 1999.

30. USACHPPM, Training Range Site Characterization and Risk Screening, Regional Range Study, Jefferson Proving Ground, IN, Aug, 2003, U.S. Army Center for Health Promotion and Preventative Medicine, Aberdeen Proving Ground, MD, 2003.

31. Pennington JC et al., Distribution and Fate of Energetics on DoD Test and Training Ranges: Interim Report 3, Technical Report, ERDC TR-03-2, U.S. Army Engineer Research and Development Center, Vicksburg, MS, 2003.

32. Walsh ME et al., Sampling for Explosives Residues at Fort Greely, Alaska. Reconnaissance Visit July 2000, Technical Report, ERDC/CRREL TR-01-15, U.S. Army Engineering Research and Development Center, Cold Regions Research and Engineering Laboratory, Hanover, NH, 2001.

33. Walsh ME et al., Range Characterization Studies at Donnelly Training Area, Alaska 2001 and 2002, ERDC/CRREL TR-04-3, U.S. Army Engineering Research and Development Center, Hanover, NH, 2004.

34. Clausen J et al., A case study of contaminants on military ranges: Camp Edwards, MA, USA, *Environ. Pollut.*, 129, 13, 2004.

35. Jenkins TF et al., Representative sampling for energetic compounds at military training ranges, *Environ. Forensics*, 6, 45, 2005.

36. Hewitt AD, personal communication, 2004.

37. Balbach H, personal communication, 2004.

38. Jenkins TF et al., Coping with spatial heterogeneity effects on sampling and analysis at an HMX-contaminated antitank firing range, *Field Anal. Chem. Technol.*, 3, 19, 1999.

39. Ampleman G et al., Evaluation of the Contamination by Explosives at Cold Lake Air Weapons Range (CLAWR), Alberta, Phase 1 Report, Technical Report TR 2003-208, Defence Research Development Canada–Valcartier, December 2003.

40. Walsh ME, Collins CM, and Racine CH, Persistence of white phosphorus particles in salt marsh sediments, *Environ. Toxicol. Chem.*, 15, 846, 1996.

41. Hays RL, Summers C, and Seitz W, Estimating Wildlife Habitat Variables, FWS/OBS-81/47, USDI Fish and Wildlife Service, Washington, DC, 1981.

42. USFWS, Standards for the Development of Habitat Suitability Index Models for Use in the Habitat Evaluation Procedures, ESM 103, U.S. Fish and Wildlife Service, Division of Ecological Services, Washington, DC, 1981, http://policy.fws.gov/ESMindex.html (accessed August 2008).

43. Steiger JW and Webb RH, Recovery of Perennial Vegetation in Military Target Sites in the Eastern Mojave Desert, Arizona, Open-File Report OF 00-355, U. S. Geological Survey, Washington, DC, 2000.

44. Turner MG, Gardner RH, and O'Neill RV, *Landscape Ecology in Theory and Practice*, Springer, New York, 2001.

45. Noss RF, A regional landscape approach to maintain diversity, *BioScience*, 33, 700, 1983.

46. Andren H, Effects of habitat fragmentation on birds and mammals in landscapes with different proportions of suitable habitat: A review, *Oikos*, 71, 355, 1994.

47. van Dorp D and Opdam P, Effects of patch size, isolation and regional abundance on forest bird communities, *Landscape Ecology*, 1, 59, 1987.

48. Rosenberg DK, Noon BR, and Meslow EC, Biological corridors: Form, function, and efficacy, *Bioscience*, 47, 677, 1997.

49. Dale VH et al., Contributions of spatial information and models to management of rare and declining species, in *Spatial Information for Land Management*, Hill MJ and Aspinall RJ, Eds., Gordon and Breach Science Publishers, Lausanne, Switzerland, 2000, pp. 159–172.

50. Carlsen TM, Cody JD, and Kercher JR, The spatial extent of contaminants and the landscape scale: An analysis of the wildlife, conservation biology, and population modeling literature, *Environ. Toxicol. Chem.*, 23, 798, 2004.

51. Suter GWII et al., *Ecological Risk Assessment for Contaminated Sites*, CRC/Lewis Press, Boca Raton, FL, 2000, chap. 9.

52. Milchunas DG, Schulz KA, and Shaw RB, Plant community structure in relation to long-term disturbance by mechanized military maneuvers in a semiarid region, *Environ. Manage.*, 25, 525, 2000.

53. Prose DV, Persisting effects of armored military maneuvers on some soils of the Mojave Desert, *Environ. Geol. Water Sci.*, 7, 163, 1985.

54. Milchunas DG, Schulz KA, and Shaw RB, Plant community responses to disturbance by mechanized military maneuvers, *J. Environ. Qual.*, 28, 1533, 1999.

55. Shaw RB and Diersing VE, Tracked vehicle impacts on vegetation at the Pinon Canyon maneuver site, CO, *J. Environ. Qual.*, 19, 234, 1990.

56. Dale V et al., Vehicle impacts on the environment at different spatial scales: Observations in west central Georgia, USA, *J. Terramechanics*, 42, 383, 2005.

57. Ayers PD, Environmental damage from tracked vehicle operation, *J. Terramechanics*, 31, 173, 1994.

58. Diersing VE, Shaw RB, and Tazik DJ, U.S. Army Land Condition-Trend Analysis (LCTA) Program, *Environ. Manage.*, 16, 405, 1992.

59. Tazik DJ et al., U.S. Army Land Condition-Trend Analysis (LCTA) Plot Inventory Field Methods, USACERL Technical Report N-92/03, U.S. Army Construction Engineering Research Laboratories, Champaign, IL, 1992.

60. Bern CM, Land Condition-Trend Analysis Installation Report: Yuma Proving Ground Arizona 1991–1992, The Center for Ecological Management of Military Lands, Colorado State University, Fort Collins, CO, 1995.

61. Green T, Land Condition-Trend Analysis Report: Yuma Proving Ground Arizona, 2003 Addendum, U.S. Army Yuma Proving Ground, Yuma, AZ, draft.
62. Cully JF and Winter SL, Evaluation of land condition trend analysis for birds on a Kansas military training site, *Environ. Manage.*, 25, 625, 2000.
63. Ackerman M, Range and Training Land Assessment spring workshops, *The ITAM (Integrated Training Area Management) Bridge*, 26, 1, 2005.
64. Lillesand TM, Kiefer RW, and Chipman JW, *Remote Sensing and Image Interpretation* (5th ed.), John Wiley and Sons, New York, 2003.
65. Underwood E, Ustin S, and Depietro D, Mapping nonnative plants using hyperspectral imagery, *Remote Sens. Env.,* 86, 150, 2003.
66. Graves M, personal communication, 2004.
67. Mann LK et al., The role of soil classification in geographic information system modeling of habitat pattern: Threatened calcareous ecosystems, *Ecosystems*, 2, 524, 1999.
68. Guan BT, Gertner GZ, and Anderson AB, Modeling Training Site Vegetation Coverage Probability with a Random Optimization Procedure: An Artificial Neural Network Approach, USACERL Technical Report 98/83, U.S. Army Corps of Engineers Construction Engineering Research Laboratories, Champaign, IL, 1998.
69. Dale VH, Fortes DT, and Ashwood TL, A landscape transition matrix approach for land management, in *Integrating Landscape Ecology into Natural Resource Management*, Liu J and Taylor W, Eds., Cambridge University Press, New York, 2002, chap. 10.
70. Childress WM, Coldren CL, and McLendon T, Applying a complex general ecosystem model (EDYS) in large-scale land management, *Ecol. Modelling*, 153, 97, 2002.
71. Sullivan PM and Anderson AB, A Methodology for Estimating Army Training and Testing Area Carrying Capacity (ATTACC) Vehicle Severity Factors and Local Condition Factors, ERDC TR-00-2, Engineer Research and Development Center, U.S. Army Corps of Engineers, Champaign, IL, 2000.
72. Leslie M et al., *Conserving Biodiversity on Military Lands: A Handbook for Natural Resource Managers*, The Department of Defense Biodiversity Initiative, Office of Deputy Undersecretary of Defense, Washington, DC, 1996.
73. Mann LK et al., Protection of biota on nonpark public lands: Examples from the U.S. Department of Energy Oak Ridge Reservation, *Environ. Manage.* 20, 207, 1996.
74. Pogue DW and Schnell GD, Effects of agriculture on habitat complexity in a prairie-forest ecotone in the Southern Great Plains of North America, *Agr. Ecosyst. Environ.*, 87, 287, 2001.
75. Krzysik AJ, Ecological Assessment of the Effects of Army Training on a Desert Ecosystem: National Training Center, Fort Irwin, California, Technical Report EN-94/07, U.S. Army Construction Engineering Research Laboratory, Champaign, IL, 1985.
76. Demarais S et al., Disturbance associated with military exercises, in *Ecosystems of Disturbed Ground*, Walker LR, Ed., Elsevier, Amsterdam, 1999, chap. 15.
77. Knouf K, personal communication, 1995.
78. McFarlane DA, Claremont College, personal communication, 2004.
79. Rotstein AH, Military coexists with endangered pronghorn on bombing range, The Associated Press State and Local Wire, http://www.globalsecurity.org/org/news/2003/031019-goldwater-range.htm, October 16, 2003 (accessed August 2008).
80. Peterson MJ, personal communication, 2005.
81. Andow DA, Baker RJ, and Lane CP, Eds., *Karner Blue Butterfly: A Symbol of a Vanishing Landscape,* Minn. Agric. Expt. Sta. Misc. Pub. 84-1994, University of Minnesota, St. Paul, 1994.
82. Wilder T, personal communication, 1998.
83. McDonald E et al., Analysis of Desert Shrubs along First-Order Channels on Desert Piedmonts: Possible Indicators of Ecosystem Condition and Historic Variation, SERDP SEED Project CS1153 Final Technical Report, Desert Research Institute, Reno, NV, 2004.

84. Smith MA, Turner MG, and Rusch DH, The effect of military training activity on eastern lupine and the Karner blue butterfly at Fort McCoy, WI, USA, *Environ. Manage.*, 29, 102, 2002.

85. Imbeau L, Drapeau P, and Mönkkönen M, Are forest birds categorised as "edge species" strictly associated with edges? *Ecography*, 26, 514, 2003.

86. Fisher SG and Woodmansee R, Issue Paper On Ecological Recovery, in *Ecological Risk Assessment Issue Papers*, EPA/630/R-94/009, U.S. Environmental Protection Agency, Washington, DC, 1994, chap. 7.

87. Booth PN et al., Evaluation of Restoration Alternatives for Natural Resources Injured by Oil Spills, Publication 304, American Petroleum Institute, Washington, DC, 1991.

88. Diersing VE et al., A user's guide for estimating allowable use of tracked vehicles on nonwooded military training lands, *J. Soil Water Conserv.*, 43, 191, 1988.

89. Efroymson RA, Nicolette JP, and Suter GWII, A Framework for Net Environmental Benefit Analysis for Remediation or Restoration of Petroleum-Contaminated Sites, ORNL/TM-2003/17, Oak Ridge National Laboratory, Oak Ridge, TN, 2003.

90. Vasek FC, Johnson HB, and Eslinger DH, Effects of pipeline construction on creosote bush scrub vegetation of the Mojave Desert, *Madroño*, 23, 1, 1975.

91. Suter GW II, Luxmoore RJ, and Smith ED, Compacted soil barriers at abandoned landfill sites are likely to fail in the long term, *J. Environ. Qual.*, 22, 217, 1993.

92. Dale V et al., Bioregional planning in Central Georgia, *Futures*, 38, 471, 2006.

93. Gilliam FS and Platt WJ, Effects of long-term fire exclusion on tree species composition and stand structure in an old-growth *Pinus palustris* (longleaf pine) forest, *Plant Ecol.*, 140, 15, 1999.

94. Conner R, Rudolph DC, and Walters JR, *The Red-Cockaded Woodpecker: Surviving in a Fire-Maintained Ecosystem,* University of Texas Press, Austin, 2001.

95. Dale VH, Beyeler, SC, and Jackson B, Understory indicators of anthropogenic disturbance in longleaf pine forests at Fort Benning, GA, USA, *Ecol. Indicators*, 1, 155, 2002.

96. Quist MC et al., Military training effects on terrestrial and aquatic communities on a grassland military installation, *Ecol. Appl.*, 13, 432, 2003.

97. Garten CT, Ashwood TL, and Dale VH, Effect of military training on indicators of soil quality at Fort Benning, GA, *Ecol. Indicators*, 3, 171, 2003.

98. Peacock AD et al., Soil microbial biomass and community composition along an anthropogenic disturbance gradient within a longleaf pine habitat, *Ecol. Indicators*, 1, 113, 2001.

99. Efroymson RA, Nicolette JP, and Suter GW II, A framework for Net Environmental Benefit Analysis for remediation or restoration of contaminated sites, *Environ. Manage.*, 34, 315, 2004.

100. Whicker FW et al., Avoiding destructive remediation at DOE sites, *Science*, 303, 1615, 2004.

101. Landon DM et al., Pronghorn use of areas with varying sound pressure levels, *Southwest. Nat.*, 48, 725, 2003.

102. Schueck LS, Marzluff JM, and Steenhof K, Influence of military activities on raptor abundance and behavior, *Condor*, 103, 606, 2001.

103. Suter GWII, A framework for assessment of ecological risks from multiple activities, *Hum. Ecol. Risk Assess.*, 5, 397, 1999.

104. Doresky J et al., Effects of military activity on reproductive success of red-cockaded woodpeckers, *J. Field Ornithol.*, 72, 305, 2001.

105. Suter GW, Norton SB, and Cormier SM, A methodology for inferring the causes of observed impairments in aquatic ecosystems, *Environ. Toxicol. Chem.*, 21, 1101, 2002.

12 Ecological Risk Assessment of Soil Contamination with Munition Constituents in North America

*Roman G. Kuperman, Ronald T. Checkai,
Mark S. Johnson, Pierre Yves Robidoux,
Bernard Lachance, Sonia Thiboutot,
and Guy Ampleman*

CONTENTS

12.1 Introduction 278
12.2 Ecological Risk Assessment in the United States of America and
     Associated Authorizations for Explosives 279
     12.2.1 CERCLA ERA Framework 279
     12.2.2 The ERAGS ERA Process 280
     12.2.3 USEPA Ecological Risk Management Principles...........................281
     12.2.4 Uniform Federal Policy for Implementing Environmental
            Quality Systems...282
     12.2.5 DoD-Specific Range Assessment Programs...................................283
     12.2.6 Types and Sources of MC on Military Ranges That May Be
            Involved in ERA 285
12.3 Tools for Screening Contaminated Soil Data 286
12.4 Tools for Assessing Risks to Wildlife...294
12.5 Ecological Risk Assessment in Canada..296
     12.5.1 Extent of Soil Contamination with Energetic Materials at
            Canadian DND Installations 298
     12.5.2 Development of Canadian Environmental Sustainability
            Indices (ESI) for Energetic Contaminants at Military Ranges
            and Training Sites...299
12.6 Conclusions and Future Outlook 301
References...303

12.1   INTRODUCTION

The importance of principles of sustainability in the management of natural resources has recently gained recognition at defense installations [1]. Challenging the attainment of the goal of sustainability is a substantially increased demand for training resources, including the corresponding increased utilization of testing and training facilities. Increased training is usually associated with increased environmental impacts at training sites due, in part, to release of munition constituents (MCs), which may jeopardize long-term sustainability of ranges and training sites. Use of explosives, propellants, and heavy metals during testing and training exercises and especially manufacturing have resulted in releases of MCs into the environment. Consequently, soil contamination with explosives and related materials is widespread at many military installations. More than 15 million acres containing elevated levels of explosives and related materials in soil have been identified in the United States, with the estimated costs of assessments and remediation of contaminated sites ranging from $8 billion to $141 billion [2]. Major Army and Air Force training ranges across Canada, including antitank, artillery, grenade, and small-arms ranges have also been extensively investigated to assess the extent of MC contamination of the environment (see Section 12.5). Residual MC has been found in many impacted areas at concentrations up to thousands of mg kg^{-1}. Moreover, mixed contamination consisting of both explosives and heavy metals that resulted from live fire procedures was generally observed at these Canadian military sites.

Contamination at manufacturing, training, and testing sites due to release of explosives may pose significant risks to military personnel, the surrounding environment, and offsite human and ecological receptors through transport of the contaminants outside of the boundaries of the ranges. Determination of the extent of chemical contamination in the environment at U.S. installations is accomplished, for the most part, through the remedial investigation. Information obtained during such investigation is used to characterize impacts of contamination on ecosystems through ecological risk assessment (ERA) procedures. The ERA is an integral part of the remedial investigation and feasibility study (RI/FS) process, which is designed to support risk management decision-making for Superfund sites [3], including those at defense installations. Although available data show that some MC can be persistent and highly mobile in the environment [4–8] (also see Chapter 2), the effects of MC on terrestrial ecological receptors have not been sufficiently investigated (for information on MC toxicity to specific groups of ecological receptors, see Chapter 3 for soil; Chapter 4 for surface water; Chapter 5 for sediment; Chapter 7 for wildlife). This presents a challenge for site managers who are required to assess environmental risks at testing and training installations, distinguish those sites that pose significant environmental risks from those that do not, quantify risk at each site, prioritize contaminated sites by the level of risk posed, and develop appropriate remedial actions and cleanup goals to secure their sustainable use. The availability of MC ecotoxicity data is particularly important at sites where threatened and endangered (T&E) species have been identified or potential for exposure of T&E species to environmental contamination exists.

The need to quantify risk to ecological receptors has provided the impetus for the generation of data and tools to aid the risk assessor. Ecotoxicological media-based benchmarks have been developed that can be used to screen risks from MC in soil. Ecological soil screening levels (Eco-SSLs) are benchmarks that have been developed by the U.S. Environmental Protection Agency (USEPA) in conjunction with stakeholders, consisting of federal, state, consulting, industry, and academic participants. Eco-SSLs have been developed for contaminants frequently found at Superfund sites, which include explosives and propellants and related soil contaminants [9]. Eco-SSLs are defined as concentrations of chemicals in soil that, when not exceeded, are theoretically protective from unacceptable harmful effects for ecological populations or communities that commonly come into contact with soil or ingest biota that live in or on soil. An Eco-SSL value is therefore a one-way test to determine whether concentration of contaminant in soil that fall within specific soil boundary conditions present an acceptable risk to ecological receptors. Exceeding an Eco-SSL value does not indicate that unacceptable risk exists; it rather indicates that additional investigation is warranted.

Terrestrial wildlife movements are such that site-specific tools are more efficiently used to refine exposure estimate. In this case, site-specific exposure estimates are used and compared with safe thresholds for toxicity, termed toxicity reference values (TRVs). Toxicity reference values for wildlife have been developed for energetic compounds. This chapter presents a brief overview of the processes used to establish these tools for ERA for explosives and related soil contaminants that are frequently of potential ecological concern at the affected military sites. This chapter also provides recommendations for use of these values in the ERA process. Investigations addressing the importance and extent of habitat disturbance as a component of the ERA process on explosives-contaminated ranges are reviewed in Chapter 11. General bioaccumulation principles and applications of the bioaccumulation factor and bioconcentration factor (BAF and BCF, respectively) concepts that are often employed in the ERA process to determine bioaccumulation potential of MC for terrestrial receptors are reviewed in Chapter 10.

12.2 ECOLOGICAL RISK ASSESSMENT IN THE UNITED STATES OF AMERICA AND ASSOCIATED AUTHORIZATIONS FOR EXPLOSIVES

12.2.1 CERCLA ERA FRAMEWORK

Ecological risk assessment is an integral part of the RI/FS process, which is designed to support risk management decision-making for Superfund sites. (The Comprehensive Environmental Response, Compensation, and Liability Act of 1980 [CERCLA, or Superfund], as amended by the Superfund Amendments and Reauthorization Act of 1986 [SARA], authorizes the USEPA to protect public health and welfare, and the environment from the release or potential release of any hazardous substance, pollutant, or contaminant.) The RI component of the process characterizes the nature and extent of contamination at a hazardous waste site and estimates risks to human health and the environment posed by contaminants at the site. The FS component of the process develops and evaluates remedial options. Thus, ERA is fundamental to the RI, and ecological considerations are also part of the FS process.

The basic guidelines for carrying out the ERA process for explosives and other materials are defined in the USEPA Guidelines for Ecological Risk Assessment [10] and Ecological Risk Assessment and Risk Management Principles for Superfund Sites [11], plus the Ecological Risk Assessment Guidance for Superfund: Process for Designing and Conducting Ecological Risk Assessments [3], the latter document frequently referred to as Ecological Risk Assessment Guidance for Superfund (ERAGS).

12.2.2 THE ERAGS ERA PROCESS

The ERAGS ERA process is defined within eight steps, with scientific-management decision points (SMDP) at the end of critical stages within the process (schematic of the format shown in Figure 12.1). Step 1 is screening-level problem formulation and toxicity evaluation. In this step, on-site biota (plants and animals) are evaluated and assessment endpoints identified (i.e., the valued ecological attributes of the site that are intended to be protected). A conceptual site model is developed that describes sources of contaminants including MC, and potential exposure pathways by which the assessment endpoints identified may come into contact with these contaminants. Step 1 concludes with the determination of whether the types of contaminants present in their current locations may pose a potential for impacting the ecosystem of the site.

Step 2 is the screening-level exposure estimate and risk calculation. The extent to which representative plants and animals are exposed to a particular explosive is estimated (usually by the maximum concentration), then compared to screening values specific for each explosive (e.g., Eco-SSL values) [9]. Steps 1 and 2 of the ERA constitute the screening level ecological risk assessment (SLERA). The SLERA portion of ERA concludes with an SMDP, which determines whether there are clearly no excesses of ecological risk (i.e., maximum concentrations do not exceed the respective screening value) and no further investigation is warranted, or exposure exceeds the selected screening value and therefore the ERA process should continue in order to better estimate the potential risks to the ecosystem posed by the presence of explosives.

When the SMDP at the end of the SLERA concludes that better estimation of the potential risks to the ecosystem posed by the presence of MC is warranted, proceeding to Steps 3 through 7 and the associated SMDP constitute the baseline ecological risk assessment (BERA). The BERA represents the transition from using preexisting site information in conjunction with published data and preestablished screening values in the SLERA, to the development and evaluation of site-specific data and information in the BERA. This step is specifically focused on determining the nature, magnitude, and extent of risk to ecological receptors posed by the presence of explosives, and the level of uncertainty associated with BERA results and evaluations. The conclusions of the BERA provide the context in which the ERA risk estimates should be evaluated for Step 8, risk management.

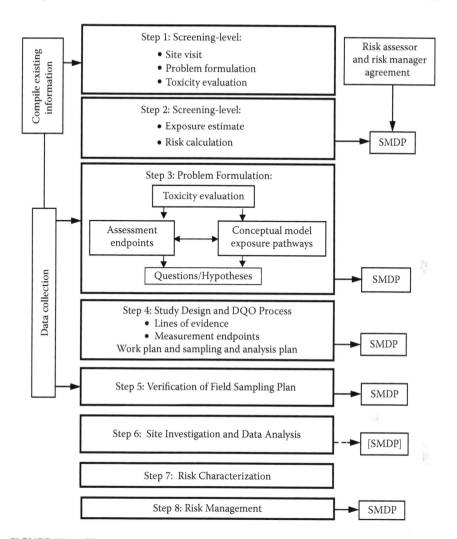

FIGURE 12.1 Eight-step ecological risk assessment process for Superfund.

12.2.3 USEPA Ecological Risk Management Principles

A critical supplement was added to the ERA process for federal facilities with the issuance of USEPA Ecological Risk Assessment and Risk Management Principles for Superfund Sites [11]. It explicitly states that the ability to make sound ecological risk management decisions is dependent upon the quality and extent of information provided in the ERA, plus it provides risk managers with six principles to consider when making their final ecological risk management decisions based on ERA. These six principles call for the following:

1. Selection of response actions that result in the recovery, or maintenance, of healthy local populations/communities of ecological receptors that are (or should be) present at the site; furthermore, in support of proper risk management, risk assessors are advised to select assessment endpoints and measures (as defined in ERAGS) that are ecologically relevant to the site and include species that are exposed and sensitive to response to site contaminants, such as explosives. In addition, if individual threatened or endangered (T&E) species or critical habitats for such species are present at a site, the Endangered Species Act may be declared an applicable or relevant and appropriate requirement (ARAR) within the ERA of the site.

2. Coordination with federal, tribal, and state natural resource trustees in the selection of ERA response actions that not only achieve levels that are protective, but also minimize residual ecological risks at sites.

3. Use of site-specific data to determine whether site releases result in acceptable risks, or to develop quantitative cleanup or management levels that are protective when unacceptable risks exist.

4. Characterization of risk to receptors from contamination by magnitude, severity, geophysical distribution, and potential for recovery.

5. Clear communication by risk managers, in collaboration with ecological risk assessors, to the public of the scientific basis and ecological relevance of the assessment endpoints used in ERA, and the measures used to determine whether risks present are acceptable or unacceptable.

6. Remediation of sites where risks from exposure to contamination outweigh those that may be created through the remediation process (e.g., temporary habitat destruction).

12.2.4 Uniform Federal Policy for Implementing Environmental Quality Systems

An authorization important for ERA, designated the Uniform Federal Policy (UFP) for Implementing Environmental Quality Systems, was recently developed under a memorandum of understanding (MOU) among the U.S. Department of Defense (DoD), USEPA, and the U.S. Department of Energy [12]. This authorization identifies the federal requirements for documenting and implementing acceptable environmental quality systems, to ensure that environmental data are of known and documented quality suitable for their intended uses including ERA. Under this authorization, environmental sampling and testing guidance that includes DoD best practices for data quality oversight of environmental sampling and testing activities has been developed, as well as instructions for preparing required quality assurance project plans (QAPP) for environmental data collection operations and the critical components for chain of custody management. All of these stipulations strengthen the requirements for the application of appropriate methodologies in Steps 3 through 7 of BERA. Moreover, use of standard methods such as those codified by the International Organization for Standardization (ISO) and USEPA are endorsed by guidance developed under the UFP. Such standard methods for explosives include ISO 22478:2006 [13] for analytical determination of concentrations of nitrotoluenes, nitramines, nitrate esters, and related compounds in drinking water, groundwater, or surface water; and USEPA Methods 8330 [14] and the adaptation 8330b [15] that describe analytical requirements for determining by high performance liquid chromatography the concentrations of nitroaromatic and nitramine explosives in water, soil, or sediment matrices.

12.2.5   DoD-Specific Range Assessment Programs

The basic methodologies of the ERA process are being applied in the ongoing Operational Range Assessment Program carried out in accordance with DoD Directive 4715.11, "Environmental and Explosives Safety Management on Operational Ranges Within the United States" [16] and Department of Defense Instruction (DODI) 4715.14, "Operational Range Assessments" [17]. The goal of this program is to support the U.S. military services' management of operational ranges to provide realistic training and testing while protecting human health and the environment. The DoD policy requires that each U.S. military service report to the appropriate regulatory agency any release, or potential threat of a release, of MC from operational ranges that pose an unacceptable risk to human health or the environment.

The U.S. Army is conducting operational range assessments at all Army installations—including Army Reserve and Army National Guard installations—in the United States and its territories. The Army is assessing approximately 424 operational range complexes, with over 11,000 individual ranges and 15 million acres. These assessments follow the Army's Operational Range Assessment Program (ORAP) conducted in two phases: a Qualitative Assessment (Phase I) and a Quantitative Assessment (Phase II). The assessment strategy for Phase I is very similar to a CERCLA (Superfund) Preliminary Assessment, or the strategy defined in the U.S. Navy Range Sustainability Environmental Program Assessment (RSEPA) Policy Implementation Manual [18]. The purpose of the Phase I assessment is to evaluate operational range areas using available data, and determine whether a potential exists for a release or a substantial threat of a release of an MC of concern (MCOC) to an off-range area that may pose an unacceptable risk to human health or the environment. MCOC are defined as those munition constituents that have the potential to migrate from a source area to an off-range receptor (human or ecological), in sufficient quantities to cause an unacceptable risk to human health or the environment.

The results of the Phase I assessment are used to identify the appropriate environmental media for sampling during Phase II investigations. The purpose of the Phase II assessments is sampling and collection of the data necessary to determine whether a potential exists for a release of an MCOC to an off-range area, or that a substantial threat exists for a release that may pose an unacceptable risk to human health or the environment. The sampling strategy in the Phase II assessments is determined by the USEPA Data Quality Objective (DQO) Process [19]. Phase II assessments will be performed on all ranges that are characterized as "Inconclusive" by Phase I. Phase II assessments will place ranges into either the "Unlikely" category or "Referred" category. The strategy for conducting Phase II assessments is based on application of the USEPA DQO systematic planning process. The application of the DQO process ensures that sampling efforts are focused on the appropriate media, and that the appropriate data are collected to determine if there is an off-range release that may pose an unacceptable risk to human health or the environment. Approximately one-third of U.S. Army installations assessed as of August 2007 in a Phase I assessment were categorized as requiring follow-up sampling [20].

In addition to assessments at the operational ranges, investigations are conducted at the Formerly Used Defense Sites (FUDS). The FUDS were transferred from DoD control prior to October 17, 1986. The FUDS program has the mandate to clean-up properties that were formerly owned, leased, possessed, or used by the U.S. Army, Navy, Air Force, or other defense agencies. The U.S. Army Corps of Engineers (USACE) has identified more than 3000 eligible properties in the FUDS inventory that require a response action, including projects within the Hazardous Toxic Radioactive Waste (HTRW) program, the Building Demolition and Debris Removal (BD/DR) program, and the Military Munitions Response Program (MMRP). More than 1300 projects have been identified as part of the MMRP [21].

The U.S. Navy began implementing its range assessment program in 2002. The Navy has separate policies for the assessment of land-based ranges and operational water ranges: the RSEPA Policy Implementation Manual [18] and Navy Policy for Conducting Operational Water Range Assessments, respectively. The Navy RSEPA process encompasses three primary components plus additional protective measures, any of which can be implemented at any time during the process. These include the range condition assessments (RCA), comprehensive range evaluations (CRE), and sustainable range oversight (SRO). Within each part, there are various detailed phases and decision points that must be followed.

The U.S. Marine Corps ensures range availability to future generations of Marines by establishing the Marine Corps Sustainable Ranges Program [22]. A key component of this program is the Range Environmental Vulnerability Assessment (REVA) Program. The REVA was developed to investigate and analyze installation and operational range encroachment relating to contamination with MC. The REVA environmental assessment focuses on MC migration that could create encroachment pressures on ranges, and assists installation and range managers in formulating encroachment control strategies and environmental investment. The scope of the REVA program includes Marine Corps operational ranges in the United States and overseas. Operational ranges include, but are not limited to, fixed ranges, live-fire maneuver areas, small-arms ranges, buffer areas, and training areas, where military munitions are known or suspected to have been used. Operational ranges exclusively used for small-arms training are being evaluated qualitatively under REVA. Data pertaining to water ranges will be collected, but the specific assessment processes for water ranges have not yet been defined.

U.S. Air Force (USAF) conducts range assessments under ORAP, which is part of the USAF Operational Range Environmental Program. The goal of USAF implementation of the ORAP program is to ensure that the natural resource infrastructures of operational ranges are capable and available to support the USAF test and training mission. It also provides USAF facilities with guidance for consistently completing a defensible assessment of potential environmental impacts to off-range receptors from military munitions used on operational ranges and range complexes. The USAF prioritized assessments of all of its air-to-ground ranges prior to other types of ranges, such as small-arms ranges (SARs), grenade and mortar ranges, and demolition and explosive ordnance disposal (EOD) ranges. The USAF air-to-ground ranges constitute the majority of the USAF operational range area. One of the reasons for prioritizing assessments of air-to-ground ranges is the size and remoteness of some of these ranges; large air-to-ground ranges in remote locations are less likely to receive the intensity of management controls that are maintained at other USAF installations. The ORAP was developed with the purpose of evaluating potential release of MC from on-range source areas to off-range receptors with the focus on human health concerns; ecological receptors were not evaluated. The USAF is, however, planning on revising its implementation of ORAP in the future to incorporate evaluation of ecological receptors in their assessments.

12.2.6 Types and Sources of MC on Military Ranges That May Be Involved in ERA

MC found primarily at impact areas and firing lines of U.S. military testing and training ranges often consist of mixtures of residues from several MC. These may include the nitroaromatic explosive compounds 2,4,6-trinitrotoluene (TNT), 2,4- and 2,6-dinitrotoluene (DNT), and trinitrophenylmethylnitramine (tetryl); nitrate esters such as nitrocellulose (NC), pentaerythritol tetranitrate (PETN), and nitroglycerin (NG); and the nitramine compounds, hexahydro-1,3,5-trinitro-1,3,5-triazine (RDX) and octahydro-1,3,5,7-tetranitro-1,3,5,7-tetrazocine (HMX). Most of these MC have been in use for decades, either as primary or secondary explosives or in propellant compositions.

Because of its worldwide use as both a military and industrial explosive, TNT is well characterized as an explosive. DNT, a byproduct of TNT production, has been used as a component of Composition C series explosives; particularly in C3, which is composed of RDX (77%), DNT (10%), MNT (mononitrotoluene 5%), TNT (4%), tetryl (3%), and NC (1%). While C3 is no longer being used as a gun projectile main charge, some stocks may still be in service. Tetryl as well as 2,4-DNT have been used in a number of explosive formulations. Besides being used as an ingredient in the Composition C series, tetryl is also the primary explosive in tetrylols (80% tetryl, 20% TNT); tetrylols were used as a base charge in detonators, but since 1980 tetryl has largely been replaced by RDX. RDX is most often used in mixtures with other MC or plasticizers, and is present in many common military explosives, such as Composition A (RDX, plus wax), Composition B (59.5% RDX, 39.5% TNT, 1% wax), and Composition C (RDX, plus plasticizer). HMX is used in nuclear devices, plastic explosives, rocket fuels, as a modifier for propellant burn rate, and in burster chargers. It is the main explosive in the formulation octol (68% HMX, 30% TNT, and 2% RDX) mostly used for the shoulder-launched antitank rockets. HMX is also formed as a byproduct in the manufacture of RDX (85:15), so it also appears as a co-contaminant with RDX. The inverse situation is also true as the process for the production of military grade HMX always leads to percentage levels of RDX as an impurity in the HMX. Nitrocellulose is a nitrate ester of cellulose, and is used in a number of military compositions especially in single-, double-, and triple-base gun propellants (e.g., the Composition C series). PETN is the most stable and least reactive of the explosive nitric esters, and one of the most powerful. PETN is used in high-efficiency detonators, detonating cords, as bursting charges in small-caliber ammunition, in land mines, and some plastic explosives.

12.3   TOOLS FOR SCREENING CONTAMINATED SOIL DATA

Screening of contaminated site soil data through the use of conservative criteria is one of the components of ERA that is conducted early in the stages of risk assessment (Figure 12.1). The primary goal of the SLERA is to shorten a long list of chemicals of potential concern to a smaller list of on-site contaminants of potential concern (COPC) on which the risk assessment will focus. Integral to an effective screening process is the availability of ecotoxicological data for the COPCs present in the soils at an investigated site. Assessment of soils at contaminated sites, including defense installations, during the SLERA is being advanced by the development of Eco-SSLs [9]. These Eco-SSL values are intended for use during Step 2 of the SLERA to iden-tify those contaminants in soil that are not of potential ecological concern, so they may be eliminated from the subsequent evaluation in a BERA.

The Eco-SSL values are derived for four groups of ecological receptors, includ-ing terrestrial plants, soil invertebrates, birds, and mammals, using published data generated from laboratory toxicity tests with different test species relevant to soil ecosystems (Figure 12.2). The Eco-SSL values have intentionally been developed to be conservative to provide confidence that contaminants that potentially present an unacceptable risk are not screened out during the SLERA process. This conser-vative nature of Eco-SSL is achieved by selecting data established in studies with upland aerobic soil types that have properties supporting high relative bioavailability of COPCs to ecologically relevant test species, using the more sensitive reproduction (for soil invertebrates) or growth (for terrestrial plants) ecotoxicological benchmarks, and by relying on the EC_{20} levels of the effect (20% reduction from controls) deter-mined in toxicity tests with analytically quantified multiple-exposure concentrations.

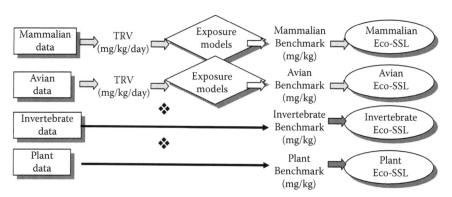

FIGURE 12.2   Derivation of Eco-SSL. ❖: Denotes directly calculated benchmarks estab-lished in definitive toxicity tests with soil invertebrates or terrestrial plants.

The Eco-SSL Workgroup, after an extensive literature review, conducted prior to release of the initial draft of the Eco-SSL Guidance in 2000 (the updated draft was released in March 2005 [9]), determined that there was insufficient information for explosives and related compounds to generate Eco-SSL values for terrestrial plants and soil invertebrates. This determination gave impetus to several ecotoxicological investigations to fill the knowledge gaps. The toxicity data developed in these investigations, including those designed to meet specific criteria [9] for establishing benchmark data for the development of Eco-SSL values for explosives and related compounds are reviewed in Chapter 3.

Sustainable use of military installations for training and testing requires the development of environmental quality criteria that can be consistently applied in different countries to gauge the ecotoxicological impacts of these operations. To overcome jurisdictional limitations of Eco-SSLs developed for U.S. Superfund sites and to make the Eco-SSL scientific approach useful internationally while preserving its core stipulations, we applied a similar methodology for developing a set of ecotoxicological tolerance values (ETVs) for several MC soil contaminants that are frequently of ecological concern at impacted military sites. The ETVs are scientifically based ecotoxicological benchmark concentrations of MC in environmental media that are not likely to exceed physiological tolerance of specific ecological populations that commonly come into contact with those environmental media or ingest biota that live in those environmental media. Thus, ETVs are theoretically protective of ecosystems from unacceptable harmful effects. Although this chapter focuses on the development of ETVs for terrestrial plants and soil invertebrates, a similar approach can be applied to developing ETVs for aquatic and sediment groups of ecological receptors.

For the purposes of developing the proposed ETVs, we selected ecotoxicological data for terrestrial plants and soil invertebrates from the list of toxicity data presented in Tables 3.2 and 3.3 of Chapter 3. These toxicity data were selected because they were established in standardized toxicity tests that were designed to comply with USEPA requirements for developing toxicity benchmarks intended for deriving Eco-SSL values (see Ref. 9 for a complete list of requirements). The selected data were developed in studies utilizing a natural aerobic upland soil, Sassafras sandy loam (SSL), which had physical and chemical characteristics supporting relatively high bioavailability of MC, including low organic matter (1.2%) and clay (17%) contents [23–31]. Furthermore, except for testing performed by Gong et al. [23], these studies included the exposures of test species to MC that were weathered and aged in soil prior to toxicity testing, in addition to corresponding separate exposures in soil freshly amended with MC. This is important because weathering and aging of MC in soil might alter the exposure of terrestrial plants and soil invertebrates because of a variety of chemical fate processes that commonly occur in soils at contaminated sites (see Chapter 2 for information on the fate and biotransformation of explosives in soil). The inclusion of weathering and aging of contaminant MC in soil in those studies more closely approximated the exposure of terrestrial receptors in the field and complied with a requirement for Eco-SSL development by USEPA [9], which was specifically undertaken for utilization at Superfund sites (locations where contaminants have been historically present). The inclusion of such a weathering-and-aging procedure also allowed the researchers to incorporate potential alterations in MC bioavailability, formation of MC transformation and degradation products, and corresponding changes in toxicity at contaminated sites into the development of toxicological benchmarks for terrestrial plants and soil invertebrates (see Chapter 3 for examples of toxicity alterations following weathering and aging of MC in soil).

We applied the selected published ecotoxicological data for terrestrial plants and soil invertebrates to developing ETVs that can be proposed as screening concentrations for the nitramine RDX; HMX; 2,4,6,8,10,12-hexanitro-2,4,6,8,10,12-hexaazaisowurtzitane (CL-20); the nitroaromatic TNT; 2,4-DNT; 2,6-DNT; and 1,3,5-trinitrobenzene (TNB) in aerobic soils. The effects of RDX, HMX, TNT, 2,4-DNT, 2,6-DNT, and TNB on the plant species alfalfa (*Medicago sativa* L.), Japanese millet (*Echinochloa crusgalli* L.), and perennial ryegrass (*Lolium perenne* L.) were determined in standardized toxicity tests [32] by Rocheleau et al. [28,29]. Similarly designed studies by Gong et al. [23] assessed the effects of CL-20 on alfalfa and perennial ryegrass in freshly amended SSL soil. The selected test species included a dicotyledonous symbiotic species (alfalfa), and two monocotyledonous species (Japanese millet and ryegrass). Thus, the selected toxicity data for benchmark derivation used representative surrogates of species that inhabit a wide range of site soils and geographical areas (i.e., are ecologically relevant). Plant growth endpoints (fresh and dry shoot mass), which were the more sensitive than seedling emergence endpoints, were used for derivation of ETVs for terrestrial plants. The selected toxicity benchmark values for plants are summarized in Table 12.1 and Table 12.2.

TABLE 12.1

Growth (Shoot Fresh Mass) Toxicity Benchmarks (EC_{20} mg kg^{-1}) for Munition Constituents (MC) Freshly Amended or Weathered and Aged in Sassafras Sandy Loam Soil for Selected Test Species

MC	Alfalfa (*Medicago sativa*)	Japanese Millet (*Echinochloa crusgalli*)	Perennial Ryegrass (*Lolium perenne*)
RDX [29]	>9740/>9537[a]	>9740/>9537[a]	>9740/>9537[a]
HMX [29]	>10411/>9341[a]	>10411/>9341[a]	>10411/>9341[a]
CL-20 [23]	>9832/NT	NT	>9832/NT
TNT	41/3[b]	33/6	94/15
2,4-DNT	11/7	3.5/3.5	11/5[b]
2,6-DNT	1.3/1.6	13/4.8[b]	18/24
TNB	38/20	16/0.3[b]	45/46

Notes: Values listed as fresh/weathered and aged exposure treatments. All values (mg kg^{-1}, dry soil mass) are based on acetonitrile extractable concentrations in soil (USEPA Method 8330). NT = not tested.

[a] Greatest concentrations tested.

[b] Statistically significant effect of weathering and aging of munition constituents in soil on toxicity based on 95% confidence intervals.

TABLE 12.2

Growth (Shoot Dry Mass) Toxicity Benchmarks (EC$_{20}$ mg kg^{-1}) for Munition Constituents (MC) Freshly Amended or Weathered and Aged in Sassafras Sandy Loam Soil for Selected Test Species

MC	Alfalfa (*Medicago sativa*)	Japanese Millet (*Echinochloa crusgalli*)	Perennial Ryegrass (*Lolium perenne*)
RDX [29]	>9740/>9537[a]	>9740/>9537[a]	>9740/>9537[a]
HMX [29]	>10411/>9341[a]	>10411/>9341[a]	>10411/>9341[a]
CL-20 [23]	>9832/NT	NT	>9832/NT
TNT	43/1.4[b]	56/11	61/13
2,4-DNT	34/15	25/6[b]	11/2[b]
2,6-DNT	2.8/0.4	11/6	26/21
TNB	62/46	43/0.7[b]	56/51

Notes: Values listed as fresh/weathered and aged exposure treatments. All values (mg kg^{-1}, dry soil mass) are based on acetonitrile extractable concentrations in soil (USEPA Method 8330). NT = not tested.

[a] Greatest concentrations tested.

[b] Statistically significant effect of weathering and aging of munition constituents in soil on toxicity based on 95% confidence intervals.

The ecotoxicological data selected for developing soil invertebrate-based ETVs that can be proposed as screening concentrations for RDX, HMX, CL-20, TNT, 2,4-DNT, 2,6-DNT, and TNB are summarized in Table 12.3. These toxicity benchmark values were established in definitive tests with the soil invertebrates earthworm *Eisenia fetida* [33], potworm *Enchytraeus crypticus* [34], and collembola *Folsomia candida* [35]. These species are representative surrogates of species that normally inhabit a wide range of site soils and geographical areas. Reproduction measurement endpoints in these tests were more sensitive compared with adult survival [24–28,31] and were consequently used for derivation of proposed soil invertebrate-based ETVs. These endpoints included cocoon production and juvenile production for earthworms, and juvenile production for potworms and collembola.

The proposed ETV for an MC–receptor pairing (e.g., CL-20–invertebrates) was calculated as the geometric mean of EC$_{20}$ benchmark toxicity values determined from the individual studies. Following the Eco-SSL derivation methodology [9], at least three toxicity data values generated under specified conditions were required to calculate an ETV. An example of the derivation process for the proposed soil invertebrate-based ETV for CL-20 is shown in Table 12.4.

The proposed ETVs developed from toxicity benchmark data for terrestrial plants and soil invertebrates are shown in Table 12.5 and Table 12.6, respectively. Separate ETVs were derived for freshly amended MC (comparable to new releases of MC into the soil environment) and for MC that were weathered and aged in soils (comparable to MC released into the environment long term). Results of toxicity tests with individual MC weathered and aged in soil (Tables 12.1 and 12.2) showed

TABLE 12.3

Reproduction Toxicity Benchmarks (EC_{20} mg kg^{-1}) for Munition Constituents (MC) Freshly Amended or Weathered and Aged in Sassafras Sandy Loam Soil for Selected Test Species

	Earthworm (*Eisenia fetida*)		Potworm (*Enchytraeus crypticus*)	Collembolan (*Folsomia candida*)
MC	Cocoon Production	Juvenile Production	Juvenile Production	Juvenile Production
RDX	1.2/19	1.6/5.0	3700/8800	28/113
HMX	2.7/>600[a]	0.4/>600[a]	>21,750/>17,500[a]	1,279[b]/1,046
CL-20	0.05/0.04	0.04/0.04	0.1/0.035[c]	0.23/0.95
TNT	38/4[c]	33/3[c]	77/37[c]	17/53[c]
2,4-DNT	31/25	44/29	19/14	10/15
2,6-DNT	14/16	9/8	37/18[c]	6/0.96
TNB	27/19	21/13	5/9	4/48[c]

Notes: Values listed as fresh/weathered and aged exposure treatments. All values (mg kg^{-1}, dry soil mass) are based on acetonitrile extractable concentrations in soil (USEPA Method 8330).

[a] Greatest concentrations tested.

[b] The EC_{20} value of 235 mg kg^{-1} was originally reported in Ref. 28. It was later revised to 1279 mg kg^{-1} by Phillips et al. [71].

[c] Statistically significant effect of weathering and aging of munition constituents in soil on toxicity based on 95% confidence intervals.

TABLE 12.4

Derivation of Proposed Soil Invertebrate-Based Ecotoxicological Tolerance Values (ETVs) for CL-20 in Freshly Amended Sassafras Sandy Loam Soil as the Geometric Mean of Reproduction Benchmarks for Selected Species

Receptor Group	EC_{20} (mg kg^{-1})	95% Confidence Intervals (mg kg^{-1})	Proposed ETV (mg kg^{-1})
Earthworm *Eisenia fetida* (cocoon production)	0.05	0.03–0.07	
Earthworm *Eisenia fetida* (juvenile production)	0.04	0.02–0.06	0.08
Potworm *Enchytraeus crypticus* (juvenile production)	0.1	0.06–0.13	
Collembola *Folsomia candida* (juvenile production)	0.23	0.12–0.34	

Note: All values (mg kg^{-1}, dry soil mass) are based on acetonitrile extractable concentrations of CL-20 in soil (USEPA Method 8330).

TABLE 12.5

Proposed Terrestrial Plant-Based Ecotoxicological Tolerance Values for Munition Constituents Determined Using Ecotoxicological Benchmarks Established in Similar Individual Definitive Toxicity Tests with Sassafras Sandy Loam Soil

Exposure Type	RDX (mg kg^{-1})	HMX (mg kg^{-1})	CL-20 (mg kg^{-1})	TNT (mg kg^{-1})	2,4-DNT (mg kg^{-1})	2,6-DNT (mg kg^{-1})	TNB (mg kg^{-1})
Freshly amended treatment	>9740, not toxic	>10411, not toxic	>9832, not toxic	52	13	8	40
Weathered-and-aged treatment	>9537, not toxic	>9341, not toxic	ND	6	6	5	9

Notes: Ecotoxicological tolerance values are based on the EC$_{20}$ values for growth toxicity endpoints (shoot dry mass) reported in Tables 12.1 and 12.2. ND = not determined (no data available).

TABLE 12.6

Proposed Soil Invertebrate-Based Ecotoxicological Tolerance Values for Munition Constituents Determined Using Ecotoxicological Benchmarks Established in Similar Individual Definitive Toxicity Tests with Sassafras Sandy Loam Soil

Exposure Type	RDX (mg kg^{-1})	HMX (mg kg^{-1})	CL-20 (mg kg^{-1})	TNT (mg kg^{-1})	2,4-DNT (mg kg^{-1})	2,6-DNT (mg kg^{-1})	TNB (mg kg^{-1})
Freshly amended treatment	21	6	0.08	36	23	13	10
Weathered-and-aged treatment	98	>600,	0.08	12	20	7	18

Note: Ecotoxicological tolerance values are based on the EC$_{20}$ values for reproduction toxicity endpoints reported in Table 12.3.

significantly increased toxicity (based on 95% confidence intervals, CI) of 2,4-DNT, 2,6-DNT, or TNB for Japanese millet; of 2,4-DNT for ryegrass; and of TNT for alfalfa and ryegrass, compared with respective toxicities in freshly amended soils [30]. Weathering and aging of individual MC in soil also affected the respective toxicities to soil invertebrates (Table 12.3). Toxicity of TNT was significantly increased (95% CI) for earthworms and potworms, and of 2,6-DNT or CL-20 for potworms, compared with respective toxicities in freshly amended soils [25–28]. In contrast, the toxicities of TNT or TNB for collembola have significantly decreased (95% CI) following weathering and aging of individual MC in soil, compared with respective toxicities in freshly amended soils [28]. These alterations in respective MC toxicities for terrestrial plants and soil invertebrates following weathering and aging in soil indicated strongly that the soil chemical environment in the experimental treatments was changing over time, similar to changes that can occur to chemical sorption and transformation in the vadose zone of the soil environments in the field. The inclusion of toxicity data for MC weathered and aged in soils in the derivation of the proposed ETVs ensures that the resulting screening values adequately represent the exposure conditions for terrestrial plants and soil invertebrates in the field.

The cyclic nitramine explosives RDX, HMX, and CL-20 did not adversely affect plants at concentrations tested; hence no plant-based ETVs could be developed for these MC. Similarly, soil invertebrate-based ETV could not be developed for HMX weathered and aged in amended soil, because the EC_{20} value was established only for one test species, *F. candida*. HMX did not adversely affect reproduction of either earthworms or potworms in weathered and aged treatments at concentrations tested. The greatest concentrations of cyclic nitramine explosives that were tested in these studies that produced no adverse effects on the test species are shown in Table 12.5 and Table 12.6 as reference values only; these values should not be considered ETVs for the respective explosives.

A review of ecotoxicological benchmark values used in this chapter for developing ETVs shows that although the majority of values were fairly uniform among test species, there were instances of high variability among EC_{20} estimates determined in toxicity bioassays. The greatest variability among EC_{20} values for plant growth was found in studies with 2,6-DNT for alfalfa and ryegrass, which ranged from 1.3 mg kg^{-1} for alfalfa fresh shoot mass to 26 mg kg^{-1} for ryegrass dry shoot mass in studies with freshly amended soil, and from 0.4 mg kg^{-1} for alfalfa dry shoot mass to 24 mg kg^{-1} for ryegrass fresh shoot mass in studies with 2,6-DNT weathered and aged in soil [30]. An even greater contrast between EC_{20} values for growth was found for alfalfa and Japanese millet exposed to TNB weathered and aged in soil. These values ranged from 46 mg kg^{-1} for alfalfa dry shoot mass to 0.3 mg kg^{-1} for Japanese millet fresh shoot mass [30]. Similar variability was found in the MC toxicity assessments with soil invertebrates exposed to RDX or HMX [24,28,31]. The greatest differences in toxicity benchmark values for phylogenetically related organisms were found between earthworm and potworm for RDX effects on reproduction endpoints, which ranged from 1.2 mg kg^{-1} for earthworm to 3700 mg kg^{-1} for potworm in studies with freshly amended soil; and from 5 to 8800 mg kg^{-1} for earthworm and potworm, respectively, in studies with RDX weathered and aged in soil. Large differences were also found between benchmark values for juvenile production by earthworm and collembola exposed to HMX, which ranged from 0.4 mg kg^{-1} for earthworm to 1279 mg kg^{-1} for collembola in studies with freshly amended soil, while no adverse effect was evident for reproduction of potworm up to and including 21,750 mg kg^{-1}, the highest HMX concentration tested. These examples of species-specific variations in toxicity data provide clear evidence in support of the use of multiple species for generating ecotoxicological benchmarks for ETV development.

The proposed ETVs can be applied to ERA at sites where terrestrial receptors may be exposed directly or indirectly to MC-contaminated soil. The selected species represent a spectrum of diverse ecological functions in soil communities: primary producers and different functional groups of soil invertebrates. The invertebrate species used in those investigations actively move through soil, thus ensuring contact with contaminants. Both terrestrial plant and soil invertebrate species tested were sensitive to a wide range of contaminants, and reflected different routes of exposure (e.g., ingestion, inhalation, dermal absorption for soil invertebrates, and uptake from soil solution for plants). All exposures were conducted under conditions of relatively high MC bioavailability in SSL soil. By deriving ETV soil screening concentrations that are protective of these receptor groups for use during ERA process, in conjunction with screening values to be developed for avian and mammalian wildlife, it is assumed that the terrestrial ecosystems will be protected from possible adverse effects associated with exposures to MC-contaminated soil.

Soil physical and chemical properties can affect the exposure of terrestrial plants and soil invertebrates to contaminants in soils (Chapter 3). Following the Eco-SSL requirements [9], the proposed ETVs may be applicable for ERA purposes in all sites where key soil constituents fall within a certain range of chemical and physical parameters. They apply to the majority of upland aerobic soils: where the pH is greater than or equal to 4.0 and less than or equal to 8.5, and the organic matter content is less than or equal to 10%. The toxicity benchmark data selected for ETV development in this chapter were established in studies using SSL soil that has characteristics supporting a relatively high bioavailability of MC, with physicochemical characteristics within the aforementioned boundaries for the specified soil properties. This ensures that the proposed ETVs for soil receptors are adequately protective for a broad range of soils within the boundary conditions described earlier.

MC can affect populations of soil organisms in different ways. These include (1) direct acute toxicity, (2) chronic toxicity such as effects on growth or reproduction, (3) indirect toxicity by altering soil structure or fertility, (4) indirect toxicity by adversely affecting nutrient and food supplies, or (5) by affecting predators and parasites. In addition, soil organisms may alter their environment changing the overall bioavailability of chemicals within the soil. No single ecotoxicological test can address all these types of effects, thus a battery of tests is required to reasonably do so. Inclusion of species from different taxonomic groups representing a range of sensitivities, which often correlate with physiologically determined modes of action and can vary among taxa, was an important consideration for selecting the test battery for establishing benchmark toxicity data for ETV development. The selected toxicity tests with representative test species have been standardized and generate reproducible, statistically valid results, which imparts a greater confidence in the data, thus generates less uncertainty associated with the decisions and recommendations that are based on the test data, including development of the proposed ETVs.

The ETVs proposed in this chapter are being developed as part of the activities of the Key Technical Area (KTA 4-32) "Development of Environmental Tolerance Values for Defense Sites Contaminated with Energetic Materials" of The Technical Cooperation Program (TTCP; www.dtic.mil/ttcp/), Weapons Technical Panel (WPN TP-4). TTCP is an international organization that collaborates in defense scientific and technical information exchange; program harmonization and alignment; and shared research activities for Australia, Canada, New Zealand, the United Kingdom, and the United States. The proposed ETVs and reports detailing their development will be provided to the WPN TP-4 for review and approval before acceptance as ETVs for RDX, HMX, CL-20, TNT, 2,4-DNT, 2,6-DNT, and TNB, and before being transitioned internationally to regulating agencies and defense organizations of Australia, Canada, New Zealand, and United Kingdom through TTCP.

12.4 TOOLS FOR ASSESSING RISKS TO WILDLIFE

With few exceptions, the chemical/physical properties of MC are such that oral exposures would constitute the predominant exposure pathway to wildlife predominantly through ingestion. Nitroaromatics, however, do show a proficiency to cross the integument [36–38]; however, dermal absorption is highly dependent upon integument type and thickness, and no models currently exist for wildlife species. In situ microcosm exposures have been conducted with salamander species in a soil matrix that may represent a worst-case exposure scenario that is likely protective of other wildlife species [37,39].

Assuming the oral route is the predominant exposure pathway, ingestion and food web models are used to calculate daily ingestion exposure (in milligram MC per kilogram body per day). Few explosives have been shown to bioaccumulate to any substantial extent in food items (e.g., animals or plants in the field; see Chapter 10), and often the predominant exposure to wildlife in modeled efforts occurs through inadvertent soil ingestion when assumptions regarding soil ingestion are made [40,41]. This point highlights the need for empirical data to describe the magnitude of soil ingestion for wildlife species.

Estimating oral exposure for wildlife species, particularly for MC, is not straightforward. Mammals and birds species can experience exposures from a wide range of locations, and MC residues at sites can be very heterogeneous [42,43]. Using a single soil concentration to represent a site such as a firing range would likely produce unreasonably inaccurate exposure estimates. Spatial exposure models have the potential to provide more accurate estimates [44–47]. One such model developed for wildlife is the Spatially Explicit Exposure Model (SEEM), and it is available for general use (http://chppm-www.apgea.army.mil/tox/herp.aspx). Another used for estimating risk to aquatic organisms in a spatial context is FISHRAND [48]. These models provide exposure estimates for a user-defined number of individuals and time periods (e.g., a season) that are compared with a toxicity reference value (TRV) or residue body burden to calculate a range of risk estimates in a population context.

Once exposure estimates are calculated, they are compared with the level of exposure deemed safe based on a review of the toxicity data. This level is often termed the TRV. There are sets of TRVs available from various sources, developed using various techniques. Examples have been developed by Oak Ridge National Laboratory and U.S. Army Center for Health Promotion and Preventive Medicine (USACHPPM), and these TRVs are available at their Web sites (www.esd.ornl.gov/programs/ecorisk/ tools.html and http://chppm-www.apgea.army.mil/tox/HERP.aspx, respectively). The most robust methods incorporate the results and parameters of a literature review, toxicity profile, and derivation rationale [49]. TRVs for many MC compounds are available within each chemical-specific Wildlife Toxicity Assessment (WTA) document via the Internet (see http://chppm-www.apgea.army.mil/tox/HERP.aspx).

Use of TRVs, which are dose-based effects values, requires modeling exposure based on soil and food EM concentrations to estimate doses. Eco-SSLs represent an alternate screening tool for wildlife that incorporates both exposure and TRVs in order to back-calculate screening-level soil concentrations (Figure 12.2). Eco-SSLs are, therefore, more readily applied within screening-level risk assessment. Mammalian- and avian-based Eco-SSL values for MC are not currently available; however, such values are planned for both TNT and RDX.

No generally accepted method for TRV derivation exists, though there are some general principles. As mentioned earlier, documentation of a robust literature review is necessary, as are transparent methods in the presentation of the toxicity information and subsequent derivation of the TRV. Toxicity benchmark value derivation for wildlife is complex in that (1) data for wildlife are rare, as laboratory models are most often used, and (2) many health-related endpoints are often inconsistently evaluated. This latter point often leads to various interpretations of the toxicity data. Issues surrounding the relevance to wildlife populations of the elevation of liver enzymes, electrolytes, or histological evaluations, for example, can be challenging and often no clear answers exist. Laboratory toxicity studies are carried out under controlled experimental conditions to evaluate the health effects occurring from exposure; however, effects that are important in an ecological context (e.g., predator avoidance behaviors, mating behaviors, foraging preferences) are rarely investigated. Therefore, it is imperative that risk assessors integrate the assistance of qualified toxicologists to use a weight-of-evidence procedure when evaluating toxic effects from exposure and determining how those data should be interpreted in TRV derivation.

Many other tools for estimating fate and transport, natural attenuation, exposure, and risk estimation, have been integrated into the Applied Risk Assessment Modeling System (ARAMS). As with the aforementioned models, ARAMS models are available as freeware on the Web site http://el.erdc.usace.army.mil/arams/. The ARAMS system is extensive, and builds on various models and databases that are contained within the system, and also pulls information from databases in real time from the Internet. These data are handled within the ARAMS system to provide reports. User-guided interfaces provide the means for operation; however, some models may require additional specific expertise. Models such as SEEM and FISHRAND can work independently or within the ARAMS system.

Any ERA risk estimate should be provided with an analysis of uncertainty. The magnitude of the uncertainty should be ascertained before decisions regarding risk management or remediation are made. The risk manager should be provided with other criteria important to the maintenance of wildlife populations during the feasibility study stage, including habitat type and relative quality, plus the impact remediation may have on the existing habitat. These factors (i.e., the magnitude of uncertainty associated with risk estimates, alterations of habitat that remediation alternatives may create, and relevance of effects from wildlife exposure to MC soil contamination) should all be considered and weighed in the feasibility study. When these factors are placed in their proper context, risk managers can then select the best course of action to balance wildlife exposure and effects of MC environmental exposure, relative to habitat changes that any remediation alternatives may create.

12.5   ECOLOGICAL RISK ASSESSMENT IN CANADA

Application of ERA at the Canadian Department of National Defense (DND) installations is limited to just a few studies [50] and was mainly used to assess the effects of specific contaminants (MC and metals) [50–52], develop ecotoxicological tools for quantifying impacts at the contaminated sites, or to validate mesocosm and biomarker approaches [53–55]. ERAs at DND installations are also conducted to protect biological diversity, to determine the maximum level of contamination that allows an acceptable protection level, and to compare remediation technologies. Published benchmark values or guidelines for MC-contaminated soil, which could be used to manage the future use of decommissioned military training sites, are not available in Canada. Soil guidelines are currently being developed to improve site assessments and to complement the results of chemical analyses at contaminated soils. Preparation of these guidelines involves the development of an ecotoxicological database that can be used for deriving environmental threshold criteria for RDX, HMX, and TNT; and more recently, for NG and 2,4-DNT. This database is intended for use by site managers involved in site characterization (chemical determinations, ecotoxicological hazard assessment), and in environmental risk assessment and management. These issues are important to the Canadian DND, as it relies on the knowledge of ecotoxicity of explosives to evaluate the impacts of defense-related activities on the environment and on human health, and to structure operational activities in a manner that ensures sustainable testing and training and the protection of the health of the military users of training ranges.

A limited number of generic criteria for soil protection and rehabilitation have been published for TNT-contaminated sites by the Quebec government, but those were developed only on the basis of human health data [56]. Canadian human health risk-based criteria for explosives are usually very conservative (below the MC analytical detection limits) and are designed to develop threshold criteria for residential, agricultural, and industrial uses; thus these criteria have limited relevance for ERA at the Canadian DND installations. Therefore, it is critical for the Canadian DND to rely on the derivation of environmental soil quality guidelines for explosives that are based on basic ecotoxicological research of explosive compounds and related MC found at testing and training ranges.

The Ministère du Développement durable, de l'Environnement et des Parcs (Department of Durable Development, Environment and Parks) of Quebec provides three levels of generic criteria (www.mddep.gouv.qc.ca/sol/terrains/politique/annexe_2_tableau_1. htm) for different substances: background levels/quantification limit (Criteria A); maximum acceptable limit for residential, recreational, and institutional sites (Criteria B); and maximum acceptable limit for commercial sites not located in a residential area and for industrial sites (Criteria C). Criteria C for soils are 0.7 mg kg^{-1} for TNT and 0.03 mg kg^{-1} for 2,6- DNT; whereas the respective Criteria C for groundwater are 120, 910, and 930 mg kg^{-1} for TNT, 2,4-DNT, and 2,6-DNT. In addition, surface water TNT criteria have been published by the Ministère de l'Environnement (Department of the Environment) of Quebec (http://www.mddep.gouv.qc.ca/eau/criteres_eau/index.htm).

The Canadian Council of Ministers of the Environment (CCME) is developing Soil Quality Guidelines (SQG) for the protection of environmental and human health in different land use categories. These CCME SQGs are being developed using "A Protocol for the Derivation of Environmental and Human Health Soil Quality Guidelines" [57]. An updated version of this protocol is under revision [58]. The CCME SQGs are derived to approximate a "no- to low-" effect level (or threshold level) based on toxicological information and other scientific data, and are generic values not intended to be applied to all contaminated sites in Canada without proper site characterization. Soil guideline values are available for different metals (e.g., As, Ba, Be, C, Cr, Cu, Hg, Mo, Pb, Sn) and organic contaminants (e.g., chlorinated benzenes, phenols, ethanes, and ethenes). To date, there are no CCME SQG for MC. However, Renoux et al. [52] proposed a preliminary Environmental Soil Quality Guideline (SQGe) for TNT of 86 mg kg^{-1} for commercial/industrial land use. The use of this value is limited to screening assessments because it is not associated with any specific receptor. Furthermore, this value requires updating to incorporate eco-toxicological data established since 2001 (when it was initially proposed). The ETVs for TNT proposed in this chapter for terrestrial plants and soil invertebrates range from 6 to 52 mg kg^{-1} (Table 12.5 and Table 12.6). The TNT concentration of 30 mg kg^{-1} (established on the basis of 50% inhibition of critical microbially mediated functions by TNT in soil) was proposed in Chapter 3. These TNT concentrations for the three major groups of soil biota suggest a downward revision of the SQGe for TNT proposed by Renoux et al. [52] may be warranted.

In collaboration with the Biotechnology Research Institute (Montreal, Quebec, Canada) of the National Research Council of Canada, the Canadian DND is developing an ecotoxicological database and military-specific Environmental Sustainability Indices (ESIs) for MC, in order to improve environmental assessments and site management at DND installations [59]. This database can be used for establishing benchmark values for site assessment, and for the assessment and validation of remedial technologies [52,60]. The database includes published toxicity data and recently established data from ecotoxicological studies that have not been published in peer-reviewed journals. Effect endpoints that are considered in this database include growth metrics for terrestrial plants and earthworms, reproduction for earthworms, soil microbial activity (see Chapter 3), and wildlife endpoints (see Chapter 7). The Canadian military-specific ESI are briefly described in Section 12.7.

In addition to soil contamination with explosives and MC-related compounds, DND sites often have elevated concentrations of metals [53,54]. Consequently, risk assessment at such sites should consider not only the presence of MC but also the potential effects of their interactions with metal co-contaminants on the toxicity to ecological receptors [61]. Robidoux et al. [53,54] reported that ecotoxicological effects (such as the effects on earthworm reproduction) in soil with contaminant mixtures could not be attributed entirely to the toxicity of MC when in the presence of elevated concentrations of metals. In contrast, reproduction toxicity in earthworms correlated with TNT concentrations in TNT-contaminated soils having low metal concentrations [62]. Findings in these and other reports show that mixtures of metals and MC in soils can present challenges for ERA and site management decisions. Resolving the effects of multiple contaminant interactions on the toxicity to ecologically relevant receptors will require additional investigations. Risk assessment of contaminant mixtures can be advanced further by developing toxicity benchmark data that consider physical and chemical properties of the soil (e.g., as in the Eco-SSL approach) to more accurately determine chemical bioavailability and toxicity at contaminated sites.

12.5.1   Extent of Soil Contamination with Energetic Materials at Canadian DND Installations

Live-fire exercises, demolition procedures, and destruction of out-of-date or faulty ammunition have produced different patterns of contamination at Canadian DND installations, and this required the development of sampling protocols to investigate spatial distribution of MC soil contaminants [63]. Multiple DND installations, including artillery, battlerun, antitank, grenade, bombing areas, demolition, and open burning–open detonation ranges have been investigated in an effort to understand the impacts of munitions used by the Canadian armed forces. Assessments of spatial distribution of contaminants have been conducted for both the MC and metals, which can be released from the ignition systems, ammunition casing, and targets during munitions testing and firing.

Canadian military training activities on ranges generally lead to low contamination with explosives. These low levels of explosives are the result of second (low-) order detonation or deflagration, malfunction of the ammunition leading to unexploded ordnance (UXO) production, munitions casing perforations leading to the release of their contents into the environment, and incomplete destruction of live ammunition using blow-in-place techniques. High-order detonation (when detonation occurs as designed) usually leads to no or low-level contamination of the area by explosives. Investigations of the open detonation of large quantities of obsolete munitions showed that very small quantities of MC were released into the environment [64]. However, open burning of MC led to greater concentrations of contaminants in the surface soils. This was observed at a gun propellant burning area where NG concentration was measured up to 80 mg kg^{-1} [65].

The M72 ammunition used on antitank ranges produced a high dud rate and high levels (up to 5,000 mg kg^{-1}) of contamination by HMX [66,67]. The M72 rockets were launched from the firing points on these ranges, hundreds of meters away from the targets. The explosive charge of these rockets was octol, a mixture of 60% HMX, 30% TNT, and 10% RDX. These rockets were powered by a double-based propellant 55% NC and 35% NG, containing 2,4-DNT. These activities resulted in contamination by high explosives in the target area and by residuals of rocket propellant, mainly NG (up to 10,000 mg kg^{-1}) and 2-4 DNT, at the firing position [68].

Distribution of MC contaminants was different on the grenade ranges [68,69]. Grenades contain TNT or Composition B (59.5% RDX, 39.5% TNT, 1% wax). Incomplete grenade detonation resulted in release of RDX and TNT at these sites. The contamination was distributed throughout the range, even at distances up to 50 m from the bunker, ranging to mg kg^{-1} levels.

The primary ammunitions used on ranges during tank maneuvers, shooting of large caliber projectiles (e.g., 105–155 mm), mortar and missile fire, and strafing contain Composition B, TNT, RDX, and in some cases, HMX. Contamination in proximity to the firing points included NG and 2,4-DNT from propellant residues expelled at the gun muzzle, and from the burning of surplus propellant charge bags. The main concerns at these sites were the presence of UXO and high concentrations of explosives (up to thousands of mg kg^{-1}) in soil next to low-order detonations, although low levels of contamination by explosives were encountered randomly at these ranges.

The art of making craters, and that of cutting through concrete, steel, or wood, are practiced at Canadian DND demolition ranges. Trigran (the name Trigran came from TRItonal GRANulated), a mixture of TNT and aluminum powder at a ratio of 80:20 containing small amounts of HMX, is mainly used for cratering, while C4, mainly composed of waxy RDX, is used for demolition. Low concentrations of TNT and RDX were generally found at these sites [65].

Finally, bombing ranges are mainly used by the air force where fighter-bombers and F-18 aircraft practice against ground targets. On Canadian DND bombing ranges, weapons such as the Sidewinder missile; CRV-7 rocket; Maverick air-to-surface guided missile; and 500, 1000, and 2000 pound bombs are used to destroy ground targets. Most of the missiles contain Composition B, while bombs typically contain tritonal. High concentrations of TNT (up to 500 mg kg^{-1}) were found in bomb drop zones, while low concentrations of TNT and RDX were found in target areas where missiles were used [70].

Overall, residual MC have been found in many impact areas and firing points at the Canadian DND training installations, at levels ranging up to thousands of mg kg^{-1}. Moreover, mixed contamination, consisting of explosives and heavy metals resulting from the use of munitions in live-firing events, was often found at these installations. Generally, low concentrations of explosive were encountered in training areas with few exceptions. These exceptions included antitank ranges, firing positions, drop zones, and in hot spots generated by low-order detonations. The greatest concentrations of explosives were found in anti-tank ranges, at both the target areas and the firing positions. When explosives were detected in soil at Canadian DND ranges, metals were consistently encountered at concentrations above background levels. Therefore, military activities on ranges that result in the release of MC may lead to a mixed-contamination pattern that should be considered during the ERA process.

12.5.2 Development of Canadian Environmental Sustainability Indices (ESI) for Energetic Contaminants at Military Ranges and Training Sites

As mentioned in Section 12.5, series of generic risk-based Environmental Sustainability Indices (ESI) values for MC in soil have been calculated using the CCME approach [57,59]. Because the use of the standard CCME industrial scenario would not be protective for wildlife (birds and mammals) present on Canadian DND testing

TABLE 12.7

Proposed Military-Specific Canadian Environmental Sustainability Indices (ESI) for Energetic Soil Contaminants at Department of National Defense Installations

Compounds	Human Health ESI	Groundwater Check	Ecological ESI	Check for Protection of Aquatic Life
TNT	41	0.15	3.7	31
RDX	250	0.24	4.7	7.6
HMX	4100	3.4	32	13
2,4 DNT	0.14	0.01	11	130
2,6 DNT	0.14	0.01	8.5	130
NG	2500	7.8	65	2.4

Notes: Units for all values are mg kg^{-1} dry soil mass. Human health ESI are noncancer or threshold based, except for 2,4- or 2,6-DNT (10^{-5} cancer risk).

and training ranges, a separate military scenario for MC exposure was developed for ecological receptors. This military scenario considers the effects on wildlife (mainly grazing herbivores and birds) but differs in the level of conservatism from the standard CCME agricultural scenario. The military-specific ESI values have been established from literature-based effect values EC_{20} and EC_{50} for microbial processes, plants, and invertebrates, and from lowest observed effect levels (LOELs) for wildlife (as described in Section 12.3 for ETV). For developing military-specific ecological ESI values, the aforementioned effect values were used instead of no observed effect concentrations or levels (NOECs or NOELs) because military activities at Canadian DND sites are more comparable with the CCME industrial land use scenario than the CCME agricultural scenario, thus allowing for a greater acceptable level of adverse effects on soil organisms. Geometric means or the 25th percentile of the distribution of effects values were used for deriving military-specific ESI for microbes, plants, and soil invertebrates. Trophic chain transfer models and an area use factor of 50% were used for deriving military-specific ESI for birds and grazing herbivores. The proposed military-specific ESI values for ecological receptors are shown in Table 12.7. The ecological ESI values for individual MC shown in this table represent the lowest concentrations established for each class of receptors. A number of conservative assumptions were used in the derivation of the ESI. For example, MC were considered to be 100% bioavailable, and degradation or binding to soil constituents were not considered. No safety factors were applied for derivation of the ESI values, in compliance with the CCME industrial land use scenario. These ESIs can facilitate management of Canadian DND sites in a sustainable manner by providing a tool that can warn the decision makers of a need for further investigations when soil concentrations of MC contaminants at training and testing ranges approach or potentially exceed ecologically sustainable levels.

The CCME protocol was also used to derive human health ESIs, using a scenario adapted for military sites. The military-specific exposure scenario was developed in collaboration with officers from the Canadian DND. Direct exposure pathways were exclusively used for derivation of human health ESI, because the groundwater is generally not used for human consumption on military testing and training sites. A "typical" human exposure scenario was established because actual exposure of soldiers to MC varies widely according to the type of testing or training activities. In the absence of analytically determined data for exposures to MC-contaminated soil-dust, or to explosives-dust directly resulting from firing activities, a 10-fold increase (relative to the CCME industrial scenario) in soil ingestion, inhalation, and dermal contact rates was arbitrarily selected for derivation of human health ESI. Exposure duration and frequency were estimated with a considerable safety margin, assuming the length of individual exposure from age 17 to 55. Additional conservative assumptions were used during the calculation of ESI values. For example, homogeneous distribution in soil, 100% bioavailability, absence of degradation over time, and no binding to soil components were assumed. Risk estimates for carcinogens and noncarcinogens were taken from the USEPA databases, and threshold effects (effects other than cancer) were considered for TNT, RDX, HMX, and NG using the USEPA reference dose published in the IRIS database (USEPA Integrated Risk Information System; http://cfpub.epa.gov/ncea/iris/index.cfm), while human health ESI values for 2,4-DNT or 2,6-DNT were calculated for an increased cancer risk of 1×10^{-5}. The resulting proposed human health ESI values are summarized in Table 12.7.

A general concern was expressed by the reviewers of the methodology for derivation of proposed human health ESIs regarding the exclusion of risk consideration for the off-site exposures to MC. Such off-site exposures may result from contaminant migration from affected areas, particularly in case of groundwater contamination. Although groundwater was not identified as an exposure pathway in the derivation of human health ESI, this media must be protected to minimize risk of exposure for individuals living around potentially contaminated Canadian DND sites. The use of reference values as warning flags, similar to those of CCME groundwater checks (Table 12.7), in conjunction with proposed ESI, can assist land managers in decision-making processes regarding protection of groundwater as drinking water sources or protection of aquatic wildlife in case of groundwater resurgence. For land management purposes, the most protective generic ESI should be the lowest value of the human or ecological health indices established for each MC. Finally, it should be noted that these proposed values require validation (for example, through risk assessment) before acceptance for land management practice at Canadian DND installations.

12.6 CONCLUSIONS AND FUTURE OUTLOOK

Available reports showed that millions of acres on defense installations in North America contain elevated levels of explosives and related materials in soil, including mixed contamination consisting of both explosives and metals. Ongoing training and testing on operational ranges further contributes to environmental impacts at these sites and may jeopardize long-term sustainability of ranges. Assessment of the extent of contamination and its impact on ecosystems at U.S. installations is accomplished, for the most part, through ERA procedures. The basic methodologies of the

ERA process are also being applied in the ongoing Operational Range Assessment Program carried out by the DoD in order to assess risks of MC migration beyond installation boundaries. Eco-SSLs have been developed by the USEPA to facilitate initial screening of contaminated soils at Superfund sites. Future Eco-SSL development will include additional selected MC. Other ERA tools include the derivation of toxicity reference values (TRVs) for wildlife and empirically derived bioaccumulation models.

Estimating oral exposure for wildlife species, particularly for MC, is challenging because mammals and birds can experience spatially heterogeneous exposures to MC residues. The use of spatial exposure models can potentially enhance the accuracy of estimates. Documentation of a robust literature review and transparency of methods used are necessary in the presentation of the toxicity information and subsequent derivation of the TRV. Risk assessors are strongly advised to seek the assistance of qualified toxicologists and to use a weight-of-evidence approach when evaluating toxic effects from exposure and determining how those data should be interpreted in TRV derivation. Any ERA risk estimate should be provided with an analysis of uncertainty. The magnitude of the uncertainty should be ascertained before any decisions regarding risk management or remediation are made, and remediation of sites should not be undertaken when cleanup causes more ecological harm than the site contamination present.

Multiple Canadian DND installations have been investigated in an effort to understand the impacts of munitions used by the Canadian armed forces. However, application of ERA at the DND installations is limited. Published benchmarks or guidelines for MC-contaminated soil are not available in Canada. Soil guidelines are currently being developed to improve site assessments and to complement the results of chemical analyses at contaminated sites. Canadian human health risk-based criteria for explosives are very conservative and are designed to develop threshold criteria for residential, agricultural, and industrial uses; thus, these criteria have limited relevance for ERA at Canadian DND installations. It is critical for the Canadian DND to establish environmental soil quality guidelines for explosives that are based on ecotoxicological research of MC found at testing and training ranges. The ongoing development of ecotoxicological databases and military-specific ESIs for MC should improve ERA and site management at DND installations. These ESI can facilitate management of Canadian DND sites in a sustainable manner by providing a tool that can inform the decision makers of a need for further investigations when soil concentrations of MC contaminants approach or potentially exceed ecologically sustainable levels.

The sustainable use of military installations for training and testing requires the development of environmental quality criteria that can be consistently applied in different countries to gauge the ecotoxicological impacts of these operations. To overcome jurisdictional limitations of Eco-SSL and ESI, developed for U.S. Superfund sites and Canadian DND sites, respectively, and to make the Eco-SSL scientific approach useful internationally, the Eco-SSL methodology was applied for developing a set of ETVs for several explosives and related soil contaminants. This chapter included development of ETVs for terrestrial plants and soil invertebrates. A similar approach, after including appropriate boundary conditions and stipulations, can be applied to developing the ETVs for aquatic and sediment groups of ecological receptors. By deriving ETVs that are protective of terrestrial plants and soil invertebrates for use during the ERA process, in conjunction with screening values to be developed for avian and mammalian wildlife, it is assumed that the terrestrial ecosystems can be protected from possible adverse effects associated with exposures to MC-contaminated soil.

Additional investigations will be required to resolve the effects of multiple contaminant interactions on the toxicity to ecologically relevant receptors. Risk assessment of contaminant mixtures can be advanced further by developing toxicity benchmark data that consider physical and chemical properties of the soil in order to more accurately determine chemical bioavailability and toxicity at contaminated sites.

REFERENCES

1. United States Army ASAIE, The Army Strategy for the Environment: Sustain the Mission—Secure the Future, Office of the Assistant Secretary of the Army for Installations and Environment, Washington, DC, 2004, www.asaie.army.mil/Public/ESOH/doc/ArmyEnvStrategy.pdf (accessed August 2008).
2. United States General Accounting Office (USGAO), Military Munitions: DOD Needs to Develop a Comprehensive Approach for Cleaning Up Contaminated Sites, Committee on Energy and Commerce, GAO-04-147, House of Representatives, Washington, DC, 2003.
3. United States Environmental Protection Agency (USEPA), Ecological Risk Assessment Guidance for Superfund: Process for Designing and Conducting Ecological Risk Assessments, EPA 540-R-97-006, Office of Solid Waste and Emergency Response, OSWER 9285.7-25; PB97-963211, Edison, NJ, June 1997.
4. Pennington JC and Brannon JM, Environmental fate of explosives, *Thermochim. Acta*, 384, 163, 2002.
5. Pennington JC et al., Distribution and Fate of Energetics on DoD Test and Training Ranges: Report 4, ERDC TR-04-4, U.S. Army Engineer Research and Development Center, Environmental Laboratory, Vicksburg, MS, 2004.
6. Pennington J C et al., Distribution and Fate of Energetics on DoD Test and Training Ranges: Interim Report 3, Technical Report, ERDC TR-03-2, U.S. Army Engineer Research and Development Center, Vicksburg, MS, 2003.
7. Pennington JC et al., Distribution and Fate of Energetics on DoD Test and Training Ranges: Interim Report 2, ERDC TR 02-8, U.S. Army Engineer Research and Development Center, Vicksburg, MS, 2002.
8. Simini M et al., Evaluation of soil toxicity at Joliet army ammunition plant, *Environ. Toxicol. Chem.*, 14, 623, 1995.
9. United States Environmental Protection Agency (USEPA), Ecological Soil Screening Level Guidance, OSWER 9285.7-55, U.S. Environmental Protection Agency, Washington, DC, November 2005, http://www.epa.gov/ecotox/ecossl/ (accessed August 2008).
10. United States Environmental Protection Agency (USEPA), Guidelines for Ecological Risk Assessment, U.S. Environmental Protection Agency, Risk Assessment Forum, Washington, DC, EPA/630/R095/002F, 1998, http://cfpub.epa.gov/ncea/cfm/recordisplay.cfm?deid=12460 (accessed August 2008).

11. United States Environmental Protection Agency (USEPA), Ecological Risk Assessment and Risk Management Principles for Superfund Sites, U.S. Environmental Protection Agency, Washington, DC, 1999, http://www.epa.gov/fedfac/documents/eco_risk_superfund.htm (accessed August 2008).

12. United States Environmental Protection Agency (USEPA), Uniform Federal Policy for Implementing Environmental Quality Systems: Evaluating, Assessing, and Documenting Environmental Data Collection/Use and Technology Programs, Intergovernmental Data Quality Task Force, EPA-505-F-03-001/DTIC ADA 395303/DOE/EH-0667, U.S. Environmental Protection Agency, Washington, DC, 2005.

13. International Organization for Standardization (ISO), *Water quality—Determination of certain explosives and related compounds: Method using high-performance liquid chromatography (HPLC) with UV detection*, ISO 22478:2006, Geneva, Switzerland, 2006.

14. United States Environmental Protection Agency (USEPA), Method 8330: Nitroaromatics and Nitramines by High Performance Liquid Chromatography (HPLC), U.S. Environmental Protection Agency, Washington, DC, 2003, http://www.epa.gov/SW-846/pdfs/8330.pdf (accessed August 2008).

15. United States Environmental Protection Agency (USEPA), Method 8330B: Nitroaromatics and Nitramines by High Performance Liquid Chromatography (HPLC), U.S. Environmental Protection Agency, Washington, DC, 2006, http://www.epa.gov/SW-846/pdfs/8330b.pdf (accessed August 2008).

16. United States Department of Defense (USDoD), Directive Number 4715.11, Environmental and Explosives Safety Management on Operational Ranges Within the United States, May 10, 2004.

17. United States Department of Defense (USDoD), Instruction Number 4715.11, Operational Range Assessments, November 16, 2004.

18. United States Department of Navy (US Navy), U.S. Navy Range Sustainability Environmental Program Assessment Policy Implementation Manual, Revision 1, Chief of Naval Operations, Environmental Readiness Division (N45), November 2006.

19. United States Environmental Protection Agency (USEPA), Guidance on Systematic Planning Using the Data Quality Objectives Process, EPA/240/B-06/001, U.S. Environmental Protection Agency, Washington, DC, February 2006.

20. SERDP and ESTCP, Army Operational Range Assessments: Summary of Use of Systematic Planning and Results, Technical Exchange Meeting on DoD Operational Range Assessment and Management Approaches, Annapolis, MD, August 7–8, 2008.

21. Lubbert RF, Transformation of the Formerly Used Defense Sites (FUDS) Program Management and Execution, Joint Services Environmental Management Conference, May 24, 2007.

22. United States Marine Corps (US Marine Corps), the Marine Corps Operational Range Assessment Program–Range Environmental Vulnerability Assessment, Headquarters Marine Corps, Washington, DC, June 2007.

23. Gong P et al., Preliminary ecotoxicological characterization of a new energetic substance, CL-20, *Chemosphere*, 56, 653, 2004.

24. Kuperman RG et al., Survival and reproduction of *Enchytraeus crypticus* (Oligochaeta, Enchytraeidae) in a natural sandy loam soil amended with the nitro-heterocyclic explosives RDX and HMX, *Pedobiologia,* 47, 651, 2003.

25. Kuperman RG et al., Weathering and aging of TNT in soil increases toxicity to potworm *Enchytraeus crypticus*, *Environ. Toxicol. Chem.,* 24, 2509, 2005.

26. Kuperman RG et al., Toxicities of dinitrotoluenes and trinitrobenzene freshly amended or weathered and aged in a sandy loam soil to *Enchytraeus crypticus, Environ. Toxicol. Chem.,* 25, 1368, 2006.

27. Kuperman RG et al., Toxicity of emerging energetic soil contaminant CL-20 to pot-worm *Enchytraeus crypticus* in freshly amended or weathered and aged treatments, *Chemosphere,* 62, 1282, 2006.

28. Kuperman RG et al., Ecological soil screening levels for invertebrates at explosives-contaminated sites: Supporting sustainability of Army testing and training, in *Proceedings of 25th Army Science Conference*, Orlando, FL, November 27–30, 2006.

29. Rocheleau S et al., Toxicity of Nitro-Heterocyclic and Nitroaromatic Energetic Materials to Terrestrial Plants in a Natural Sandy Loam Soil, Technical Report No. ECBC-TR-351, U.S. Army Edgewood Chemical Biological Center, Aberdeen Proving Ground, MD, 2005.

30. Rocheleau S et al., Phytotoxicity of nitroaromatic energetic compounds freshly amended or weathered and aged in sandy loam soil, *Chemosphere,* 62, 545, 2006.

31. Simini M et al., Reproduction and survival of *Eisenia fetida* in a sandy loam soil amended with the nitro-heterocyclic explosives RDX and HMX, *Pedobiologia,* 47, 657, 2003.

32. American Society for Testing and Materials (ASTM), Standard Guide for Conducting Terrestrial Plant Toxicity Tests; E 1963-98, ASTM International, West Conshohocken, PA, 1998.

33. International Organization for Standardization (ISO), *Soil Quality—Effects of Pollutants on Earthworms (Eisenia fetida)—Part 2: Determination of Effects on Reproduction*, ISO 11268-2, ISO 11268-2:1998, International Organization for Standardization, Geneva, Switzerland, 1998.

34. International Organization for Standardization (ISO), *Soil Quality—Effects of Pollutants on Enchytraeidae (Enchytraeus sp.)—Determinations of effects on reproduction and survival*, ISO/CD 16387, International Organization for Standardization, Geneva, Switzerland, 2005.

35. International Organization for Standardization (ISO), *Soil Quality—Inhibition of Reproduction of Collembola (Folsomia candida) by Soil Pollutants*, ISO 11267, International Organization for Standardization, Geneva, Switzerland, 1999.

36. Reifenrath WG et al., Percutaneous absorption of explosives and related compounds: An empirical model of bioavailability of organic nitro compounds from soil, *Toxicol. Appl. Pharmacol.,* 182, 60, 2002.

37. Johnson MS et al., Bioaccumulation of 2,4,6-trinitrotoluene and polychlorinated biphenyls through two routes of exposure in a terrestrial amphibian: Is the dermal route significant? *Environ. Toxicol. Chem.,* 18, 873, 1999.

38. Johnson MS et al., Effects of 2,4,6-trinitrotoluene in a holistic environmental exposure regime on a terrestrial salamander, *Ambystoma tigrinum*, *Toxicol. Path.,* 28, 334, 2000.

39. Johnson M et al., Toxicologic and histopathologic response of the terrestrial salamander *Plethodon cinereus* to soil exposures of 1,3,5-trinitrohexahydro-1,3,5-triazine, *Arch. Environ. Contam. Toxicol.,* 47, 496, 2004.

40. Beyer WN, Connor EE, and Gerould S, Estimates of soil ingestion by wildlife, *J Wildl. Manage.,* 58, 375, 1994.

41. Johnson MS et al., Are songbirds at risk from lead at small arms ranges? An application of the spatially explicit exposure model (SEEM), *Environ. Toxicol. Chem.,* 26(10), 2215, 2007.

42. Jenkins TF et al., Characterization of Explosives Contamination at Military Firing Ranges, Technical Report ERDC TR-01-5, Engineering Research and Development Center, U.S. Army Corps of Engineers, Cold Regions Research and Engineering Laboratory, Hanover, NH, 2001.

43. Pennington JC et al., Distribution and Fate of Energetics on DoD Test and Training Ranges, ERDC TR-04-4, U.S. Army Corps of Engineers, Engineer Research and Development Center, Vicksburg, MS, 2004.

44. Freshman JS and Menzie CA, Two wildlife exposure models to assess impacts at the individual and population levels and the efficacy of remedial actions, *Human Ecol. Risk Assess.*, 3, 481, 1996.

45. Wickwire WT et al., Incorporating spatial data into ecological risk assessments: The spatially explicit exposure module (SEEM) for ARAMS, in *Landscape Ecology and Wildlife Habitat Evaluation: Critical Information for Ecological Risk Assessment, Land-Use Management Activities and Biodiversity Enhancement Practices*, ASTM STP 1458, Kapustka LA et al., Eds., American Standards for Testing and Materials International, West Conshohocken, PA, 2004, 297.

46. Hope BK, Generating probabilistic spatially-explicit individual and population exposure estimates for ecological risk assessments, *Risk Analysis*, 20, 573, 2000.

47. Hope BK, A case study comparing static and spatially explicit ecological exposure analysis methods, *Risk Analysis*, 21, 1001, 2001.

48. Menzie-Cura & Associates Inc., FISHRAND Probabilistic Fish Bioaccumulation Model, User's Manual, version 1.0, prepared for the Environmental Laboratory of the U.S. Army Engineer Research and Development Center, Vicksburg, MS, 2003.

49. United States Army Center for Health Promotion and Preventive Medicine (USACHPPM), Standard Practice for Wildlife Toxicity Reference Values, Technical Guide 254, U.S. Army Center for Health Promotion and Preventive Medicine, Aberdeen Proving Ground, MD, USA, 2000.

50. Robidoux PY et al., Ecotoxicological risk assessment of explosives contaminated sites, in *Environmental Analysis of Contaminated Sites: Toxicological Methods and Approaches*, Sunahara GI et al., Eds., John Wiley and Sons, UK, 2002, 335.

51. Talmage SS et al., Nitroaromatic munition compounds: Environmental effects and screening values, *Rev. Environ. Contam. Toxicol.*, 161, 1, 1999.

52. Renoux AY et al., Derivation of environmental soil quality guidelines for 2,4,6-trinitrotoluene using the CCEM approach, *Human Ecol Risk Assess.*, 7, 1715, 2001.

53. Robidoux PY et al., Toxicity assessment of contaminated soils from an antitank firing range, *Ecotoxicol. Environ. Saf.*, 58, 300, 2004.

54. Robidoux PY et al., Assessment of an antitank firing range using *Lumbricus terrestris* and *Eisenia andrei* in mesocosms and laboratory studies, *Ecotoxicology*, 13, 603, 2004.

55. Robidoux PY et al., Ecotoxicological risk assessment of an explosives-contaminated site, *Can. Tech. Rep. Fish. Aquat. Sci.*, 2331, 34, 2000.

56. Ministère de l'Environnement, Politique de protection des sols et de réhabilitation des terrains contaminés, 1999.

57. Canadian Council of Ministers of Environment (CCME), A Protocol for the Derivation of Environmental and Human Health Soil Quality Guidelines, Draft, Winnipeg, Manitoba, 1996.

58. Canadian Council of Ministers of Environment (CCME), A Protocol for the Derivation of Environmental and Human Health Soil Quality Guidelines, Winnipeg, Manitoba, 2006.

59. Robidoux PY et al., Development of Ecological and Human Health Guidelines for Energetic Materials to Ensure Training Sustainability of Canadian Forces, Final Report. Applied Ecotoxicology Group, Biotechnology Research Institute, National Research Council, Montreal, Canada, prepared for National Defence, Directorate Environmental Protection, Ottawa, Canada, June 2006.

60. Sunahara GI et al., Laboratory and field approaches to characterize the ecotoxicology of energetic substances, in *Environmental Toxicology and Risk Assessment: Science, Policy and Standardization—Implications for Environmental Decisions*, Tenth Volume STP 1403, Greenberg BM et al., Eds., American Society for Testing and Materials, West Conshohocken, PA, 2001, 293.

61. Robidoux PY et al., Effects of heavy metals and explosive polynitro-organic mixtures on terrestrial plants and earthworms, *SETAC 25th Annual Meeting Abstract Book*, Portland, OR, 2004.

62. Robidoux PY et al., Evaluation of tissue and cellular biomarkers to assess 2,4,6-tri-nitrotoluene (TNT) exposure in earthworm: Effect-based assessment in laboratory studies using *Eisenia andrei*, *Biomarkers*, 7, 306, 2002.

63. Thiboutot S, Ampleman G, and Hewitt A, Guide for Characterization of Sites Contaminated with Energetic Materials, ERDC-CRREL TR-02-01, February 2002.

64. Ampleman G et al., Study of the Impacts of OB/OD Activity on Soils and Groundwater, at the Destruction Area in CFAD Dundurn, DREV Report R-9827, December 1998.

65. Ampleman G et al., Evaluation of the Soils Contamination by Explosives at CFB Chilliwack and CFAD Rocky Point, DREV Report TR-2000-103, November 2000.

66. Thiboutot S et al., Characterization of Antitank Firing Ranges at CFB Valcartier, WATC Wainwright and CFAD Dundurn, DREV R-9809, October 1998.

67. Jenkins T et al., Coping with spatial heterogeneity effects on sampling and analysis at an HMX contaminated antitank firing range, *Field Anal. Chem. Technol.*, 3, 19, 1999.

68. Thiboutot S et al., Environmental Conditions of Surface Soils and Biomass Prevailing in the Training Area at CFB Gagetown, New Brunswick, DRDC-Valcartier TR 2003-152, October 2003.

69. Ampleman G et al., Evaluation of the Impacts of Live Fire Training at CFB Shilo, DRDC-Valcartier TR 2003-066, April 2003.

70. Ampleman G et al., Evaluation of the Contamination by Explosives in Soils, Biomass and Surface Water at Cold Lake Air Weapons Range (CLAWR), Alberta, Phase I Report, DRDC-Valcartier TR 2003-208, December 2003.

71. Phillips CT et al., Toxicity Benchmark Determinations for RDX or HMX Freshly Amended into Five Soils, The 2007 SERDP Partners in Environmental Technology Technical Symposium & Workshop, Washington, DC, December 4–6, 2007.

13 Closing Remarks

Guilherme R. Lotufo, Geoffrey I. Sunahara,
Jalal Hawari, and Roman G. Kuperman

The primary focus of this book has been to provide a comprehensive review, summary, and interpretation of decades of research on the fate and ecotoxicological effects of energetic materials (EMs) in terrestrial and aquatic environments, and to identify the strengths and limitations of the available scientific information. This book was structured to first discuss analytical tools available to detect and identify explosives and their degradation products in the environment; second, to bring to light the current knowledge available on the ecological impact of these chemicals; and, finally, to highlight the knowledge gaps and define the future research needs.

Environmental contamination with EMs has been adequately characterized in many industrialized countries, including the United States, Canada, and Germany. However, information on the spatial distribution, quantities, and bioavailability of EMs present at former manufacturing plants, disposal sites, and at active military training areas (i.e., Army, Navy, and Air Force ranges), including both terrestrial and underwater sites around the globe, is insufficient. This knowledge gap precludes an accurate evaluation of the local and global environmental impacts of the presence of EMs in aquatic and terrestrial habitats. We hope that continuing progress in the assessments of contaminated sites and the advancements in the basic research of the fate and transport of EMs, as well as the toxicological effects on ecological receptors, will fill the remaining knowledge gaps.

Research on the environmental fate of EMs reviewed here showed that degradation of explosives and their transformation products is limited in the vadose soil zone, which can explain EM persistence in terrestrial and inland aquatic habitats. The fate of EMs in surface soil and their migration into groundwater is governed by the physicochemical properties of the EMs and their interaction with the soil constituents (mainly clay and organic matter). The relatively soluble nitroaromatic compounds (such as TNT), and the less soluble but highly mobile nitramines (such as RDX), have been frequently detected in groundwater; whereas less soluble and less mobile compounds, such as HMX, are typically restricted to the soil surface and vadose zone. Studies reviewed in this book also showed that EMs can vary widely in their propensity to undergo transformation under various environmental conditions, with daughter products co-occurring with parent compounds at field sites. Therefore, the scope of future studies should include identification of EM transformation products and pathways leading to their formation, the mechanisms of their interaction with soil constituents, and the ecotoxicity of these transformation products, in order to unequivocally relate the potential hazards of individual EMs to specific ecological receptors. Additional investigations will be required to better understand the mechanisms controlling fate and bioavailability of nitroaromatic EMs in the environment; and to explain the wide variability in their effective concentrations in different soil or sediment types, and under different exposure conditions.

There is a need for improved in situ monitoring tools that can provide a more comprehensive field data for use in environmental risk assessment and management. Advanced analytical methods and soil sampling techniques should be employed to ensure sufficient detection of trace amounts of EMs, and the wide range of their respective breakdown products at contaminated sites. Presently, the analysis of EMs in the environment is based mostly on the U.S. Environmental Protection Agency (USEPA) Methods 8330 and 8330B. However, because the recovery of these EMs and their transformation products depends on the matrix (e.g., soil, sediment, or tissue) properties, we think that a new generation of predictive models should be developed to more effectively account for the effects of field parameters, including organic matter and clay contents, on the fate of EMs in the environment.

Substantial progress in EM ecotoxicology over the last decades has generated a database sufficient to provide a general understanding of the potential hazards associated with the presence of EMs in terrestrial and aquatic environments. Nonetheless, ecotoxicological data for the transformation products of many explosives are insufficient or nonexisting, which justifies continuing support for basic toxicity investigations using ecologically relevant taxonomic groups. Studies reviewed in this book have established ecotoxicological benchmarks for EM effects on a variety of ecological receptors, including microbes, and ecologically relevant plant, invertebrate, and vertebrate species. The majority of the investigated aquatic invertebrates, terrestrial plants, and soil microbes were relatively insensitive to the effects of cyclic nitramine explosives RDX, HMX, and CL-20. However, soils amended with RDX, HMX, or CL-20 were highly toxic to the reproduction of earthworms, whereas CL-20 was much more toxic than HMX to enchytraeids. Studies reviewed in this book showed that TNT can selectively and adversely affect certain components of the soil microbial community and disrupt the biogeochemical cycles of carbon and nitrogen. Plant and soil invertebrate species were adversely affected by exposure to nitroaromatic EMs at environmentally realistic concentrations, and the magnitude of the effects was influenced by the soil type and test species. Weathering and aging of several EMs in soil altered their respective toxicities to plants and soil invertebrates by either increasing or decreasing the exposure effects, depending on the species and soil type tested. These findings emphasize the importance of including weathering and aging of EMs in soil in toxicity testing protocols.

Aqueous toxicity of nitroaromatic EMs has been determined for a variety of freshwater and marine species. RDX was acutely toxic to several species of fish and aquatic invertebrates at relatively high concentrations, while HMX did not elicit mortality at its aqueous solubility level. Exposure of benthic invertebrates to sediments spiked with RDX or HMX did not elicit lethal toxicity, even at exceedingly high concentrations. Therefore, those nitramine EMs pose minimal risk for fish and aquatic invertebrates at contaminated sites. The concentration values that caused negative effects were generally lower and more variable for nitroaromatic EMs compared to nitramine EMs, as determined in studies with fish and invertebrates using either aqueous or sediment exposures. Comparison of toxicity values determined in laboratory tests with the limited information on EM concentrations in field sites suggests a low risk from exposure to EMs for fish and aquatic invertebrates.

Determining the relationships among soil or sediment constituents and EM toxicity is critical for improving the utility of models used in the ecological risk assessment (ERA) process. This will require new studies with multiple soil and sediment types that also include greatly expanded ranges of individual properties. Those investigations will greatly reduce the level of uncertainty, thereby decreasing remedial costs associated with the use of limited ecotoxicological data in site management by transitioning from reliance on conservative assumptions to include the use of predictive models.

Energetic materials have long been known as toxic and mutagenic to mammalian species, including humans. Cytological studies reviewed in this book unequivocally demonstrate the involvement of oxidative stress in the toxicity of nitroaromatic explosives, and strongly indicate the need for identifying protein targets of the EM toxic action using the proteomic and other molecular approaches, as well as the characterization of the mode of their binding to DNA. Measurement of the effects at the organism level by using biochemical and molecular responses as markers of exposure or toxicity are being evaluated as a screening approach to characterizing sites contaminated with EMs. Promising diagnostic tools for exposure detection and for chemical toxicity screening are being developed in the field of ecotoxicogenomics, the genomewide analysis of gene expression to study the effect of toxicants on ecological receptors [1,2]. Expression profiling used in transcriptomics, proteomics, and metabolomics can provide important clues on the modes of toxic action of EMs. However, none of these techniques can currently be considered routine for establishing the toxicological benchmarks that are typically used in the environmental regulation of EMs (i.e., survival, growth, and reproduction).

While sufficient evidence indicates the toxicity of many EMs to soil and aquatic invertebrates and to fish occurs at concentrations expected or reported for contaminated sites, toxicological studies have also indicated that only exceedingly high concentrations cause adverse effects on most wildlife species. Additional toxicity data and more accurate risk assessment techniques will be required to more accurately characterize the EM exposure risk to further improve contaminated site management decisions.

In addition to the mixtures of parent EMs and their transformation products, contaminated terrestrial, benthic, and aquatic sites typically contain other classes of chemicals, such as heavy metals associated with munitions. Limited information is available regarding the chemical and biological interactions among EMs and of EMs with other chemicals such as heavy metals. Some of these contaminants have been present at legacy sites for decades, whereas other chemicals may pose future issues in view of emerging environmental contaminants from new technologies, such as the nanotechnologies and nano-enabled technologies.

The use of nanomaterials (i.e., engineered particles or objects with at least one dimension less than 0.1 μm in size) in the research, development, testing and evaluation of new weapon systems and platforms has been ongoing for at least a decade, and the basic understanding of the ecotoxicological effects of nanosized materials is increasing rapidly [2–4]. Due to their high specific surface area and chemical reactivity, these chemicals can be used in energetic formulations to increase the energy and flame temperature in rocket propellants and underwater weapons. Currently, the use of energetic nanomaterials is relegated to research and testing, and includes, for example, combinations of RDX and nanosized metals such as nano-aluminum. Because the use of these materials is still in the testing phase, studies of their potential for transport, fate, and ecological effects in the environment are highly desirable. Likewise, the toxicological interactions of nanomaterials and EMs will require further research.

Physical modification of habitat resulting in diminished availability of resources appears to be a greater contributing factor to the reported declines in wildlife populations on military installations compared to the chemical contamination. Although highly desirable, studies of the relative and cumulative contributions of stressors present at military installations (e.g., chemical contaminants; detonation craters; range clearing practices; and habitat loss, gain, or fragmentation) have not been commonly reported in the available literature.

Overall, this book represents the up-to-date compilation and critical review of available information on environmental fate and ecotoxicological effects of explosives, propellants, and related EMs. The readers will undoubtedly benefit from the summaries of ecotoxicological data prepared by the chapters' authors who have also identified the key knowledge gaps and outlined the directions for future research. Despite the availability of a large body of laboratory-derived ecotoxicological data for EMs, the successful application of these data for accurate and predictive risk assessment at contaminated sites will require a better understanding of the mechanisms of toxic action, the effects of exposure media properties on chemical bioavailability to terrestrial and aquatic receptors, multispecies or community-level assessments (e.g., microcosms, mesocosms, terrestrial model ecosystems, etc.) and in situ field validation studies, and the development of more advanced models for the fate and transport of EMs in the environment. Continuous research to address these challenges will provide contaminated-site stakeholders with better tools to gauge the ecotoxicological impacts at legacy sites and the impacts of operations that involve the use of explosives and propellants, ultimately promoting the sustainable use of testing and training ranges and successful contamination management.

REFERENCES

1. Ankley GT et al., Eds., *Genomics in Regulatory Toxicology: Applications and Challenges*, CRC Press, Boca Raton, FL, 2007.
2. Gong P et al., Toxicogenomic analysis provides new insights into molecular mechanisms of the sublethal toxicity of 2,4,6-trinitrotoluene in *Eisenia fetida, Environ. Sci. Technol.*, 41, 8195, 2007.
3. Handy RD, Owen R, and Valsami-Jones E, The ecotoxicology of nanoparticles and nanomaterials: Current status, knowledge gaps, challenges, and future needs, *Ecotoxicology*, 17, 315, 2008.
4. Klaine SJ et al., Nanomaterials in the environment: Behavior, fate, bioavailability, and effects, *Environ. Toxicol. Chem.*, 27, 1825, 2008.

Index

A

Abiotic transformations, 8–13
  cyclic nitramines, 11–13
    alkaline hydrolysis, 11–12
    photolysis, 12–13
    reduction by iron, 13
  nitroaromatic compounds, 8–11
    alkaline hydrolysis, 8–10
    photolysis, 11
    reduction by iron, 11
  transformation pathways, 8–13
    cyclic nitramines, 11–13
    nitroaromatic compounds, 8–11
Acetobacterium malicum, 19
Acrosiphonia coalita, 148
2-ADNT, *see* 2-Amino-4,6-dinitrotoluene
4-ADNT, *see* 4-Amino-2,6-dinitrotoluene
ADNTs, *see* Aminodinitrotoluenes
ADR, *see* Adrenodoxin reductase
Adrenodoxin (ADX), 217
Adrenodoxin reductase (ADR), 217
ADX, *see* Adrenodoxin
AFLP, *see* Amplified fragment length
    polymorphism
Agaricus eastivalis, 16
Agelaius phoeniceus, 164
Agency for Toxic Substance Disease Registry
    (ATSDR), 158
Agrobacterium sp., 19
Agrocybe praecox, 16
Air-to-ground combat training ranges, 255
ALD, *see* Approximate lethal dose
Allium cepa root tip micronucleus assay, 193
Alternaria sp., 16
Ambystoma tigrinum, 142, 169, 246
Americamysis bahia, 81, 86, 102
2-Amino-4,6-dinitrotoluene (2-ADNT), 8, 65,
    147
4-Amino-2,6-dinitrotoluene (4-ADNT), 8, 14,
    65, 119, 147, 219
Aminodinitrotoluenes (ADNTs), 6, 36, 38, 89,
    120, 151
  elimination rate of, 145
  genotoxicity and, 180
  glyco-conjugates, 147
  wildlife species and, 158
2-Amino-6-nitrotoluene (2-A-6-NT), 125
Aminonitrotoluenes (ANTs), 89

Aminophenols (APs), 98
Ammodramus henslowii, 263
Ammonium perchlorate (AP), 166
Ampelisca abdita, 124, 126, 127
Amplified fragment length polymorphism
    (AFLP), 199
Anabaena flos-aquae, 103, 104, 105
Anas platyrhynchos, 167
2-A-6-NT, *see* 2-Amino-6-nitrotoluene
Antilocapra americana sonoriensis, 266
Antitank rocket ranges, 258
ANTs, *see* Aminonitrotoluenes
AP, *see* Ammonium perchlorate
Aplexa hypnorum, 81
Applicable or relevant and appropriate
    requirement (ARAR), 282
Applied Risk Assessment Modeling System
    (ARAMS), 295
Approximate lethal dose (ALD), 163
APs, *see* Aminophenols
Aquatic organisms, bioconcentration,
    bioaccumulation, and
    biotransformation of explosives and
    related compounds in, 135–155
  aquatic animals, 136–145
    bioaccumulation from diet, 142–143
    bioconcentration, 136–141
    body distribution, 142
    toxicokinetics and biotransformation,
      143–145
  BCF values for fish and aquatic invertebrates,
    137
  elimination rate of ADNTs, 145
  elimination of TNT, 144
  parent compound BCF, 137, 141
  phototrophic organisms, 145–151
    microalgae, 145–146
    nonvascular freshwater macrophytic
      algae, 150–151
    nonvascular marine macrophytes,
      148–150
    vascular aquatic plants, 146–147
  phytophotolysis, 147
  phyto-treatment of explosives-contaminated
    water, 136
  recommendations, 151–152
  TNT removal from seawater by tissue
    cultures of marine macroalgae, 149

313

Aquatic toxicology, 77–115
    acute and chronic values derived as water
        quality criteria for protection of
        aquatic life, 108
    criteria and screening benchmarks, 107–109
    cyclic nitramines, 105–106
        diethylene glycol dinitrate, 107
        hexahydro-1,3,5-trinitro-1,3,5-triazine,
            105
        hexanitrohexaazaisowurtzitane, 106
        nitrocellulose, 106
        nitroglycerin, 106–107
        nitroguanidine, 106
        octahydro-1,3,5,7-tetranitro-1,3,5,7-
            tetrazocine, 105
        pentaerythritol tetranitrate, 107
    disposal of munitions into ocean, 77
    explosives, propellants, and related
        chemicals, 79–107
    nitroaromatic compounds, 79–105
        nitrobenzenes, 79–83
        nitrophenols, 93–98
        nitrotoluenes, 83–93
        photoactivation, 99–105
        2,4,6-trinitrophenylmethylnitramine,
            98–99
    ordnance compounds, toxicity of, 100
    pink water, 77
    QSAR studies, 93
    recommendations, 109–110
    toxicity of reduction products of DNT
        isomers, 92
    toxicity of TNT reduction products and their
        isomers, 90–91
    water quality parameters, 89
Arabidopsis thaliana, 69, 199
ARAMS, *see* Applied Risk Assessment
        Modeling System
ARAR, *see* Applicable or relevant and
        appropriate requirement
Arbacia punctulata, 81, 83, 87, 89, 98, 103, 128
Army Training and Testing Carrying Capacity
        model, 265
Artemia salina, 86, 95, 98
Artificial soil, 36
Asellus militaris, 102, 106
Aspergillus terrus, 16
ATSDR, *see* Agency for Toxic Substance
        Disease Registry
Avena sativa L., 52

B

Bacillus sp., 15
BAFs, *see* Bioaccumulation factors
Basal respiration (BR), 37
Baseline ecological risk assessment (BERA), 280

BCF, *see* Bioconcentration factor
BCNU, see 1,3-Bis-(2-chloroethyl)-1-nitrosourea
BERA, *see* Baseline ecological risk assessment
Bioaccumulation factors (BAFs), 142, 229
Bioconcentration, *see* Aquatic organisms,
        bioconcentration, bioaccumulation,
        and biotransformation of explosives
        and related compounds in; Terrestrial
        systems, bioconcentration,
        bioaccumulation, and
        biomagnification of nitroaromatic and
        nitramine explosives in
Bioconcentration factor (BCF), 136, 229
Biomimetic devices, use of to assess explosives
        bioavailability, 129–131
Biotic transformations, 13–21
    cyclic nitramines, 18–21
        aerobic degradation, 20–21
        anaerobic degradation, 18–20
    nitroaromatic compounds, 13–18
        DNT biotransformation, 18
        picric acid biotransformation, 18
        TNT biotransformation, 13–17
1,3-Bis-(2-chloroethyl)-1-nitrosourea (BCNU),
        221
BR, *see* Basal respiration
Brachionus calyciflorus, 87
Brassica rapa Metzg, 52
Bromus mollis L., 52
Bufo americanus, 169

C

Canadian Council of Ministers of the
        Environment (CCME), 297
Carassius auratus, 137
Carbon cycle, effects of nitroaromatic
        compounds on, 43–44
Catharanthus roseus, 147, 237, 238
CCME, *see* Canadian Council of Ministers of the
        Environment
Central nervous system (CNS), 163
CERCLA, *see* Comprehensive Environmental
        Response, Compensation, and
        Liability Act of 1980
Ceriodaphnia dubia, 81, 85, 89, 90, 101, 105, 110
China Lake 20 (CL-20), 36
Chinese hamster ovary (CHO) cells, 185
Chironomus tentans, 102, 106, 123, 124, 125,
        132, 137
Chloramphenicol, 40
Chlorella pyrenoidosa, 79, 81, 83, 87, 98
CHO cells, *see* Chinese hamster ovary cells
Citrobacter freundii, 18, 19
CL-20, *see* 2,4,6,8,10,12-Hexanitro-2,4,6,8,10,12-
        hexaazaisowurtzitane
Clitocybe odora, 16

Clostridium
  acetobutylicum, 69
  bifermentans, 18
  sp., 19
Clupea harengus, 94
CMC, *see* Criterion maximum concentration
CNS, *see* Central nervous system
Colinus virginianus, 163, 165
Comprehensive Environmental Response,
          Compensation, and Liability Act of
          1980 (CERCLA), 279
Comprehensive range evaluations (CRE), 284
Contaminants of potential concern (COPC), 286
COPC, *see* Contaminants of potential concern
Coprinus comatus, 16
Corynebacterium sp., 19
Crangon septemspinosa, 95
Crassostrea
  gigas, 91
  virginica, 95
CRE, *see* Comprehensive range evaluations
Criterion maximum concentration (CMC), 107
Cryptotis parva, 160
Cyclic nitramines, 11–13, 18–21, 105–106
  aerobic degradation, 20–21
  alkaline hydrolysis, 11–12
  anaerobic degradation, 18–20
  diethylene glycol dinitrate, 107
  hexahydro-1,3,5-trinitro-1,3,5-triazine, 105
  hexanitrohexaazaisowurtzitane, 106
  nitrocellulose, 106
  nitroglycerin, 106–107
  nitroguanidine, 106
  octahydro-1,3,5,7-tetranitro-1,3,5,7-
          tetrazocine, 105
  pentaerythritol tetranitrate, 107
  photolysis, 12–13
  reduction by iron, 13
Cygnus olor, 167
Cyperus esculentus, 147, 235
Cyprinodon variegatus, 80, 85, 94, 137
Cyprinus carpio, 80, 84, 94, 137, 143

D

2,4-DANT, *see* 2,4-Diamino-4-nitrotoluene
DANTs, *see* Diaminonitrotoluenes
Daphnia
  carinata, 81, 85, 95
  magna, 66, 79, 80, 85, 90, 92, 94, 99, 101, 110
Data Quality Objective (DQO) Process, 283
Deep sea contamination, *see* Aquatic toxicology
DEGDN, *see* Diethylene glycol dinitrate
Dehydrogenase activity (DHA), 37
Denitrification, 42
Department of National Defense (DND)
          (Canada), 296

Depleted uranium (DU) tank penetrator rounds,
          257
Desert vegetation, recovery of, 268
Desulfovibrio sp., 19
Detonation craters, 260
DHA, *see* Dehydrogenase activity
2,4-Diamino-4-nitrotoluene (2,4-DANT), 119,
          120, 247
Diaminonitrotoluenes (DANTs), 6, 14, 89, 120,
          181
Diethylene glycol dinitrate (DEGDN), 107
3,5-Dinitroaniline (3,5-DNA), 65, 120
Dinitroaniline herbicides, 68
1,3-Dinitrobenzene (1,3-DNB), 78
2,4-Dinitrophenol (2,4-DNP), 129
Dinitrophenols (DNPs), 93
2,4-Dinitrotoluene (2,4-DNT), 8, 9, 11, 36, 239,
          257
2,6-Dinitrotoluene (2,6-DNT), 36, 71, 117, 120,
          125
Dinitrotoluenes (DNTs), 6, 182
  biotransformation, 18
  ERA involvement at military ranges, 285
  USEPA classification, 191
  wildlife species and, 158
Dinophilus gyrociliatus, 81, 83, 86, 95, 102, 105,
          110, 129
N,N′-Diphenyl-*p*-phenylene diamine (DPPD),
          221
Dipodomys merriami, 266
3,5-DNA, *see* 3,5-Dinitroaniline
1,3-DNB, *see* 1,3-Dinitrobenzene
DND, *see* Department of National Defense
2,4-DNP, *see* 2,4-Dinitrophenol
DNPs, *see* Dinitrophenols
2,4-DNT, *see* 2,4-Dinitrotoluene
2,6-DNT, *see* 2,6-Dinitrotoluene
DNTs, *see* Dinitrotoluenes
DNX, *see* Hexahydro-5-nitro-1,3-dinitroso-
          1,3,5-triazine
DoD, *see* U.S. Department of Defense
DPPD, see *N,N′*-Diphenyl-*p*-phenylene diamine
DQO Process, *see* Data Quality Objective
          Process
Drosophila melanogaster, 170
Dugesia dorotocephala, 99
DU tank penetrator rounds, *see* Depleted
          uranium tank penetrator rounds

E

Echinochloa crusgalli, 46, 242, 288
Ecological dynamics simulation modeling
          (EDYS), 265
Ecological risk assessment (ERA), 278, 311
Ecological Risk Assessment Guidance for
          Superfund (ERAGS), 280, 282

Ecological soil screening levels (Eco-SSLs), 279, 286
Eco-SSLs, *see* Ecological soil screening levels
Eco-SSL Workgroup, 287
Ecotoxicological tolerance values (ETVs), 287
EDYS modeling, *see* Ecological dynamics simulation modeling
EGDN, *see* Ethyleneglycoldinitrate
Eglin Air Force Base, 266
Eisenia
    andrei, 53, 56, 57, 59, 61, 64, 66, 244
    fetida, 53, 54, 55, 60, 64, 289
Elodea canadensis, 146
EMH, *see* Extramedullary hematopoiesis
EMs, *see* Energetic materials
Enchytraeus
    albidus, 52, 54, 58, 61, 63, 65
    crypticus, 53, 54, 55, 56, 57, 58, 59, 60, 62, 63, 66, 289
Endangered Species Act, 282
Energetic materials (EMs), 1, 36, 309
    aqueous toxicity of nitroaromatic, 310
    effects on wildlife species, 311
    microbial transformation of, 65
    need for research, 312
    toxicity mechanisms, 67
    transformation products, 309
Enterobacter cloacae, 15, 19, 220
Entosyphum sulcatum, 81, 83, 96
Environment, fate and transport of explosives in, 5–33
    adsorption coefficients of explosive contaminants in natural and model systems, 23
    abiotic transformations, 8–13
    biotic transformations, 13–21
        cyclic nitramines, 18–21
        nitroaromatic compounds, 13–18
    cyclic nitramines, 18–21
        aerobic degradation, 20–21
        anaerobic degradation, 18–20
    future outlook, 25
    military explosives classification, 7–8
    nitroaromatic compounds, 13–18
        DNT biotransformation, 18
        picric acid biotransformation, 18
        TNT biotransformation, 13–17
    sorption processes and bioavailability, 21–25
        cyclic nitramines, 24–25
        nitroaromatic compounds, 22–23
    transformation pathways, 8–13
Environmental Sustainability Indices (ESIs), 297, 299
EOD ranges, *see* Explosive ordnance disposal ranges
Eohaustorius estuarius, 123, 124

ERA, *see* Ecological risk assessment
ERAGS, *see* Ecological Risk Assessment Guidance for Superfund
EROD, *see* Ethoxyresorufin-*O*-deethylase
Escherichia coli, 19, 97, 179, 180, 183, 220
ESIs, *see* Environmental Sustainability Indices
Ethoxyresorufin-*O*-deethylase (EROD), 145
Ethyleneglycoldinitrate (EGDN), 6, 9
ETVs, *see* Ecotoxicological tolerance values
Explosive ordnance disposal (EOD) ranges, 285
Explosives Ordnance Demolition Ranges, 255
Extramedullary hematopoiesis (EMH), 159

F

FAD, *see* Flavin adenine dinucleotide
Festuca arundinacea Schreb., 52
FETAX, *see* Frog Embryo Teratogenesis Assay–*Xenopus*
FISHRAND model, 294, 295
Flavin adenine dinucleotide (FAD), 213
Flavin mononucleotide (FMN), 213
FMN, *see* Flavin mononucleotide
Folsomia candida, 53, 57, 58, 63, 65, 289, 292
Fomes fomentarius, 164
Food web models, 228
Formerly Used Defense Sites (FUDS), 284
Fort Benning Military Reservation, 269, 271
Fort Sill Military Reservation, 266
Frog Embryo Teratogenesis Assay–*Xenopus* (FETAX), 169, 170
FUDS, *see* Formerly Used Defense Sites
Fungi, biodegradation of TNT, 16

G

Gallus domesticus, 167
Gammarus
    fasciatus, 102, 106
    minus, 101
    pseudolimnaeus, 102
Genotoxicity to explosives, 177–209
    Allium cepa root tip micronucleus assay, 193
    application to ecotoxicology, 195–198
    AZT formation, 184
    data gaps and future directions, 201–202
    definition of genotoxicity, 177
    in vitro mammalian cell-based systems used to assess mutagenic potential, 186
    in vitro methodologies, 178–189
        cyclic nitramines and their products, 183, 188
        dinitrotoluenes and their products, 182–183, 187–188
        mammalian cell line–based genotoxicity assays, 185–189

mammalian cell line testing of
environmental samples, 189
microorganism-based mutagenicity
assays, 178–185
mixtures and environmental samples,
184–185
other compounds, 183–184, 188–189
2,4,6-trinitrotoluene and its products,
180–182, 185–187
in vivo methodologies, 189–194
cyclic nitramines and their products,
192–193
dinitrotoluenes and their products, 191–192
mammalian in vivo genotoxicity
investigations, 190–193
other compounds, 193
other investigations, 193–194
2,4,6-trinitrotoluene and its products,
190–191
K-strategy, 197
new approaches for assessing genotoxicity of
explosives, 198–199
nitroglycerin, 193
organism repair of DNA damage, 197
quantitative structure–activity relationship
models, 194
random amplified polymorphic DNA
technique, 198
Salmonella assay, 178, 195
structure–activity relationships, 194–195
summary for genotoxicity testing of
explosives compounds, 200
Tradescantia stamen hair mutation (Trad-
SHM) assay, 193
Trad-SHM assay, see *Tradescantia* stamen
hair mutation assay
USEPA classification, 191
Vicia faba root tip micronucleus assay, 193
Geographic information system (GIS) model,
265
GIS model, *see* Geographic information system
model
Glutathione reductase (GR), 217
Glutathione S-transferases (GSTs), 69
GR, *see* Glutathione reductase
GSH, *see* Reduced glutathione
GSSG, *see* Oxidized glutathione
GSTs, *see* Glutathione S-transferases

H

Habitat disturbance at explosives-contaminated
ranges, 253–276
Air-to-ground combat training ranges, 255
Army Training and Testing Carrying
Capacity model, 265

confounding effects of multiple stressors,
269–271
contamination at antitank rocket ranges, 258
desert vegetation, recovery of, 268
detonation craters, 260
distributional heterogeneity of residues of
energetic compounds, 259
double-base propellant, 257
effects of explosives ranges on terrestrial
wildlife habitat variables, 261
firing tests, 256
foot traffic, 269
Fort Benning Military Reservation, 269, 271
Fort Sill Military Reservation, 266
fragmentation effects, 263
geographic information system model, 265
habitat alteration on military installations,
263
habitat disturbance, 260
habitat fragmentation, 262
habitat suitability and connectivity, 261–263
Habitat Suitability Index models, 262
impact areas, 259
Integrated Natural Resources Management
Plans, 253–254
Jefferson Proving Ground, 256
Kansas Army National Guard Training
Facility, 264
Military Munitions Rule, 255
munitions ranges, 254–256
physical modification of habitat, 271
phytotoxicity, recovery from, 268
quantifying habitat change, 263–265
resiliency and recovery, 267–268
responses of species to disturbance, 265–267
Snake River Birds of Prey National
Conservation Area, 270
spatial distribution of contamination and
phytotoxicity, 257–260
spatial scale of physical effects, 260
Thornwaite Moisture Regions, 267
threatened species on DoD installations, 265
turf grass monocultures, 269
urban development, 270
U.S. Air Force bombing ranges, 260
U.S. Department of Defense Directive, 255
U.S. Fish and Wildlife Service habitat
models, 261
vegetation removal, 262
white phosphorus-based smoke rounds, 259
Yuma Proving Ground, 264, 267
Habitat Suitability Index models, 262
Hazardous Toxic Radioactive Waste (HTRW)
program, 284
Hb, *see* Hemoglobin
Helianthus annuus, 46, 240

Hemoglobin (Hb), 164
Henry's law constant, 21
Hexagenia bilinata, 102
Hexahydro-3,5-dinitro-1-nitroso-1,3,5-triazine
    (MNX), 13, 61, 71, 147, 192
Hexahydro-5-nitro-1,3-dinitroso-1,3,5-triazine
    (DNX), 13, 71
Hexahydro-1,3,5-trinitro-1,3,5-triazine (RDX,
    TNX), 6, 9, 11, 36, 78, 105, 117, 157
  abiotic degradation pathways of, 12
  accumulation of in herbivore, 246
  adsorption coefficients, 24
  biodegradation of initiated by denitration
    under aerobic and anaerobic
    conditions, 20
  biodegradation of initiated by reduction
    to nitroso derivatives followed by
    α-hydroxylation, denitration, or
    reduction, 21
  biodegradation of by microorganisms, 19
  carcinogenicity, 232
  concentration in plant tissue, 241
  degradation pathway, 232
  earthworm toxicity and, 310
  effects of on indigenous soil, 37
  effects of on terrestrial plants, 45
  embryo hatching success and, 105
  ERA involvement at military ranges, 285
  genotoxicity and, 178, 183
  losses of radioactivity and, 127
  mononitroso degradation product of, 192
  photolysis of, 105
  plant absorption of, 239
  resistance to degradation, 12
  terrestrial plant sensitivity to, 70
  terrestrial systems, 239–243
  wildlife species and, 158
2,4,6,8,10,12-Hexanitro-2,4,6,8,10,12-
    hexaazaisowurtzitane (CL-20),
    6, 9, 36
  biodegradation of by microorganisms, 19
  earthworm toxicity and, 310
  terrestrial plant sensitivity to, 70
High melting explosive (HMX), 257, *see also*
    Octahydro-1,3,5,7-tetranitro-1,3,5,7-
    tetrazocine
High performance liquid chromatography
    (HPLC), 99, 140, 189
HMX, *see* Octahydro-1,3,5,7-tetranitro-1,3,5,7-
    tetrazocine
Hordeum vulgare L, 45
HPLC, *see* High performance liquid
    chromatography
HTRW program, *see* Hazardous Toxic
    Radioactive Waste program

Hyalella azteca, 81, 85, 89, 90, 101, 123, 124, 132
Hydra littoralis, 103, 105

I

Ictalurus punctatus, 80, 84, 101, 105, 137, 143
In situ monitoring tools, need for improved, 310
Integrated Natural Resources Management
    Plans, 253–254
Integrated Risk Information System (IRIS), 190
International Organization for Standardization
    (ISO), 282
IRIS, *see* Integrated Risk Information System
ISO, *see* International Organization for
    Standardization

J

Jefferson Proving Ground, 256
Joliet Army Ammunition Plant, 40

K

Kansas Army National Guard Training Facility,
    264
Key Technical Area (KTA), 1, 293
Klebsiella pneumoniae, 18, 19
KTA, *see* Key Technical Area
Kupffer cells, 163

L

LAAP, *see* Louisiana Army Ammunition Plant
Lactuva sativa, 45, 242
Land Condition Trend Analysis (LCTA)
    program, 263–264
LCTA program, *see* Land Condition Trend
    Analysis program
Lemna
  minor, 87
  perpusilla, 46
Lepidium sativum L., 52
Lepomis macrochirus, 80, 84, 94, 101, 105, 109,
    142
Leptocheirus plumulosus, 123, 124, 128
Leuciscus idus, 137
LOEC, *see* Lowest observed effect concentration
Lolium perenne, 46, 242, 288
Louisiana Army Ammunition Plant (LAAP), 24
Lowest observed effect concentration (LOEC),
    37, 122, 257, 300
Lumbriculus variegatus, 86, 102, 137, 143
Lumbrokinase-3, 70
Lupinus perennis, 266–267
Lycaecides melissa samuelis, 266
Lytechinus variegatus, 99

M

Mammalian cell cytotoxicity, mechanisms of, 211–226
  concentrations of nitroaromatic explosives, 214
  enzymatic reactions of nitroaromatic explosives, 213–220
    formation of stable metabolites, 218–220
    free radical reactions, 213–218
  enzymatic reactions of other explosives, 222
  NADPH:cytochrome P-450 reductase, 213
  NAD(P)H:quinone oxidoreductase, 219
  nitroaromatic explosives, 221–222
  oxidative stress, 221
  oxygen-insensitive nitroreductases, 220
  structural formulae of nitroaromatic explosives and their metabolites, 212
Mammalian cell line-based genotoxicity assays, 185–189
  cyclic nitramines and their products, 188
  dinitrotoluenes and their products, 187–188
  mammalian cell line testing of environmental samples, 189
  2,4,6-trinitrotoluene and its products, 185–187
  other compounds, 188–189
Mammalian in vivo genotoxicity investigations, 190–193
  cyclic nitramines and their products, 192–193
  dinitrotoluenes and their products, 191–192
  other compounds, 193
  TNT and its products, 190–191
Maneuver impact miles (MIMs), 265
Marine Corps Sustainable Ranges Program, 284
MC, *see* Munition constituents
MCOC, *see* MC of concern
MED, *see* Minimum effective dose
Medicago sativa, 46, 242
MEDINA, *see* Methylenedinitramine
Meisenheimer complex, 8, 10
Memorandum of understanding (MOU), 282
Methemoglobinemia, 211
Methylenedinitramine (MEDINA), 18, 24
Methylobacterium sp., 15
Microbial efficiency, definition of, 43
Microcystis aeruginosa, 79, 81, 83, 87, 96, 103, 104, 105
Microorganism-based mutagenicity assays, 178–185
  cyclic nitramines and their products, 183
  dinitrotoluenes and their products, 182–183
  mixtures and environmental samples, 184–185
  other compounds, 183–184
  2,4,6-trinitrotoluene and its products, 180–182

Microtus ochrogaster, 246
Military explosives classification, 7–8
Military Munitions Rule, 255
MIMs, *see* Maneuver impact miles
Minimum effective dose (MED), 68
MNX, *see* Hexahydro-3,5-dinitro-1-nitroso-1,3,5-triazine
Model(ing)
  Applied Risk Assessment Modeling System, 295
  Army Training and Testing Carrying Capacity model, 265
  ecological dynamics simulation modeling, 265
  FISHRAND, 294, 295
  food web models, 228
  geographic information system model, 265
  Habitat Suitability Index models, 262
  quantitative structure–activity relationship models, 194
  Spatially Explicit Exposure Model, 294
  U.S. Fish and Wildlife Service habitat models, 261
Morganella morganii, 18, 19
MOU, *see* Memorandum of understanding
Mucor mucedo, 16
Munition constituents (MCs), 278, *see also* Soil contamination, ecological risk assessment of with munition constituents (North America)
  of concern (MCOC), 283
  Environmental Sustainability Indices for, 297
Munitions ranges, 254–256, *see also* Habitat disturbance at explosives-contaminated ranges
Mus musculus, 160, 161
Mycobacterium
  sp., 15
  vaccae, 15
Myriophyllum
  aquiticum, 147, 237
  spicaticum, 237

N

NACs, *see* Nitroaromatic compounds
NADPH:cytochrome P-450 reductase, 213
Nanomaterials, ecotoxicological effects of, 311
National Institute for Environmental Health Sciences (NIEHS) database, 187
Navicula pelliculosa, 103, 105
NB, *see* Nitrobenzene
NC, *see* Nitrocellulose
NDAB, *see* 4-Nitro-2,4-diazabutanal
Neanthes arenaceodentata, 122, 123, 124, 128, 132

Nematoloma frowardii, 16
Neuronal nitric oxide synthase (nNOS), 68
Neurospora crassa, 16, 179
NFA, *see* Nitrogen fixation activity
NG, *see* Nitroglycerin
NIEHS database, *see* National Institute for
    Environmental Health Sciences
    database
NIEHS Pathology Working Group, 192
Nitella spp., 150, 151
Nitocra spinipes, 86, 99, 102, 105
Nitroaromatic compounds (NACs), 8–11, 13–18,
    22, 79–105
  alkaline hydrolysis, 8–10
  chronic toxicity of, 64
  DNT biotransformation, 18
  ecological consequences of soil
      contamination with, 44–45
  effects of on carbon cycle, 43–44
  effects of on nitrogen cycle, 41–43
  effects of on soil invertebrates, 62–64
  effects of on soil microbial community,
      39–41
  microbial toxicity of, 38–45
  nitrobenzenes, 79–83
  nitrophenols, 93–98
  nitrotoluenes, 83–93
  photoactivation, 99–105
  photolysis, 11
  phytotoxicity of, 46–52
  picric acid biotransformation, 18
  reduction by iron, 11
  TNT biotransformation, 13–17
  2,4,6-trinitrophenylmethylnitramine, 98–99
Nitrobenzene (NB), 80–82, 137
Nitrocellulose (NC), 6, 9, 109, 285
4-Nitro-2,4-diazabutanal (NDAB), 11, 24
Nitrogen cycle, effects of nitroaromatic
    compounds on, 41–43
Nitrogen fixation activity (NFA), 37
Nitroglycerin (NG), 6, 9, 106, 193, 285
Nitroguanidine (NQ), 106
Nitro organic compounds (NOCs), 6, 25
Nitrophenols (NPs), 84–88, 93, 94–97
2-Nitrotoluene (2-NT), 125
5-Nitro-1,2,4-triazol-3-one (NTO), 212
nNOS, *see* Neuronal nitric oxide synthase
NOCs, *see* Nitro organic compounds
NOEC, *see* No observed effect concentration
No observed effect concentration (NOEC), 37,
    122, 300
Notopterus notopterus, 94
NPs, *see* Nitrophenols
NQ, *see* Nitroguanidine
2-NT, *see* 2-Nitrotoluene
NTO, *see* 5-Nitro-1,2,4-triazol-3-one

O

OB/OD, *see* Open burning or detonation
Octahydro-1,3,5,7-tetranitro-1,3,5,7-tetrazocine
    (HMX), 6, 9, 11, 36, 78, 105, 117
  adsorption coefficients, 24
  biodegradation of by microorganisms, 19
  earthworm toxicity and, 310
  effects of on terrestrial plants, 45
  ERA involvement at military ranges, 285
  genotoxicity and, 178, 183
  mammalian cell cytotoxicity and, 222
  microbial toxicity of in soil, 38
  resistance to degradation, 12
  terrestrial plant sensitivity to, 70
  terrestrial systems, 243–244
  wildlife species and, 158
Odocoileus virginiana, 245
OECD, *see* Organization for Economic
    Cooperation and Development
OECD artificial soil, 36, 244
OM, *see* Organic matter
Onchorhynchus mykiss, 80, 101, 145
Onobrychis viciifolia Scop., 46
Onychomys torridus, 266
Open burning or detonation (OB/OD), 6
Operational Range Assessment Program
    (ORAP), 282, 283, 284
ORAP, *see* Operational Range Assessment
    Program
ORCs, *see* Ordnance related compounds
Ordnance compounds, toxicity of, 100–104
Ordnance related compounds (ORCs), 69
Organic matter (OM), 38, 44
Organization for Economic Cooperation and
    Development (OECD), 36, 234
Oryctologus cuniculus, 162
Oryzias latipes, 80
Oxidized glutathione (GSSG), 217

P

PA, *see* Picric acid
Packed cell volume (PCV), 164
PAHs, *see* Polycyclic aromatic hydrocarbons
Paratanytarsus
    dissimilis, 102
    parthenogeneticus, 86, 102
PAT, *see* Production or lot acceptance testing
PCBs, *see* Polychlorinated biphenyls
PCV, *see* Packed cell volume
Penicillium frequentans, 16
Pentaerythritol tetranitrate (PETN), 6, 9, 107,
    222
  ERA involvement at military ranges, 285
  uses, 286
Perognathus longimembris, 266

Peromyscus leucopus, 159, 161, 245
PETN, *see* Pentaerythritol tetranitrate
Phalaris arundinacea, 146
Phanerochaete chrysosporium, 16, 19, 20
Phaseolus vulgaris L., 52
Phospholipid fatty acids (PLFAs), 40
Photoactivation, 99
Phytophotolysis, 147
Picoides borealis, 268
Picramic acid, 129
Picric acid (PA), 6, 9, 18, 93, 126
PICT, *see* Pollution-induced community tolerance
Pimephales promelas, 80, 84, 89, 90, 92, 94, 100, 105, 109, 137, 142
Pink water, 77
Pinus palustris, 268
Plethodon cinereus, 169, 170
PLFAs, *see* Phospholipid fatty acids
PNA, *see* Potential nitrification activity
Poecilia reticulata, 79, 80, 85
Pollution-induced community tolerance (PICT), 41
Polychlorinated biphenyls (PCBs), 228
Polycyclic aromatic hydrocarbons (PAHs), 99
Populus deltoids × nigra, 243
Portieria hornemannii, 148
Potential nitrification activity (PNA), 37
Production or lot acceptance testing (PAT), 256
Providencia rettgeri, 18, 19
Pseudokirchneriella subcapitata, 87, 89, 91, 92, 103, 105, 109, 145
Pseudomonas
    aeruginosa, 15
    fluorescens, 15
    pseudoalcaligens, 15
    putida, 19, 79, 82, 83, 93, 97
    savastanoi, 15
    sp., 14, 15, 19

Q

QAPP, *see* Quality assurance project plans
QSAR models, *see* Quantitative structure–activity relationship models
QSARs, *see* Quantitative structure–activity relationships
Quality assurance project plans (QAPP), 282
Quantitative structure–activity relationship (QSAR) models, 194
Quantitative structure–activity relationships (QSARs), 171

R

Random amplified polymorphic DNA (RAPD) technique, 198
Range condition assessments (RCA), 284

Range Environmental Vulnerability Assessment (REVA) Program, 284
Ranges, explosives-contaminated, *see* Habitat disturbance at explosives-contaminated ranges
Range Sustainability Environmental Program Assessment (RSEPA), 283
RAPD technique, *see* Random amplified polymorphic DNA technique
RBC, *see* Red blood cell count
RCA, *see* Range condition assessments
RCW, *see* Red-cockaded woodpeckers
RDX, *see* Hexahydro-1,3,5-trinitro-1,3,5-triazine
Red blood cell count (RBC), 164
Red-cockaded woodpeckers (RCW), 268, 271
Reduced glutathione (GSH), 218, 221
Regional Simulator (RSim), 270
Remedial investigation and feasibility study (RI/FS) process, 278
Restriction fragment length polymorphism (RFLP), 199
REVA Program, *see* Range Environmental Vulnerability Assessment Program
RFLP, *see* Restriction fragment length polymorphism
Rhizoctonia solani, 16
Rhodococcus
    erythropolis, 15
    rhodochrous, 69
    sp., 19, 20
RI/FS process, *see* Remedial investigation and feasibility study process
Royal demolition explosive (RDX), 105, 257, *see also* Hexahydro-1,3,5-trinitro-1,3,5-triazine
RSEPA, *see* Range Sustainability Environmental Program Assessment
RSim, *see* Regional Simulator

S

Saccharomyces cerevisiae, 179
SAGE, *see* Serial analysis of gene expression
Salmo
    gairdneri, 84, 94, 101, 105
    salar, 94
Salmonella typhimurium, 178, 179
SARA, *see* Superfund Amendments and Reauthorization Act of 1986
SARs, *see* Structure–activity relationships
Sassafras sandy loam (SSL), 38, 193, 233, 287
Sceloporus
    occidentalis, 168, 169
    undulatus, 168
Scenedesmus
    obliquus, 81, 87
    quadricauda, 79, 81, 83, 87, 93, 96

Schizopera knabeni, 86, 92, 93, 95
Sciaenops ocellatus, 80, 85, 94, 101, 105
Scientific-management decision point (SMDP), 280
Screening level ecological risk assessment (SLERA), 280, 286
Sediments, fate and toxicity of explosives in, 117–134
    differences in species sensitivity, 125
    effects of explosives on reproduction and growth of invertebrates, 128
    fate and lethal toxicity of explosives in sediment exposures, 119–128
        cyclic nitramines, 127–128
        2,4,6-trinitrophenol, 126–127
        2,4,6-trinitrophenylmethylnitramine, 127
        trinitrotoluene and related compounds, 2,6-dinitrotoluene, 125–126
    measured sum concentration of TNT, 120
    methodology for amending explosives to sediments, 118–119
    nominal concentration of TNT measured after mixing into spiked sediments, 121
    renewal of overlying waters, 121
    research recommendations, 131–133
    toxicity of explosives determined for aquatic invertebrates using spiked sediment exposures, 123–124
    toxicity of porewater extracted from sandy or fine-grained spiked sediments, 130
    toxicity of porewater extracted from sediments spiked with explosives, 128–129
    use of biomimetic devices to assess explosives bioavailability, 129–131
SEEM, *see* Spatially Explicit Exposure Model
Selenastrum capricornutum, 66, 87, 89, 91, 92, 103, 145
SERDP, *see* Strategic Environmental Research and Development Program
Serial analysis of gene expression (SAGE), 68
Serratia marcescens, 15, 19
SHE cells, *see* Syrian hamster embryo cells
Shewanella sp., 19
Sigmodon hispidus, 159
SIR, *see* Substrate-induced respiration
SLERA, *see* Screening level ecological risk assessment
Small-arms ranges, 285
SMDP, *see* Scientific-management decision point
Snake River Birds of Prey National Conservation Area, 270
Soil, artificial, 36
Soil bacteria, 37

Soil contamination, ecological risk assessment of with munition constituents (North America), 277–307
    baseline ecological risk assessment, 280
    Canada, 296–301
        ammunitions used on ranges during tank maneuvers, 299
        antitank ranges, 298
        bombing ranges, 299
        Environmental Sustainability Indices for energetic contaminants as military ranges and training sites, 299–301
        extent of soil contamination with energetic materials at DND installations, 298–299
        grenade ranges, 298
        Trigran, 299
        UXO production, 298
    derivation of Eco-SSL, 286
    derivation of ETVs for CL-20 in freshly amended SSL soil, 290
    Ecological Risk Assessment Guidance for Superfund, 280
    ecological risk assessment procedures, 278
    ecological soil screening levels, 279, 286
    effects of soil properties on terrestrial plants, 293
    Endangered Species Act, 282
    Formerly Used Defense Sites, 284
    future outlook, 301–303
    growth toxicity benchmarks for MC freshly amended in SSL soil, 288
    Hazardous Toxic Radioactive Waste program, 284
    Marine Corps Sustainable Ranges Program, 284
    MC of concern, 283
    Operational Range Assessment Program, 283
    oral exposure for wildlife species, 294
    Range Environmental Vulnerability Assessment Program, 284
    remedial investigation and feasibility study process, 278
    reproduction toxicity benchmarks for MC freshly amended in SSL soil, 290
    scientific-management decision point, 280
    screening level ecological risk assessment, 280
    Superfund, ecological risk assessment process, 281
    terrestrial plant-based ETVs for MC determined using ecotoxicological benchmarks, 291
    The Technical Cooperation Program, 293
    threatened and endangered species, 278
    tools for assessing risks to wildlife, 294–296

tools for screening contaminated soil data, 286–294

toxicity benchmark value derivation for wildlife, 295

United States and associated authorizations for explosives, 279–286

    CERCLE ERA framework, 279–280

    DoD-specific range assessment programs, 283–285

    ERAGS ERA process, 280

    types and sources of MC on military ranges that may be involved in ERA, 285–286

    Uniform Federal Policy for Implementing Environmental Quality Systems, 282–283

    USEPA ecological risk management principles, 281–282

    USAF Operational Range Environmental Program, 284

Soil organisms, effects of energetic materials on, 35–76

    cellular protein degradation, 70

    cytochrome-containing nitrite reductase, 42

    denitrification, 42

    dinitroaniline herbicides, 68

    ecotoxicological benchmarks for energetic materials established in standardized single-species toxicity tests with terrestrial plants, 47–51

    effects on soil invertebrates, 52–65

        acute toxicity of nitroaromatic compounds, 62–64

        chronic toxicity of nitroaromatic compounds, 64

        effects of cyclic nitramines, 52–62

        effects of nitroaromatic compounds, 62–64

        effects of perchlorate, 64–65

    effects on soil microorganisms, 37–45

        ecological consequences of soil contamination with nitroaromatic compounds, 44–45

        effects of cyclic nitramines on microbial activity in soil, 37–38

        effects of nitroaromatic compounds on carbon cycle, 43–44

        effects of nitroaromatic compounds on nitrogen cycle, 41–43

        effects of nitroaromatic compounds on soil microbial community, 39–41

        microbial toxicity of nitroaromatic compounds, 38–45

    effects on terrestrial plants, 45–52

        effects of cyclic nitramines, 45–46

        phytotoxicity of nitroaromatic compounds, 46–52

effects of weathering and aging energetic materials in soil on toxicity to soil organisms, 65–66

estimates of microbial efficiency, 43

future outlook, 70–71

mechanisms of toxicity, 67–70

ordnance related compounds, 69

serial analysis of gene expression, 68

stimulating effects of energetic materials, 66–67

substrate-induced respiration, 43

triazine herbicides, 46

ubiquitin activating enzyme, 70

Soil Quality Guidelines (SQG), 297

Solid phase microextraction fibers (SPMEs), 131

Spatially Explicit Exposure Model (SEEM), 294

SPMEs, *see* Solid phase microextraction fibers

SQG, *see* Soil Quality Guidelines

SRO, *see* Sustainable range oversight

SSL, *see* Sassafras sandy loam

Staphylococcus sp., 15

Stenotrophomonas maltophilia, 19

Strategic Environmental Research and Development Program (SERDP), 6

Streptomyces chromofuscus, 15

Structure–activity relationships (SARs), 194

Sturnus vulgaris, 164

Substrate-induced respiration (SIR), 37

Superfund, ecological risk assessment process, 281

Superfund Amendments and Reauthorization Act of 1986 (SARA), 279

Sustainable range oversight (SRO), 284

Syrian hamster embryo (SHE) cells, 187

T

Tanytarsus dissimilis, 86

TAT, *see* 2,4,6-Triaminotoluene

TATB, *see* 1,3,5-Triaminotrinitrobenzene

Terrestrial systems, bioconcentration, bioaccumulation, and biomagnification of nitroaromatic and nitramine explosives in, 227–252

    environmental degradation and transformation of energetic compounds, 230–233

        nitroaromatics, 230–232

        nitramines, 232–233

    food web models, 228

    OECD artificial soil, 244

    plants, 235–244

        HMX, 243–244

        nitroaromatics, 235–239

        RDX, 239–243

    plant test BAFs, 242

polychlorinated biphenyls in fat, 228
properties of tested soils, 237
RDX concentration in plant tissue, 241
soil biota, 233–235
    development and application of BAF
        values for ecological risk assessment
        process, 234–235
    effects of soil properties on
        bioaccumulation of energetic
        compounds, 233–234
    effects of weathering and aging
        on energetic compounds
        bioaccumulation, 233
soil invertebrates, 244–245
summary, 247
sunflower BAF, 241
terrestrial animals, 245–247
tissue TNT-derived radiocarbon, 238
USEPA early seedling growth test, 242
yellow nutsedge grown in hydroponic media,
    236
T&E species, *see* Threatened and endangered
    species
Test ranges, firing tests, 256
Tetrahymena pyriformis, 81, 87, 96
4,5,6,7-Tetranitrobenzimidazol-2-one (TNBO),
    212
1,3,6,8-Tetranitrocarbazole (TNC), 212
Tetryl, *see* 2,4,6-Trinitrophenylmethylnitramine
The Technical Cooperation Program (TTCP),
    1, 293
Thioredoxin (Trx), 217
Threatened and endangered (T&E) species, 278,
    282
Tigriopus californicus, 90
TNB, *see* 1,3,5-Trinitrobenzene
TNBO, *see* 4,5,6,7-Tetranitrobenzimidazol-2-one
TNC, *see* 1,3,6,8-Tetranitrocarbazole
2,4,6-TNP, *see* 2,4,6-Trinitrophenol
TNT, *see* 2,4,6-Trinitrotoluene
TNT⁻, *see* 2,4,6-Trinitrobenzyl anion
TNX, *see* Hexahydro-1,3,5-trinitroso-1,3,5-
    triazine
Toxicity reference values (TRVs), 279, 294, 302
Tradescantia micronucleus bioassay (Trad-
    MCN), 68, 193
Tradescantia stamen hair mutation (Trad-SHM)
    assay, 193
Trad-MCN, see *Tradescantia* micronucleus
    bioassay
Trad-SHM assay, see *Tradescantia* stamen hair
    mutation assay
Trametes versicolor, 16, 69
2,4,6-Triaminotoluene (TAT), 11, 89, 186
1,3,5-Triaminotrinitrobenzene (TATB), 7, 9
Triazine herbicides, 46

Trifolium repens, 199
Trigran, 299
1,3,5-Trinitrobenzene (TNB), 6, 11, 36, 78, 117,
    184
2,4,6-Trinitrobenzyl anion (TNT⁻), 10
2,4,6-Trinitrophenol (2,4,6-TNP), 6, 93, 126
2,4,6-Trinitrophenylmethylnitramine (tetryl) 6,
    78, 98, 117, 127
2,4,6-Trinitrotoluene (TNT), 6, 9, 36, 78, 117, 157
    biotransformation, 13–17
        aerobic and anaerobic conditions, 17
        fish, 136
        fungi, 16
    environmental degradation, 230
    ERA involvement at military ranges, 285
    explosives-contaminated ranges, 257
    genotoxicity and, 178, 185
    mammalian cell cytotoxicity and, 21
    measured sum concentration of, 120
    reduction products, toxicity of, 90–91
    schematic toxicological impacts of on soil
        microbial community, 39
    sensitivity of Gram-positive organisms to, 40
    soil microbial endpoints and, 44
    sorption of, 22
Triticum aestivum L., 52
TRVs, *see* Toxicity reference values
Trx, *see* Thioredoxin
TTCP, *see* The Technical Cooperation Program
Tubifex tubifex, 123, 131, 132, 137, 143, 144
Turf grass monocultures, 269

U

Ubiquitin activating enzyme, 70
UDS assay, *see* Unscheduled DNA synthesis
    assay
Ultraviolet (UV) radiation, 93
Ulva fasciata, 81, 83, 88, 92, 93, 96, 104, 105,
    110, 129
Unexploded ordnance (UXO), 1, 118, 151, 260,
    269, 298
Unscheduled DNA synthesis (UDS) assay, 187,
    189, 191
Urban development, habitat disturbance and, 270
USACHPPM, *see* U.S. Army Center for Health
    Promotion and Preventive Medicine
USAF, *see* U.S. Air Force
U.S. Air Force (USAF), 284
    bombing ranges, 260
    Operational Range Environmental Program,
        284
U.S. Army Center for Health Promotion and
    Preventive Medicine (USACHPPM),
    294
U.S. Department of Defense (DoD), 202, 265
    Directive, 255

memorandum of understanding, 202
Operational Range Assessment Program, 282
U.S. Department of Energy, memorandum of
understanding, 282
U.S. Environmental Protection Agency
(USEPA), 191, 279
artificial soil, 36
Data Quality Objective Process, 283
DNT classification, 191
early seedling growth test, 242
Ecological Risk Assessment and Risk
Management Principles for Superfund
Sites, 281
ecological soil screening levels, 279
EM analysis, 310
Guidelines for Ecological Risk Assessment,
280
memorandum of understanding, 282
requirements for developing toxicity
benchmarks, 287
USEPA, *see* U.S. Environmental Protection
Agency
U.S. Fish and Wildlife Service habitat models,
261
U.S. Marine Corps, Marine Corps Sustainable
Ranges Program, 284
U.S. Navy, Range Sustainability Environmental
Program Assessment Policy
Implementation Manual, 283, 284
UV radiation, *see* Ultraviolet radiation
UXO, *see* Unexploded ordnance

V

Vibrio fischeri, 82, 83, 88, 91, 97, 99, 104, 179
Vicia faba root tip micronucleus assay, 193
Vireo atricapillus, 266

W

Water quality criteria (WQC), 107
Weapons Technical Panel (WPN), 293
White phosphorus-based smoke rounds, 259
Wildlife species, toxicity of energetic
compounds to, 157–175
amphibians, 169–171
artillery ranges, sizes of, 157
birds, 163–166, 166–168
extramedullary hematopoiesis, 159
FETAX assay, 169, 170
mammals, 158–163
reptiles, 168
Wildlife Toxicity Assessment (WTA) document,
295
WPN, *see* Weapons Technical Panel
WQC, *see* Water quality criteria
WTA document, *see* Wildlife Toxicity
Assessment document

X

Xanthine oxidase, 217, 219
Xanthomonas maltophilia, 19
Xenopus
laevis, 84, 89, 90
sp., 169

Y

YPG, *see* Yuma Proving Ground
Yuma Proving Ground (YPG), 264, 267

Z

Zea mays, 242